2nd Edition
Brand new content and
assessment features

Plumbing

NVQ and Technical Certificate Level 2

D1334283

...nemann.co.uk
Free online support
✓ Useful weblinks
...our online ordering

01865 888118

Heinemann

Heinemann is an imprint of Pearson Education Limited, a company incorporated in England and Wales, having its registered office at Edinburgh Gate, Harlow, Essex, CM20 2JE. Registered company number: 872828

www.heinemann.co.uk

Heinemann is a registered trademark of Pearson Education Limited

Text © JTL 2008

First published 2005
Seventh edition published 2008

12 11
10

British Library Cataloguing in Publication Data
A catalogue record for this book is available from the British Library

ISBN 978 0 435402 50 1

Designed by HL Studios
Typeset by HL Studios
Original illustrations © Pearson Education Limited 2008
Illustrated by HL Studios
Cover design by GD Associates
Cover photo/illustration © Pearson Education Limited/Gareth Boden
Printed in China (CTPSC/10)

Acknowledgements

Every effort has been made to contact copyright holders of material reproduced in this book. Any omissions will be rectified in subsequent printings if notice is given to the publishers.

The websites used in this book were correct and up-to-date at the time of publication. It is essential for tutors to preview each website before using it in class so as to ensure that the URL is still accurate, relevant and appropriate. We suggest that tutors bookmark useful websites and consider enabling students to access them through the school/college intranet.

Contents

Acknowledgements

JTL would like to thank the following people for their contribution to this book: John Thompson (lead author), Mike Phoenix, Keith Arrell, Dave Bowers, Mark Brown, Karen Easton, Julian Hodges and Keith Powell.

The authors and publishers would like tot hank the following for permission to reproduce copyright material:

JTL, SummitSkills, The Chartered Institute of Plumbing and Heating Engineering, City & Guilds, EMTA Awards Ltd, WRAP

Photo acknowledgements

The author and publisher would like to thank the following individuals and organisations for permission to reproduce photographs:

Alamy, p243, p341, p435, p439; Alamy/Brand X Pictures, p95; Alamy/Brian Harris, p310 (bottom); Alamy Images/Comstock Images, p291 (top); Alamy Images/Ingram Publishing, p278; Alamy Images/Niall McDiarmid, p281 (middle); Alamy Images/ Photofusion/Steve Morgan, p433; Alamy Images/Sciencephoto, p118; Alamy Images/The Photolibrary Wales, p310 (top); Arco, p38 (gloves); Art Directors and TRIP/Chris Parker, p77; Art Directors and TRIP/Helene Rogers, p38 (boots), p105, p290 (top); Construction Images, p445; Constructionphotography.com, p1, p17, p49, p75, p185, p398; Constructionphotography.com/Tom Lee, p448; Corbis/Jim Craigmyle, p458; Dreamstime/Chris Green, p27; Fernox-Cookson Electronics, p306 (bottom); Jake Fitzjones/Redcover.com, p281 (bottom); GAH Group, p287; Getty nrf, p2, p309; Hepworth Building Products, p161 (all); Institue of Heating and Plumbing Engineers, p235; Lead Sheet Association, p401, p402 (all), p405 (all), p407 (all), p409 (all), p411 (all), p425, p426 (all), p429; MorgueFile/Ronnie Bergeron, p118; Pearson Education Ltd/Jules Selmes, p65; Rex Features/Image Source, p92; Rex Features/ Voisin/Phanie, p89; Science Photo Library/Shelia Terry, p186; Ginny Stroud-Lewis, p37 (all), p38 (bottom), p40 (all), p100, p285, p291 (middle, bottom), p312.

All others Pearson Education Ltd/Gareth Boden.

Introduction

How to use this book

This book has been designed with you in mind. It has a dual purpose:

1 To lead you through the Level 2 MES Plumbing qualifications, providing background information and technical guidance.

2 To provide a future reference book that you will find useful to dip into long after you have gained your qualification.

The book has been specially produced to assist you in completing your Plumbing 6129 Scheme and NVQ qualification. The schemes are applicable in England, Wales and Northern Ireland and are designed to set a quality standard for learning and training while attending college/a learning centre or as a mature experienced worker wishing to update qualifications.

The book details the various sections that you will undertake to complete the Level 2 Units from the Plumbing 6129 Technical Certificate Scheme, all of which support the full NVQ. Each chapter concludes with job knowledge checks which are essential parts of the overall qualification.

The book is also a key reference document for you to consult in support of your plumbing work. Your tutor will further explain the content of the various sections to you when you attend an approved assessment centre.

The chapters are in a similar format to the units of the City & Guilds assessment, making it easier to complete a section with a successful assessment before moving on to the next.

Qualifications

1 Technical Certificate: this is the job knowledge and training part of the qualification for new entrants.

2 NVQ: this in full means National Vocational Qualification, which is the collection of evidence that you have done the work in the real workplace.

Together they give the full qualification for an operative to be able to work with a company under supervision.

For those of you who are already engaged in the trade and have been for some time, you can simply do the tests of the Technical Certificate to accompany the NVQ workplace evidence. Either way, this book has the content you require.

How this book can help you

There are other key features of this book which are designed to help you make progress and reinforce the learning that has taken place. Such features are:

- **Photographs:** easy-to-follow sequences of key operations
- **Illustrations:** clear drawings, many in colour, showing essential information about complex components and procedures
- **Margin notes:** short helpful hints to aid you to good practice
- **Tables, bullet points and flowcharts:** easy-to-follow features giving information at a glance
- **On-the-job scenarios:** typical things that happen on the job – what would you do?
- **Safety tips and Remember:** margin notes to emphasise important points
- **Did you know?:** useful information on things you always wondered about
- **End-of-chapter knowledge checks:** test yourself to see if you have absorbed all the information; are you ready for the real test?
- **Glossary:** clear definitions and explanations of those strange words and phrases.
- **Candidate CD-ROM:** contains Progress check questions and answers and Knowledge check answers so you can test your knowledge of the material

Why choose plumbing?

Plumbing is a very satisfying and rewarding industry to join, with a variety of work on a range of systems. There is always a demand for workers who have the skills, knowledge and qualifications.

Why this book?

Because it is structured to give all the basic information required to gain the relevant qualification and set you on course for an exciting career in a buoyant industry. Well done for choosing such a good start!

There are another two similar qualifications at Level 3 and there's another book like this to match!

Additional Resources

Your tutor can access additional free resources at www.heinemann.co.uk. Resources cover the new employment rights and responsibilities material included in this edition, as well as resources for heating and ventilation and refrigeration and air conditioning.

Free Resources can be accessed at:

www.heinemann.co.uk/Series/FEAndVocational/PlumbingNVQAndTechnical
CertificatesLevels2And3/FreeResources/FreeResources.aspx

chapter 1

Industry and qualification overview

OVERVIEW

This chapter will introduce you to what you can expect when you start to work in the plumbing industry. This chapter includes:

- **An overview of the plumbing industry**
    - what is plumbing?
    - the size of the plumbing industry

- **Key industry organisations and bodies**

- **Plumbing training for new entrants**
    - qualification levels
    - plumbing apprenticeship schemes
    - an overview of plumbing qualifications

- **Career opportunities**

An overview of the plumbing industry

This section will explore what the job of a plumber really involves and should prepare you for what to expect on site. It will also provide information about the structure of the plumbing industry and some of its key organisations.

What is plumbing?

Generally, plumbing satisfies the basic needs of people to keep clean, warm and healthy by providing heating, hot- and cold-water systems and sanitation 365 days a year, 24/7.

A plumber's job usually includes the **installation**, **service** and **maintenance** of a wide range of systems such as:

- cold water, including underground services, to a dwelling
- hot water
- heating systems fuelled by gas, oil or solid fuel
- sanitation (or above-ground drainage), including the installation of baths, wash-hand basins, WCs and sinks
- sheet lead weathering
- rainwater runoff and drainage
- drainage
- associated electrical systems.

A qualified plumber could install, service and maintain all the various components contained within those systems, for example:

- pipe materials, fittings, fixings, controls
- heat exchangers, boilers, radiators
- pumps, accelerators and motorised/isolating valves
- storage vessels, cylinders, cisterns
- sanitary appliances, baths, WCs, etc.
- domestic appliances including washing machines and dishwashers
- cabling and electrical components
- sheet weathering, aprons, gutter backs, step flashings, soakers, lead slates
- gas appliances (natural or LPG), boilers, water heaters, cookers and fires.

Some well-used tools of the trade!

Find out

How would you describe plumbing? How would your friends describe it? Jot down some thoughts and ideas

Definition

Installation – fitting a new system

Service – regular checking of a system to ensure it is still working properly

Maintenance – carrying out any necessary repairs to worn or broken parts; keeping appliances and systems in good working order

Safety tip

The safety equipment required for a particular job should always be used. Get into this habit at the beginning of your training

Plumbers work in many and varied work locations. Plumbing companies tend to specialise in either industrial/commercial work – factories, office blocks – or domestic properties, such as houses and flats. Some companies specialise in certain types of system installation or maintenance, such as central-heating systems, others undertake the full range of activities.

A competent plumber should be able to:

- work safely at all times
- plan the job and agree a schedule of work with the customer
- prepare the work location, including ensuring adequate access, protecting customer's property and making sure all the tools, materials and equipment are available
- mark out, measure and work out the installation requirements
- fabricate, position and fix system components
- pre-commission (including testing), commission and decommission systems
- service and maintain system components
- work effectively with customers, workmates and other site visitors
- work in an environmentally friendly manner
- promote the products and services of the plumbing business.

A plumber also needs to be able to read and interpret drawings, specifications and manufacturers' catalogues – plumbing is not just about unblocking drains!

Find out

What range of plumbing activities does the company you work for provide?

What skills and knowledge do you think a plumber needs in order to be able to carry out the job competently?

Figure 1.1 Essential knowledge

The size of the plumbing industry

The plumbing industry contains approximately 25,000 businesses. About 80 per cent of these are classified as **micro-businesses**. Recent research indicates that the industry employs around 110,000 people.

You will find that smaller firms tend to concentrate on domestic plumbing, covering installation, servicing and maintenance. Some firms specialise in installation only, such as new house build. Larger firms tend to be more multi-disciplinary, providing a plumbing service to their customer, but also covering electrical installation and heating and ventilating in domestic, industrial and commercial premises. Some will also cover industrial and commercial gas work.

Plumbing has always been a mobile industry, moving from one site to another, and the trend these days is more and more towards working further afield, particularly if you work for a larger firm.

The structure of a business will vary depending on the size of the firm. In a micro-business, the owner/manager or sole trader will be responsible for every aspect of running the business, such as getting the work, pricing the job, ordering the materials, organising the job, doing the job and customer care, as well as managing the finances of the business. In a larger firm, some of the responsibilities of running the business are shared among the management team and technical support staff such as designers and estimators.

Key Organisations in the Plumbing Industry

SummitSkills

SummitSkills is the recognised Sector Skills Council for the building services engineering sector comprising the plumbing, electrotechnical, heating and ventilating, refrigeration and air conditioning industries.

Summit SKILLS

SummitSkills looks after the training and development needs of the sector, as well as the National Occupational Standards that are used to develop plumbing qualifications. The board of SummitSkills is made up from senior representatives from industry. It includes employer representatives from trade associations such as the Association of Plumbing and Heating Contractors (APHC), the Heating and Ventilating Contractors Association (HVCA) and the Electrical Contractors Association (ECA), as well as trade union representatives from Unite and other industry sectors.

Association of Plumbing and Heating Contractors

The principal trade association for the plumbing industry is the Association of Plumbing and Heating Contractors (APHC). As a trade association the organisation provides a number of services to its employer members including technical support, legal advice, representation to Government on industry-related matters, regular updates on the latest industry developments through its own magazine, and raising the profile of the membership to potential customers, among other services. The organisation's membership is made up from businesses ranging from sole traders to larger organisations, including several plcs.

Unite

The principal trade union for the plumbing sector is Unite. Unite represents the interests of the employees of businesses in terms of such issues as pay and conditions, health and safety, and advice and guidance on employment issues. Unite is the largest union in the United Kingdom, with over 2 million members in the public and private sectors.

JTL

JTL is the leading training provider of apprenticeship services to the building services engineering sector in England and Wales (electrotechnical, plumbing and heating and ventilating). Apprenticeships are managed and administered by a managing agent. There are a number of these in the UK, but at this time the leading training provider in plumbing is JTL.

JTL is one of the top three training providers in the country. It was jointly established by the Electrical Contractors Association (ECA) and the union Unite in 1990 and it arranges and manages the training and funding for the apprentices by liaising with them and their employers. The company operates across a number of regions across England and Wales. Through its network of training officers, JTL monitors and assesses the progress of its apprentices and gives guidance and advice throughout the apprenticeship.

The role of a managing agent is to find suitable young people for apprenticeship employment with plumbing contractors; they will process all the applications for the apprenticeship scheme and conduct selection tests at various locations. Agents produce careers brochures that they send out to careers advisers. Agents also distribute information to employers, explaining their apprenticeship schemes and encouraging them to take on an apprentice.

The managing agent then oversees the arrangements in colleges or training centres that apprentices attend for the off-the-job learning part of their apprenticeship. JTL, for example, have training officers who review and monitor the progress of apprentices in the workplace and at college, and who are technically qualified to undertake the NVQ assessments in the workplace.

The Chartered Institute of Plumbing and Heating Engineering

The Chartered Institute of Plumbing and Heating Engineering (CIPHE) is the professional body for plumbers. The institute is a registered charity with the prime objective of improving the science, practice and engineering principles of plumbing in the public interest. The Institute provides a professional home for qualified owner managers, sole proprietors, journeymen, designers, lecturers, inspectors and consultants. All corporate members (Fellows, Members and Associates) agree to abide by a five point Code of Professional Standards, which requires them to act professionally, competently and responsibly. A Professional Standards Committee meets monthly to consider possible contraventions of the Code.

Did you know?

The CIPHE has a student category of membership specifically for apprentices and trainees

Joint Industry Board for Plumbing and Mechanical Engineering Services

The Joint Industry Board for Plumbing and Mechanical Engineering Services (JIB for PMES) is the registration body for the plumbing industry in England and Wales. It is affiliated to the Construction Skills Certification Scheme (CSCS) registration scheme. The JIB-PMES comprises the APHC and the Construction Confederation, representing plumbing employers, and Unite, the trade union representing employees.

The JIB-PMES is owned and administered by employer (APHC and Construction Confederation) and operative (Unite) representatives.

The JIB-PMES has three main functions:

1. The administration of holiday credit/benefit schemes for participant employers and operatives.

2. The setting of terms and conditions of employment of operatives working for participant employers.

3. As an affiliate of the CSCS registration scheme, registering and providing cards to operatives employed in plumbing and mechanical engineering services and related sectors.

Did you know?

The CSCS scheme is a competence card scheme recognising a basic level of site safety awareness on construction sites

Council of Registered Gas Installers

The Council of Registered Gas Installers (CORGI) was founded in 1970 to protect the public from the dangers of unsafe gas installations. CORGI is the sole body approved by the Health and Safety Executive (HSE) to maintain an up-to-date register of qualified gas installers. All businesses undertaking gas work must be registered with CORGI by law. CORGI also sets out to promote the core values of professional service and high standards in safety. The organisation provides member companies with key services such as technical advice and periodic updating via its monthly magazine.

City & Guilds

City & Guilds offers learners over 500 qualifications in 28 industry areas so that they can learn skills that equip them to fulfil their career ambitions or enrich their leisure time. City & Guilds qualifications are valued by employers because they are developed in conjunction with key industry bodies – so they are always relevant and up to date. City & Guilds awards both the technical certificate and NVQ programmes in plumbing.

EMTA Awards Ltd

EMTA Awards Ltd (EAL) is a leading UK awarding body for NVQs and their Scottish equivalents (SVQs).

EMTA Awards Limited
Part of the **SEMTA** *Group*

Established with the sole purpose of helping industry invest in its future through vocational qualifications, and as a sister company to SEMTA (Sector Skills Council for Science, Engineering and Manufacturing Technologies) EAL has expertise in meeting the qualification needs of the engineering and technology sector. Building on its reputation for high quality awards, it has broadened its range of NVQs and SVQs, enabling people in many other sectors and occupational areas to benefit from its experience.

EAL currently awards qualifications in the NVQ programme in plumbing.

Did you know

All operatives working on gas systems have to demonstrate correct applications of their ACS competencies through regular work assessments by CORGI

Knowledge check 1

1 Is the fitting of a new bathroom suite normally regarded as a maintenance or an installation activity?

2 Approximately what percentage of businesses in the UK plumbing industry employ between one and four operatives?

3 Which organisation maintains a statutory register of gas installation businesses?

4 Which organisation is the employers' trade association for the plumbing industry in England and Wales?

5 Which organisation determines the pay and conditions of plumbing operatives in the plumbing industry in England and Wales?

Plumbing training for new entrants

New entrant training is loosely defined as the training of operatives who have little or no plumbing and heating background training/skills at the point of commencing a plumbing qualification. This is distinct from an existing worker in the industry.

New entrants tend to fall into two categories:

- Those who are eligible to participate in the Government's apprenticeship scheme (typically this will be those in the 16–25 year age bracket), and

- Those whose eligibility falls outside the apprenticeship scheme (typically the over-25s).

A brief overview of the qualification levels

Before we consider the various new entrant schemes it is worth looking at the plumbing and heating qualification levels, as they will be referred to at length throughout this section.

- **Level 2** – a Level 2 qualification is for a craft operative who is to work on-site under minimum supervision on a range of relatively straightforward work activities (this may be someone who works in a team but takes direction as to the work to be carried out).

- **Level 3** – a Level 3 qualification is for a craft operative who is to work on-site under their own initiative on a range of straightforward and complex work activities. Supervision of the work of others may be a part of the job role.

Remember

It is a Level 3 qualification that you'll require to be fully qualified in plumbing, to take responsibility for your own work or possibly become self-employed

Plumbing apprenticeship schemes

The plumbing apprenticeship schemes are Government recognised and funded. They consist of:

- apprenticeship schemes taking the apprentice through to completion of a Level 2 Plumbing Qualification

- advanced apprenticeship schemes taking the apprentice through to completion of a Level 3 Plumbing Qualification.

Key features of the apprenticeship schemes:

- Apprentices entering the scheme usually have employed status (employed by the company).

- Specific entry requirements must be met:
  - minimum 4 GCSEs at Grade D or above in mathematics, English, science and craft, design and technology
  - comprehensive application form completed
  - an initial assessment undertaken (to assess the prospective apprentice's ability to complete the programme)
  - an assessment for colour deficiency if progressing to Level 3, as electrical work on central-heating controls is undertaken.

- A training agreement is established between the apprentice, employer and the funding agency. The funding agency in England is the Learning and Skills Council (LSC) and in Wales it is the Education and Learning Wales (ELWa).

- An individual learning plan must be developed for the apprentice detailing the key learning targets across phases of the apprenticeship programme that the apprentice must aim to complete.

- The apprentice must receive a comprehensive induction to the job both at work and at college.

Apprenticeship outcomes

To successfully complete an apprenticeship in plumbing the apprentice must achieve the following:

Apprenticeship programme

- NVQ Level 2 qualification (offered by EAL or City & Guilds)

- Level 2 Certificate in Basic Plumbing Studies (offered by City & Guilds)

- Key skills units in Communication (Level 2) and Numeracy (Level 2)

Advanced apprenticeship programme

- NVQ Level 3 qualification (offered by EAL or City & Guilds)

- Level 3 Certificate in Domestic Plumbing Studies (offered by City & Guilds)

- Key skills units in Communication (Level 2) and Numeracy (Level 2)

About key skills

The apprenticeship schemes require you to obtain a key skills certificate in Communication and Application of Number at Level 2. Achieving the key skills units involves developing a portfolio of evidence as you work through a topic, and completing a final test. You can be exempt from the end test if you have certain GCSE results. You will normally be required to achieve key skills certification early in the plumbing programme. The key skills units are designed to ensure that you have the adequate skills to communicate with people and can carry out essential calculations.

An overview of plumbing qualifications

The responsibility for the development, content and structure of plumbing qualifications is that of SummitSkills, the recognised Sector Skills Council (SSC) for the building services engineering sector.

Plumbing qualifications are then offered by EAL and City & Guilds across England and Wales. It is these two organisations that are responsible for developing the assessment tests and processes which candidates must complete, as well as the approval of colleges/training centres to offer the range of plumbing qualifications.

The qualifications in the plumbing industry are produced from a set of National Occupational Standards (NOS). National Occupational Standards are a description of what a competent plumber needs to do when carrying out work on plumbing systems, including:

- planning to install or maintain systems
- installing systems
- commissioning and decommissioning systems
- maintaining systems.

The operation of qualifications across all industries is now fundamentally different from what it used to be 20 years ago. A new entrant is essentially required to complete two distinct qualifications at each qualification level:

- a technical certificate (currently offered by City & Guilds) which is the college taught element of the programme.
- NVQ (currently offered by EAL or City & Guilds) which is a programme of assessment activities carried out in the actual workplace.

To be recognised as a competent plumber in the industry you must successfully complete either the NVQ Level 2 or Level 3 qualification. Completion of the technical certificates alone is not recognised as a completed plumbing qualification.

The aims of the two qualifications which are run in tandem is to provide a detailed national standard for the basic learning carried out in the college or centre by the new entrant, and then confirm that he/she has the necessary skills (and experience) to put what has been learned into practice in the workplace through the NVQ programme.

Key content of the plumbing qualifications

Plumbing Technical Certificate – Level 2 Certificate in Basic Plumbing Studies

Job knowledge

Candidates on the Level 2 programme are required to complete the following knowledge units:

- Safety in Plumbing Activities
- Key Plumbing Principles
- Common Plumbing Processes
- Cold-Water Systems
- Domestic Hot-Water Systems
- Sanitation Systems
- Central-Heating Systems (Pipework Only)
- Electrical Supply and Earth Continuity
- Sheet Lead Weathering
- Environmental Awareness in Plumbing
- Effective Working Relationships in the Plumbing Industry.

As a completion to each unit the candidate must sit (and successfully pass) a knowledge assessment test before moving on to the next unit. In addition, all apprentices on an apprenticeship programme must complete a unit covering Employment Rights and Responsibilities in the Workplace.

Practical

Practical work carried out in the college or centre is based on training being provided, leading to the completion of a range of basic plumbing skills tests devised by the plumbing industry, including pipework installation to exacting dimensional tolerances.

The range of Level 2 practical tasks which must be successfully completed in the college or centre include:

- health and safety tasks, for example safe use of ladders
- copper pipework tasks – cutting, fixing, bending, jointing and testing
- plastic pipework tasks – cutting, fixing, jointing and testing
- low carbon steel pipework tasks – cutting, fixing, bending, jointing and testing
- sheet lead weathering – abutments and chimney weathering sets
- airing cupboard installation – cistern and cylinder
- bathroom installation – bathroom suite, combi jig and radiators

Remember

Completing a plumbing technical certificate isn't just about completing a range of assessment tests, comprehensive training is required to complete those tests

- decommissioning systems

- routine maintenance of simple components – taps and valves, etc.

Only when all the knowledge tests and practical tasks have been successfully completed is the Level 2 Certificate in Basic Plumbing Studies awarded.

Plumbing Technical Certificate – Level 3 Certificate in Domestic Plumbing Studies

Before progressing on to Level 3 all the Level 2 knowledge tests and practical tasks must be completed.

Job knowledge

Candidates on the Level 3 programme are required to complete the following knowledge units:

- Systems Planning assignment

- Complex Cold-Water Systems

- Complex Domestic Hot-Water Systems (including Unvented Systems)

- Complex Sanitation Systems

- Central-Heating Systems – (Boilers and Controls)

- Domestic Gas Supply Systems (Natural or LPG)

- Improvement of Business Products and Services in the Plumbing Industry.

Practical

The practical tasks at Level 3 conclude with the installation of a boiler, associated controls, radiators and unvented system in an airing cupboard situation. The range of Level 3 practical tasks which must be successfully completed in the college or centre include:

- unvented systems tasks – installation, decommissioning, maintenance and fault diagnostics on systems

- heating controls – installation and maintenance tasks (including controls wiring and methods of filling and venting)

- gas boiler tasks – installation, decommissioning, maintenance and fault diagnostics on boilers

- gas space heater tasks – installation, decommissioning, maintenance and fault diagnostics on space heaters

- gas cooker tasks – installation, decommissioning, maintenance and fault diagnostics on cookers

- gas water heater tasks – decommissioning, maintenance and fault diagnostics on water heaters

- commissioning above-ground sanitation systems, macerator WCs and boosted showers.

Again, only when all the knowledge tests and practical tasks have been successfully completed is the Level 3 Certificate in Domestic Plumbing Studies awarded.

Plumbing National Vocational Qualifications (NVQs) – Levels 2 and 3

With new entrants, the Plumbing NVQ runs alongside the college based technical certificates at Level 2 or 3. The NVQ tasks conducted on-site are the concluding aspect of the qualifications path and require candidates to prove that they can carry out the tasks effectively in the workplace.

In order to complete an NVQ:

- An assessor from the awarding body approved college, or training centre, must review work being carried out, or that has been previously carried out by the candidate.

- The employing company must appoint a workplace recorder (this is usually the candidate's immediate work supervisor). His/her role is to work with the assessor to review the candidate's work confirming that it is to an acceptable company standard. If the workplace recorder does not believe that the work is to an acceptable standard then the work will have to be repeated and the qualification will not be achieved until such time as it is.

Tasks that the candidate must complete in the workplace include:

Did you know

A candidate is the term used to describe an apprentice or trainee who has registered with the awarding body to undertake a plumbing qualification

Level 2

- health and safety tasks ranging from safe use of ladders to safety of the work location itself
- preparatory work tasks – preparing the work location to accept systems components, for example cutting and lifting floorboards
- installation tasks, a minimum of two of the following tasks to be completed:
  - cold-water system (including a bathroom suite)
  - hot-water system (including a bathroom suite)
  - complete sanitation system
  - central-heating system (excluding fuel supply)
- decommissioning tasks – a minimum of two of the following tasks to be completed:
  - cold-water system (either permanent or temporary)
  - hot-water system (either permanent or temporary)
  - complete sanitation system (either permanent or temporary)
  - central-heating system (either permanent or temporary)
- simple maintenance tasks including component replacement and diagnosing faults

Level 3

- improvement tasks – dealing with basic customer care issues and making recommendations on improvements to customer care

- planning tasks (planning to undertake work and co-ordinating it once it is underway), a minimum of two of the following tasks to be completed:
  - cold-water system (including a bathroom suite)
  - hot-water system (including a bathroom suite)
  - complete sanitation system
  - central-heating system

- commissioning tasks – a minimum of two of the following tasks to be completed:
  - complete cold-water system
  - complete hot-water system
  - complete sanitation
  - complete central-heating system.

In addition, at Level 3 the candidate must successfully complete the following industry tests through a recognised scheme provider:

- unvented hot-water systems

- water supply (water fittings) regulations assessment

- gas ACS scheme (core gas safety and boilers as a minimum).

Only when all the workplace tasks have been completed and the relevant technical certificate has been obtained will the new entrant be qualified as a plumber.

Career opportunities in the plumbing industry

You are probably only just starting out your career as a plumber if you are reading this book, and getting to the end of your craft qualifications is currently your main target. However, your career need not end when you've completed those qualifications. There are many and varied options available, including:

- running your own business

- teaching or training

- working in a large plumbing business – managing staff, tendering for work, etc.

The chart on the following page provides some idea of the career options available together with supporting qualifications.

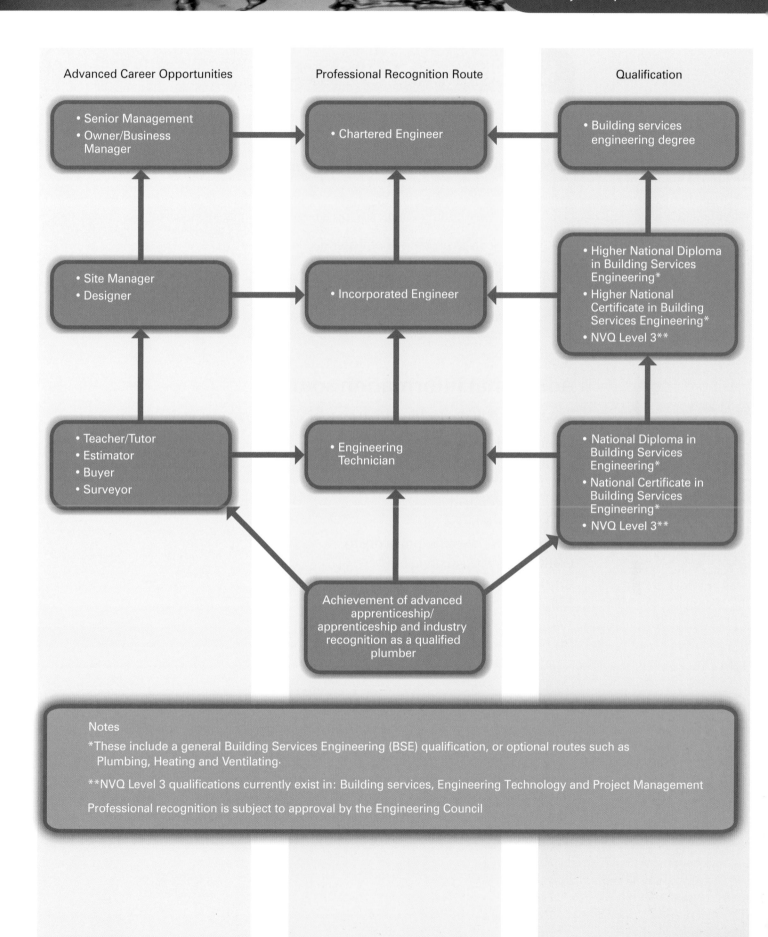

Figure 1.2 Career development paths

Knowledge check 2

1 What is the recommended plumbing qualification for running a small business?

2 Is a plumbing technical certificate completed in the workplace or in a training centre?

3 What three components make up an advanced apprenticeship in plumbing?

4 Is work on an unvented hot-water system included in Level 2 or Level 3 practical task activities?

5 What type of qualification will a senior manager/designer in a large building services business normally require?

Additional information sources

Sources of additional information on topics in this chapter include:

SummitSkills – www.summitskills.org.uk
APHC – www.aphc.co.uk
UNITE – www.unitetheunion.org.uk
JTL – www.jtlimited.co.uk
IPHE – www.iphe.org.uk
JIB for PMES – www.jib-pmes.org.uk
CORGI – www.trustcorgi.com
City and Guilds – www.city-and-guilds.co.uk
EAL – www.eal.org.uk

chapter 2

Health and Safety

OVERVIEW

Health and Safety forms a very important part of your everyday working life. This chapter provides the knowledge and learning to meet the requirements of Level 2 – Health and Safety – a unit required to complete the City & Guilds Certificate: Domestic Plumbing Studies.

This chapter will include sections on the following topics:

- Personal safety responsibilities and the safety of others
- Regulations
- Risk assessments
- Personal safety
- Access to work
- Fire safety and emergencies
- Electric shock and first aid
- Control of substances hazardous to health (COSHH)
- Safe use of LPG and lead working

Health and Safety is a very important aspect of your job. There are a number of legal Health and Safety requirements that must be met by you and your employer. These have been put in place to keep you, your fellow workers and all site visitors free from accidents. Having a good grasp of Health and Safety requirements will also help you to be aware of the potential hazards when working on-site and in the plumbing workshop.

Every year in the construction industry tens of thousands of building-site accidents are reported to the Health and Safety Executive (HSE), some of which will be fatal. These figures are not intended to put you off your new career but to make sure you understand that accidents do happen if you do not follow Health and Safety guidance.

The main body that establishes and implements the Health and Safety laws is the Health and Safety Executive (HSE).

During this chapter, you will learn more about the regulations designed to keep you safe at work. You will also learn about the various hazards you may face, safe methods of working and actions to be taken in the event of accidents.

Personal safety responsibilities and the safety of others

Health and Safety responsibilities

The rules that govern your own and your employer's Health and Safety responsibilities come directly from the Health and Safety at Work Act (HASAWA) 1974. This crucial piece of legislation will be referred to frequently throughout this chapter. Under the HASAWA, you, as an employee, can be prosecuted for breaking safety laws. You are legally bound to cooperate with your employer to ensure your company complies with the requirements of the Act. It is the legal responsibility of all workers to take reasonable care of their own Health and Safety and to ensure that they act in a responsible manner so as not to endanger other workers or members of the public. Some of the most important aspects of the HASAWA are detailed below.

- In domestic premises, the occupier is normally covered by insurance (Occupier's Liability) for visitors such as friends. When you are working on domestic premises, however, you as the plumber become responsible for the health, safety and welfare of the occupant, because it is your place of work.

- It is an offence under the Act to misuse or interfere with equipment provided for your health and safety or that of others. Employers provide you with safety equipment and should instruct you in the use of it. Your employer also has a duty to ensure that the safety equipment provided is kept in good condition, but they cannot cater for human error or negligence. You should always use Health and Safety equipment correctly, for the purpose it is provided for.

- Substances such as oil, grease, cutting compounds, paints and solvents are hazardous if spilled on the floor. Items such as off-cuts of pipe, cables, tools and even food are also dangerous if left underfoot. You have a responsibility both to yourself and to others to keep the workplace hazard free.

What do you need to do?

You should be beginning to get an idea of some of your Health and Safety responsibilities. Your precise legal obligations are listed below. The information may seem obvious, but it is very important that you stick to these obligations throughout your apprenticeship and working career.

Your legal responsibilities

1 Take reasonable care at work of your own Health and Safety and that of others who may be affected by what you do or do not do.

2 Do not intentionally or recklessly interfere with or misuse anything provided for your Health and Safety.

3 Cooperate with your employer on Health and Safety matters. Assist your employer in meeting their statutory obligations.

4 Bring to your employer's attention any situation you think presents a serious and imminent danger.

5 Bring to your employer's attention any weakness you might spot in their Health and Safety arrangements.

What does your employer need to do?

Your employer also has important Health and Safety responsibilities. You should know what your employer's legal responsibilities are for keeping you safe at work. The HASAWA requires your employer to ensure, so far as is reasonably practicable, your health, safety and welfare at work. The matters to which this duty extends include:

1 the provision and maintenance of plant and systems of work that are safe and without risk to health

2 safety in the use, handling, storage and transport of articles and substances

3 the provision of information, instruction, training and supervision as necessary to ensure the Health and Safety at work of employees

4 the provision of access to and exit from the workplace that is safe and without risk

5 the provision of adequate facilities and arrangements for welfare at work.

The HASAWA also says that employers have other duties towards employees in companies that employ five or more members of staff. Such companies must have a Health and Safety policy statement. These will be examined in more detail during the section on Regulations.

Ensuring personal safety and the safety of others
Personal hygiene

It is in your own interests to keep yourself as hygienic as possible while at work, but this may be hard at times, given the nature of some plumbing jobs! To ensure good personal hygiene you should:

- keep your overalls as clean as possible and ensure they are washed regularly
- wash your hands thoroughly before contact with food
- avoid washing with solvents (e.g. white spirits), which can cause dermatitis.

Get into the habit of using barrier cream before starting a job. Barrier cream fills the pores of the skin with a water-soluble antiseptic cream, so that when you wash your hands the dirt and germs are removed with the cream. This should be reapplied each time you wash your hands.

Accident prevention

The best way to reduce the risk of accidents is to try to remove the cause. It is important that, wherever possible, the workplace has clearly defined passageways, good lighting, ventilation, reduced noise levels and non-slip floorings.

Another potential hazard is removed by locking dangerous substances in approved locations so that people will not be tempted to interfere with them, thus causing danger to themselves and others.

Storing materials correctly will also reduce the risk of accidents. Lengths of tube and pipe should be stored horizontally to minimise the risk of them falling.

Hazards that cannot be removed or reduced can be guarded against. For example:

- safety guards and fences can be put on or around machines
- safe systems of work can be introduced
- wearing safety goggles, safety helmets and safety shoes can be made standard practice
- other protective clothing can be supplied for your personal protection such as ear defenders, respirators, eye protection and overalls.

Personal attitudes to safety training

It is essential that you adopt a positive attitude towards safety training, not only for your own welfare but also to ensure the safety of your workmates and the general

Remember

Your employer must provide protective clothing, free of charge, when the work process requires its use public. To do this, you must make a positive decision to act and work in a careful and responsible manner and to put into practice what you have learned. You must make yourself aware of your company's Health and Safety policies/procedures and act accordingly. You need to know what dangers can occur, what protection is available, how to use it and how to prevent accidents.

Personal habits such as drinking alcohol and drug abuse can render a worker a hazard not only to himself but also to others. It stands to reason that evidence of alcohol and drug abuse will not indicate a positive personal attitude towards safety training.

Regulations

The Health and Safety at Work Act (HASAWA) 1974

This Act was a milestone in bringing in laws dealing with the health, safety and welfare of persons at work. Duties are placed on all persons connected with Health and Safety at work, whether as employers, employees, self-employed workers, manufacturers or suppliers of plant and materials. Protection is also given to members of the public affected by the activities of persons at work.

In addition to setting out the basic Health and Safety requirements of employers and employees, the HASAWA also gives more detailed guidance to employers on aspects such as Health and Safety policy statements, which are required for companies employing five or more members of staff.

Additional HASAWA employer requirements

The Health and Safety policy statement must be brought to the attention of all members of staff; arrangements for ensuring this must be in place. Under the HASAWA employers also have a duty to:

- carry out an assessment of risks associated with all the company's work activities
- identify and implement control measures
- inform employees of the risks and control measures
- periodically review the assessments
- record the assessment if more than five persons are employed.

Employers must be prepared to consult a safety representative, if one is appointed by a recognised trade union, about matters affecting their employees' health and safety. Employers must also (if requested in writing by any two safety representatives) establish a safety committee within three months of the request being made.

Construction (Design and Management) Regulations

The Construction (Design and Management) (CDM) Regulations are the main set of Health and Safety regulations which apply to safety on construction sites. The regulations require that notification is provided to the HSE if a 'construction project' is likely to:

- last longer than 30 days or
- involve more than 500 person days of construction work.

The Regulations place responsibilities on 'dutyholders', including:

- Clients – anyone having construction or building work carried out as part of their business.
- CDM co-ordinators – have to be appointed to advise the client on projects that last more than 30 days or involve 500 person days of construction work. The CDM co-ordinator's role is to advise the client on Health and Safety issues during the design and planning phases of construction work.
- Designers – those who, as part of their work, prepare design drawings, specifications, bills of quantities and the specification of articles and substances. This could include architects, engineers and quantity surveyors.
- Principal contractors – have to be appointed for projects which last more than 30 days or involve 500 person days of construction work. The principal contractor's role is to plan, manage and co-ordinate Health and Safety while construction work is being undertaken. The principal contractor is usually the main, or managing contractor, for the work.
- Contractors – a business involved in construction, alteration, maintenance or demolition work.
- Workers – anyone who carries out work during the construction, alteration, maintenance or demolition of a building or structure. A worker could be, for example, a plumber, electrician, scaffolder, painter, etc.

The main aim of the Regulations is to ensure that all parties involved in a construction project work together in a manner that will ensure that the work is properly co-ordinated, so providing a safe environment for the work carried out.

The CDM Regulations 2007 have now incorporated a number of previously separate regulations, in particular the Construction (Health, Safety and Welfare) Regulations, which have now been superseded. Other areas covered under CDM include:

- safe places of work
- good order and site security
- stability of structures
- demolition or dismantling
- explosives
- excavations

- cofferdams and caissons
- reports of inspections
- energy distribution installations
- prevention of drowning
- traffic routes
- vehicles
- prevention of risk from fire, etc.
- emergency procedures
- emergency routes and exits
- fire detection and fire-fighting
- fresh air
- temperature and weather protection
- lighting
- welfare facilities – sanitary and washing facilities, drinking water, changing/rest facilities

Reporting of Injuries, Diseases and Dangerous Occurrences Regulations (RIDDOR)

The Reporting of Injuries, Diseases and Dangerous Occurrences Regulations (RIDDOR) lay down the requirements for the employer reporting the following to the Health and Safety Executive:

- Injuries – fatalities (including members of the public) or injuries resulting in three days off work
- Diseases – if a doctor advises that an employee is suffering from a work-related disease listed under RIDDOR
- Dangerous occurrence – something that happened that could have resulted in a reportable injury, for example the collapse of an excavation.

Details of reportable accidents should be forwarded by the employer to the HSE on form F2508. If a fatality is being reported then this must take place within 10 days.

The HSE, based on the report form forwarded, will make a decision on the level of investigation and subsequent action required which may include legal action for breaches of health and safety regulations.

A firm should maintain an **accident book** for recording detail of all types of injury that occur while carrying out work, however minor. The accident book is used by the employing firm to monitor the level of accidents that occur and establish whether any additional safety controls are required to do the work.

Definition

Accident book – a document where detail of all accidents is recorded no matter how minor

If an accident occurs (whether it is RIDDOR reportable or not), an employee may be required to complete an accident report form. This should preferably be completed by the injured employee as close to the accident occurring as possible. If completed by a third party then the employee must carefully check that the details are a clear identification of the events that took place. Once signed an accident report could be used in legal proceedings!

Fire Precautions Act

This Act applies to a range of commercial properties including factories, shops and offices. Employer duties include safety in relation to fire hazards, both from the work processes and activities carried out, as well as general fire safety in the workplace. Employers must carry out a fire risk assessment. The risk assessment is to enable employers to identify and then take steps to eliminate, reduce or control safety risks (including risks from fire) to make sure that no one gets hurt or becomes ill. Where more than 5 people are employed at a property then the property will require a written fire risk assessment.

General fire precautions that an employer must provide include:

- means of detecting and giving warning in case of fire
- escape routes
- fire-fighting equipment
- training of staff in fire safety.

Health and Safety (Safety Signs and Signals) Regulations

The Health and Safety (Safety Signs and Signals) Regulations (1996) provide a standardised approach for displaying safety signs in the workplace. They require employers to provide specific safety signs whenever there is a risk that has not been avoided or controlled by other means, for example by engineering controls or safe systems of work. Where a safety sign would not help to reduce that risk, or where the risk is not significant, there is no need to provide a sign.

- They require, where necessary, the use of road traffic signs within workplaces to regulate road traffic.
- They also require employers to:
  - maintain the safety signs which are provided by them
  - explain unfamiliar signs to their employees and tell them what they need to do when they see a safety sign.

The Regulations apply to all places and activities where people are employed, but exclude signs and labels used in connection with the supply of substances, products and equipment or the transport of dangerous goods.

The Electricity at Work Regulations (1989)

The Electricity at Work Regulations, which came into force on 1 April 1990, lay down general Health and Safety requirements regarding electricity at work for employers, self-employed persons and employees.

The Regulations are made under the HASAWA 1974 and, broadly speaking, impose a duty upon every employer and self-employed person to comply with the provisions of the Regulations. They also impose a duty on every employee to cooperate with their employer so far as is necessary to enable the Regulations to be complied with.

As a plumber, you will need to be especially careful when dealing with electricity. The Electricity at Work Regulations provide the full safety standards requirement; more will be said about the practical use of electricity on site in a later section.

The Regulations refer to a person as a 'duty holder' in respect of systems, equipment and conductors. These Regulations are therefore statutory, and consequently penalties can be imposed on those people found guilty of malpractice or misconduct.

Definitions relevant to persons who have responsibilities under these Regulations

1 Employer

For the purpose of the Regulations, an employer is any person or company that (a) employs one or more individuals under a contract of employment or apprenticeship, or (b) provides training under the schemes to which the HASAWA applies.

2 Self-employed

A self-employed person is an individual who works for gain or reward other than under contract of employment, whether or not he or she employs others.

3 Employee

Regulation 3(2) (b) reiterates the duty placed on employees by the HASAWA.

This clause places duties on employees equivalent to those placed on employers and self-employed persons where these matters are within their control. This will include apprentices – like you – who will be considered as employees under these Regulations.

Remember

Can you remember your responsibilities under the HASAWA?

This arrangement recognises the level of responsibility that many employees in the plumbing trade are expected to take on as part of their job. The 'control' that they exercise over the electrical safety in any particular circumstances will determine to what extent they hold responsibilities under the Regulations to ensure that the Regulations are complied with.

A person may find that they are responsible for causing danger to arise elsewhere in an electrical system, at a point beyond their own area of work. This situation may arise, for example, if you are working on a circuit while somebody else is working in a different room on that same circuit. This is obviously a dangerous situation. Because such circumstances are 'within [your] control', the effect of Regulation 3 is to bring responsibilities for compliance with the rest of the Regulations to you, thus making you a duty holder.

4   Absolute/reasonably practicable

Duties in some of the Regulations have a qualifying term: 'reasonably practicable'. Where qualifying terms are absent, the requirement in the Regulation is said to be absolute. The meaning of 'reasonably practicable' has been well established in law. The interpretations below are given only as a guide.

5   Absolute

If the requirement in a Regulation is 'absolute' – for example, if the requirement is not qualified by the words 'so far as is reasonably practicable' – the requirement must be met regardless of cost or any other consideration.

6   Reasonably practicable

Someone who is required to do something 'so far as is reasonably practicable' must think about the amount of risk of a particular work activity or site and, on the other hand, the costs in terms of the physical difficulty, time, trouble and expense which would be involved in taking steps to reduce the risks to Health and Safety of a particular work process. For example, in your own home you would expect to find a fireguard in front of a fire to prevent young children from touching the fire and being injured. This is a cheap and effective way of preventing accidents and would be a reasonably practicable situation.

If the cost or technical difficulties of taking certain steps to prevent those risks are very high, it might not be reasonably practicable to take those steps.

In the context of the Regulations, where the risk is often that of death from electrocution, and where the necessary precautions are often very simple and cheap – for example, insulating surrounding cables – the level of duty to prevent that danger approaches that of an absolute duty.

The Provision and Use of Work Equipment Regulations

The Provision and Use of Work Equipment Regulations (PUWER) require risks to a person's health and safety, from equipment that they use at work, to be prevented or controlled. Work equipment includes all tools (power or hand) and equipment such as threading machines and bending machines. In general terms, the Regulations require that equipment provided for use at work is:

- suitable for the intended use

- safe for use, maintained in a safe condition and, in certain circumstances, inspected to ensure this remains the case

Remember

The safe condition of hand tools such as hammers and spanners is covered by PUWER

- used only by people who have received adequate information, instruction and training
- accompanied by suitable safety measures, for example protective devices, markings, warnings.

Manual Handling Operations Regulations

The Manual Handling Operations Regulations lay down requirements for moving loads by hand, which could involve pushing, pulling, lowering the load etc. The following control measures are identified in the Regulations.

- Avoid hazardous manual handling operations so far as is reasonably practicable, for example by using a mechanical lifting aid such as a sack trolley to move the load.

- Make a suitable and sufficient risk assessment of any hazardous manual handling operations that cannot be avoided.

- Reduce the risk of injury from those operations so far as is reasonably practicable, providing such controls as training staff in safe or kinetic lifting techniques.

A risk assessment for manual handling will include taking into account the following five factors:

- the task
- the load
- the working environment
- individual capability (of the person carrying out the lifting)
- other factors, for example use of protective clothing.

Did you know?

The use of abrasive wheels is now covered by PUWER – all staff using such equipment need to be properly trained, informed and supervised

Did you know?

PUWER has laid down new requirements for the use of cartridge fixing devices based on a risk assessment of the work and proper training of staff – any age restriction detailed in previous legislation has been removed

A sack trolley is an example of a mechanical lifting device

Personal Protective Equipment at Work Regulations

Personal protective equipment (PPE) is defined in the Personal Protective Equipment at Work Regulations as all equipment (including clothing affording protection against the weather) which is intended to be worn or held by a person at work and which protects him against one or more risks to health or safety, for example safety helmets, gloves, eye protection, high visibility clothing, safety footwear and safety harnesses. The provision of respiratory and hearing protection is the subject of separate legislation.

The main requirement of the PPE at Work Regulations is that personal protective equipment is to be supplied and used at work wherever there are risks to health and safety that cannot be adequately controlled in other ways. The Regulations also require that PPE:

- is properly assessed before use to ensure it is suitable
- is maintained and stored properly
- is provided with instructions on how to use it safely
- is used correctly by employees.

Under the Regulations, PPE must be supplied by the employer on a free of charge basis to the employee.

Lifting Operations and Lifting Equipment Regulations

The Lifting Operations and Lifting Equipment Regulations aim to reduce risks to people's health and safety from lifting equipment (such as cranes, pulleys, ropes, slings etc.) provided for use at work. Generally, the Regulations require that lifting equipment provided for use at work is:

- strong and stable enough for the particular use and marked to indicate safe working loads
- positioned and installed to minimise any risks
- used safely, i.e. the work is planned, organised and performed by competent people
- subject to an ongoing thorough examination and, where appropriate, inspection by competent people.

Work at Height Regulations

The Work at Height Regulations apply to all work at height where there is a risk of a fall liable to cause personal injury. The Regulations place duties on employers, the self-employed, and any person who controls the work of others. The Regulations require duty holders to ensure:

- all work at height is properly planned and organised
- all work at height takes account of weather conditions that could endanger health and safety
- those involved in work at height are trained and competent
- the place where work at height is done is safe
- equipment for work at height is appropriately inspected
- the risks from fragile surfaces are properly controlled
- the risks from falling objects are properly controlled.

Control of Substances Hazardous to Health Regulations

The Control of Substances Hazardous to Health Regulations (COSHH) require employers to control exposure to hazardous substances to prevent ill health. Employers have to protect employees, and others who may be exposed, by complying with COSHH. To comply with COSHH an employer must:

- assess the risks
- decide what precautions are needed
- prevent or adequately control exposure
- ensure that control measures are used and maintained
- monitor the level of exposure to the substance
- carry out appropriate health surveillance where required
- prepare procedures to deal with accidents and emergencies
- ensure employees are adequately informed, trained and supervised.

Control of Asbestos at Work Regulations

Key requirements of the Control of Asbestos at Work Regulations are:

- Asbestos insulating materials or linings of white, blue and brown asbestos must only be removed/worked on by licensed contractors.
- On identifying high risk white, blue or brown asbestos materials, work must be stopped in order that a risk assessment may be conducted in relation to working with/or working in the proximity of the asbestos material.
- Work with asbestos material must be the subject of a risk assessment with appropriate control measures provided, for example on-site washing facilities.
- Workers who may encounter asbestos in the workplace (including plumbers) are required to attend mandatory asbestos awareness training sessions.
- Asbestos materials must be properly disposed of in approved packaging (preventing the spread of any fibres) and asbestos materials must be properly disposed of at approved waste sites.
- Operatives undertaking work in the proximity of asbestos or in the removal of asbestos materials must be properly informed, instructed and trained.

Work with asbestos cement-based materials does not present as high a risk. This may be carried out by non-licensed contractors, provided a risk assessment is undertaken, that effective control measures are used to prevent the spread of any material and that workers are protected, for example by the use of effective protective clothing and respiratory protective equipment. The HSE website at www. hse.gov.uk/asbestos includes a series of useful guidance notes on safety measures associated with the work where contact with asbestos cement-based materials may occur.

Did you know?

Hazardous substances can be in the form of solids, liquids, dusts, fumes, vapours, gases or micro-organisms that may present a risk to health

Did you know?

There are three types of asbestos, commonly known as chrysotile, amosite and crocidolite (white, blue and brown respectively)

Remember

Breathing in asbestos dust can lead to chronic (long-term) illness fatal diseases such as mesothelioma (cancer of the inner chest lining), lung cancer or asbestosis (scarring of the lung)

Did you know?

Asbestos is more likely to be found in properties built before 1980

Control of Lead at Work Regulations

The Control of Lead at Work Regulations requires the following to be put in place:

- Duties - employers and employees alike are responsible for protecting themselves and other persons on premises where leadwork is being carried out or who are likely to be exposed to lead from that work.

- Training - operatives should be provided with proper information, instruction and training on the safe use of lead.

- Assessment of risk of exposure - if there is a risk of exposure to lead in a form which may be inhaled or ingested, the level of risk must be assessed and the findings properly recorded. Correct working practices and proper controls should be established and appropriate protective and safety equipment provided before the work proceeds.

- Control measures - the employer and employee must ensure, as far as is reasonably practical, that all measures are taken to restrict and control exposure.

- Protective clothing - each employee should be provided with and should wear adequate protective clothing.

- Respiratory equipment and ventilation - suitable masks or respiratory equipment must be provided and used where there is a risk of exposure to airborne lead dust or fumes.

- Washing facilities and canteen areas - adequate washing and changing facilities should be provided. Washing of the hands is essential before eating, drinking or smoking and before leaving the workplace. Food and drink should not be consumed in any area where leadwork is being carried out. Washing of hands and face and changing of contaminated clothes should take place before entering canteen areas.

- Spread of contamination - both employer and employee should take such steps as are reasonably practicable to prevent lead contamination from spreading beyond the workplace or storage area.

- Waste and scrap - old lead sheet being replaced should be removed with care. Scrap and dust should be taken to approved collectors to facilitate recycling. Vehicles transporting scrap and dust should be thoroughly washed and cleaned after use.

- Lifting lead - lead is heavy. Proper lifting equipment and additional staff should be available and always used to ensure the safety of operatives lifting and moving lead sheet.

- Medical surveillance - the blood lead level of operatives regularly working with lead should be monitored at regular intervals.

- Maintenance records - it is the duty of the employer to provide and maintain adequate records which show details of risk assessments, information and training provided, precautionary measures taken, medical surveillance and ventilation and respiratory equipment provided.

Knowledge check 1

1   A company is required to have a written policy statement when employing more than how many operatives?

2   Which is the main piece of health and safety legislation dealing with work on large construction sites?

3   Which item of health and safety legislation lays down controls when using or applying solvents?

4   Under the Control of Lead at Work Regulations what two control measures could be applied to minimise the effect of lead fumes when lead welding indoors for prolonged periods?

5   Who is responsible for providing PPE under the Personal Protective Equipment at Work Regulations?

FAQ

Do I really need to know all these Health and Safety Regulations?

Yes. You have a responsibility to work within the bounds of these Regulations. They are there to ensure your safety and the safety of others.

Risk assessments

Risk assessments are a kind of checking system designed to keep you safe at work. The aim of risk assessments is to make you think about the possible hazards you may face while performing a specific working task. At this stage of your apprenticeship you will not be required to write risk assessments, but you must have a working knowledge of them and understand their purpose.

What is risk assessment?

A risk assessment is nothing more than a careful examination of what, in your work, could cause harm to you or other people. It enables you to weigh up whether you have taken enough precautions or could do more to prevent any harm. The main aim of a risk assessment is to ensure that no one gets hurt or becomes ill.

Your employer needs to decide whether a hazard is significant, and whether it is covered by satisfactory precautions so that the risk is small. Your employer is also legally bound to assess the risks in your workplace. Legislation that governs the requirements of risk assessment is contained within the Management of Health and Safety at Work Regulations.

On the job: Risk assessment

Jane, a new apprentice, was carrying out a risk assessment with the senior plumber. Jane spotted some obvious hazards where a number of nails were sticking out from upturned floorboards. The plumber told Jane not to bother about the nails because no one would be silly enough to tread on them. Jane knew that loads of casual labourers would be clearing the building in the morning.

1 What should Jane do in this situation?

2 Has the senior plumber contravened any Health and Safety legislation by not completing the risk assessment correctly?

3 If so, what legislation has been contravened?

Generic risk assessment

Hazard and risk identification

In order to assess the potential of a risk, processes are usually undertaken to 'rank' hazardous operations. The example that follows is from a model developed by the British Plumbing Employers' Council (BPEC) specifically for the plumbing industry. It is based on a number of factors, which are presented on a generic risk-assessment form. These are used to assess any risk associated with the plumbing industry.

Here is an example of a generic risk-assessment form for working with specialist tools.

TASK	
MANUAL HANDLING OF LOADS SPECIALIST TOOLS	

APPLICATION OF EQUIPMENT	**APPLICATION OF SUBSTANCE**
PIPE-BENDING MACHINES, STILSONS, ROPES, LEAD DRESSERS, BENDING SPRINGS, BLOCK AND TACKLE, SPANNERS ETC.	N/A

ASSOCIATED HAZARDS
RISK OF MUSCLE STRAINS
RISK OF SPRAINS
RISK OF MUSCULO-SKELETAL INJURY

LIKELIHOOD	**CONSEQUENCE**	**RISK FACTOR**
3	3	9

RISK EXPOSURE	**SAFEGUARDS HARDWARE**
EMPLOYEES	NIL

CONTROL MEASURES
1 SPECIFIC TRAINING AND INSTRUCTION TO EMPLOYEES – KINETIC LIFTING
2 INDIVIDUAL ASSESSMENT TO BE PERFORMED FOR ALL TASKS
3 WORKPLACE INSPECTIONS CONDUCTED AT 3 MONTHLY INTERVALS
4 RANDOM SAFETY INSPECTIONS
5 SUITABLE AND SUFFICIENT PERSONAL PROTECTIVE EQUIPMENT
6 MEDICAL SCREENING FOR STAFF AT RISK

Figure 2.1 Risk- assessment form

Most of the form is self-explanatory, but we want to focus here on the following items from the form:

- likelihood of accident occurring

- maximum consequences (severity) of an accident

- risk factor (likelihood of accident occurring × maximum consequences (severity) of an accident)

Likelihood of accident occurring

This is assessed from the following table:

Likelihood	Scale value
No likelihood	0
Very unlikely	1
Unlikely	2
Likely	3
Very likely	4
Certainty	5

Figure 2.2

Maximum consequences of an accident

This is assessed from the following table:

Injury or loss	Scale value
No injury or loss	0
Treated by first aid	1
Up to 3 days off work	2
Over 3 days off work	3
Specified major injury	4
Fatality	5

Figure 2.3

Risk factor

Risk factors are then calculated using a simple formula:

Likelihood × Consequence = Risk

The outcome of the likelihood of an accident occurring and the maximum consequences should it happen, will reveal a risk factor of between 1 and 25.

- a figure between 1 and 6 = minor risk, can be disregarded but closely monitored

- a figure between 8 and 15 = significant risk, requires immediate control measures

- a figure between 16 and 25 = critical risk, activity must cease until risk is reduced.

Other factors contained on the assessment form are discussed below.

Risk exposure

Risk exposure describes the individuals or groups of people that may be affected by the work activity or process. Control measures must take account of all those people.

Safeguards hardware

Safeguards hardware describes the in-built safety features of work equipment. For example, on powered machines this would include machine guards or trip switches. In this particular example safeguards hardware is nil.

Control measures

These describe the additional safeguards that underpin your arrangements. Where these are identified they must be followed through, and a record kept of any outcomes.

The list below shows some of the most common risks you will face during your plumbing career.

Checklist of possible hazardous operations

- Working with electrical-powered plant:
  - power transformers
  - extension cables
  - plugs and sockets
  - portable tools
  - electric arc welders
  - threading machines
  - large hammer drills
  - specialist equipment
  - fixed equipment
- Working with non electrically-powered plant:
  - hand tools
  - specialist tools
  - pneumatic tools
  - hydraulic tools
- Working with non-powered tools:
  - specialist equipment
- Manual handling of loads:
  - specialist tools
  - general lifting

- Working with hazardous substances
- Working in excavations:
  - physical conditions
  - interruption of gas, water or electrical services
- Working with powered industrial trucks:
  - fork-lift trucks
  - dump trucks
  - JCBs/tractors
  - tail-lift vehicles
  - road vehicles
- Working with highly flammable liquids
- Working with liquefied petroleum gases
- Working with lead:
  - manual handling
  - lead hygiene

- Working at heights:
  - ladders
  - scaffolds
  - ropes/harnesses
- Working with demolitions
- Working with electrical installations
- Controlling work with fumes
- Controlling work with noise
- Controlling work with dust
- Working with asbestos
- Working within vessels
- Working in confined spaces
- Clearing of hazardous waste
- Cartridge-fixing devices
- Working on suspended timber floors

Other health and safety systems

Method statements

Firms produce method statements for high-risk activities such as working at height to line a flue system. The method statement draws together the findings and control measures of all the risk assessments associated with the activity such as working at heights, working with flue liner materials, working with mortar mixes, etc., into one simple document. This document is given to employees, co-contractors, etc. to provide guidance on how the work activity is going to be carried out.

Permit to work systems

A permit to work system is used as a safety control for very high-risk or potentially fatal activities. The permit to work is essentially a fixed checklist of tasks that must be completed before an activity is carried out. The permit to work will usually be countersigned by a supervisory level company official. A permit to work system could apply to high-risk activities such as working with gas heating equipment in potential fire risk areas, working on electrical equipment or working in confined spaces such as sewers.

Personal safety

Safety signs

During your plumbing career you will spend time on various building/construction and other work sites. All sites are littered with potential dangers. Any site you enter will have safety signs. Safety signs are designed to keep visitors and workers safe, pass on useful information (regarding exits, fire extinguishers, etc.) and to warn personnel about the possible dangers they may be exposed to during their time on site.

There are four categories of safety sign:

1 mandatory signs
2 prohibition signs
3 warning signs
4 information signs.

> ### Find out
>
> Find out what the common warning, information and prohibition signs found in construction environments are. You will need to remember these. When looking at the signs, try to make a mental note of which group they fit into

	Prohibition signs	**Mandatory signs**	**Warning signs**	**Information or safe condition signs**
Shape:	Circular	Circular	Triangular	Square or rectangular
Colour:	Red borders and cross bar. Black symbols on white background	White symbol on blue background	Yellow background with black border and symbol	White symbols on green background
Meaning:	Shows what must NOT be done	Shows what must be done	Warns of hazard or danger	Indicates or gives information on safety provision
Example:	No smoking	Wear eye protection	Danger electric shock risk	First aid kit

Figure 2.4 Safety signs

General introduction to 'PPE'

The guidelines that govern protective clothing are set out in the Personal Protective Equipment at Work Regulations. You will usually hear colleagues refer to personal protective equipment as 'PPE', which is defined as all equipment designed to be worn or held to protect against a risk to health or safety, and includes most types of protective clothing and equipment such as eye, hand, foot and head protection.

Your employer has a duty to assess and, where possible, minimise the risks you face at work, although some jobs you will be asked to complete will carry a degree of risk. In these cases, PPE will be used as a last resort to keep you as safe as possible. Your employer should ensure:

Did you know?

Eye protection that is used to guard against the threat created by flying objects should include lenses that are of the high impact resistant type

- all PPE is provided free of charge

- PPE is suitable for its purpose

- information on procedures to maintain, clean and replace damaged PPE is freely available

- storage for PPE is provided

- PPE is maintained in an efficient state and is in good repair

- that PPE is properly used.

Eye protection

Your eyes are two of your most precious possessions. You need them for learning and for earning, because without your eyesight you probably could not perform your present job! Your eyes are the parts of your body that are most vulnerable to injury at work.

Every year, thousands of workers suffer eye injuries, which result in pain, discomfort, lost income and even blindness. Following safety procedures correctly and wearing eye protection can prevent these injuries. There are many types of eye protection equipment available – for example:

- safety glasses

- safety goggles

- welding goggles.

Foot protection

Although this refers to items to protect your toes, ankles, and feet from injury, wearing the correct footwear can also protect your whole body from injury, for example from electric shock.

Safety glasses

Safety goggles

Accidents to the feet have serious consequences and can result in pain and suffering, disability and loss of work and income.

Protective footwear can help prevent injury and reduce the severity of the injuries that do occur. The basic universal form of foot protection is the safety shoe, which will usually include some kind of metal toe protection, rubber soles and sturdy leather uppers. The photograph shows typical foot protection.

Safety boots

Hand protection

This involves the protection of two irreplaceable tools – your hands, which you use for almost everything: working, playing, driving, eating, etc. Unfortunately hands are often injured. Almost 1 in 4 work-related injuries happen to hands and fingers.

One of the most common problems – other than cutting, crushing or puncture wounds – is dermatitis. Dermatitis is an inflammation of the skin normally caused by the hands coming into contact with irritating substances.

Gloves used for general work

This skin irritation may be indicated by sores, blisters, redness or dry, cracked skin that can easily become infected.

To protect your hands from irritating substances, you need to keep them clean by regular washing using approved cleaners. Make good use of barrier creams where provided, and wear appropriate personal protection (usually a strong pair of gloves) when required. The type of glove used depends on the work being carried out:

Specialist gloves used for working with chemicals

- prevention of cuts: general purpose gloves
- working with hazardous substances (an example could be the contents of a toilet bowl): rubber-type gloves.

Head protection

Head protection is important because it guards your most vital organ: your brain. A head injury can disable a person for life or

Safety helmet

even be fatal. One study revealed that over 80 per cent of industrial head injuries occurred as a result of people not wearing 'hard hat' protection.

On building sites the risk from falling objects is very real. There are also other potential traps, such as objects (e.g. scaffolding poles) at head height, just waiting for you to walk into them!

On the job: Head protection

Fazal is a first-year apprentice who is starting his first day on a construction site. The plumber in charge gives Fazal a hard hat to wear. Fazal notices that the hat has a crack down one side and reports this to his supervisor. The supervisor tells Fazal that he must wear this hat because there are no more in the store and he needs to get to work.

1   What should Fazal do in this situation?

2   If an object hit Fazal on the head and he was injured because of the crack in the hat, who would be at fault for this injury?

You need to wear your safety helmet whenever it is required, and you should ensure it is worn correctly.

Safety tip

Always wear your safety helmet the correct way round – despite what your workmates might be doing with theirs!

Checklist for safety helmets

✔ Adjust the fit of your safety helmet so that it is comfortable

✔ All straps should be snug but not too tight

✔ Do not wear your helmet tilted or back to front

✔ Never carry anything inside the clearance space of a hard hat, for example cigarettes, playing cards

✔ Never wear an ordinary hat under a safety helmet

✔ Do not paint your safety helmet as this could interfere with electrical protection or soften the shell

✔ Handle it with care: do not throw it or drop it, etc.

✔ Regularly inspect and check the helmet for cracks, dents or signs of wear.

✔ Check the strap for looseness or worn stitching and also check your safety helmet is within its specified 'use-by' date

To keep safe, use your head.

- Know the potential hazards of your job and what protective gear to use.
- Follow safe working procedures.
- Take care of your protective headgear.
- Notify your supervisor of unsafe conditions and equipment.
- Get medical help promptly in the case of head injury.

Ear protection

Earplugs

You may have to wear ear protection when you are carrying out a noisy job, for example using a large hammer drill to bore a hole through an external wall for an overflow pipe. Ear protection ranges from simple earplugs to earmuffs.

Ear defenders

Respiratory protection

There are many types of respiratory protection device: each is used to guard against a particular circumstance. In plumbing, respiratory protective devices are used primarily to guard against the following:

- Breathing in dusts, mists or fibre particles, for example in a loft space where loft insulation particles can cause irritation or during the removal of a solid-fuel fireback boiler where the dust and soot created can require the use of a respirator.

Respirator

- Breathing in dangerous gases, for example in situations where you have to undertake lead welding indoors or where there is no mechanical extraction system. In these situations, a respirator would typically be used.

- Occasionally a plumber may have to enter a sewer system, which may contain harmful sewer gases – this is a dangerous situation where a plumber may be required to wear full breathing apparatus but only after special training.

The consequence of not protecting yourself is the strong likelihood of lung problems in later life which could include potentially fatal conditions such as lung cancer and emphysema.

Dust mask

Tools and equipment

Safety tip

Remember that the filters on respirators require regular replacement.

Full breathing apparatus requires specialist training in its use

You will use a variety of tools and pieces of equipment as a plumber. All of them are potentially dangerous if misused or neglected. Practical instruction in the proper use of tools and equipment will form part of your training on-site and in your training centre. Tools and equipment are covered in greater detail in Chapter 5, *Common plumbing processes.*

Hand tools and manually operated equipment are often misused. You should always use the right tool for the job. Here are some guidelines to follow:

- Never use a hammer on a tool with a wooden handle (e.g. wood chisel) as you may damage the wooden handle and release dangerous splinters. If the handle splits as you hit it, the hammer could slip off and damage your hand.

- Cutting tools, saws, drills, etc. must be kept sharp and in good condition. As a plumbing apprentice, you will frequently be asked to use cutting tools such as hacksaws and wood saws: you should ensure that the blades are always fitted properly and are sharp. Hacksaw teeth should be pointing in the forward direction of cut. After use, guards should be fitted where possible.

- Handles should be properly fitted to tools such as hammers and files, and should be free from splinters. Hammerheads should be secured correctly using metal or wooden wedges. 'Mushroom heading' of chisels is also a dangerous condition which can lead to serious eye injury. Unprotected file tangs present a serious danger of cuts and puncture wounds.

- It is particularly important to check that the plugs and cables of hand-held electrically operated power tools are in good condition. Frayed cables and broken plugs should be replaced. Electric power tools of 110 or 230 volts must be PAT **tested** in accordance with your employer's procedures. It is good practice to check all electrical equipment and test labels to ensure they are in safe working order.

- Other common items of equipment, e.g. barrows, trucks, buckets, ropes and tackle, etc. are all likely to deteriorate with use. If they are damaged or broken they will sooner or later fail in use and may cause an accident. Unserviceable tools and equipment should not be used. They should be repaired or replaced and the unsafe equipment removed from the site.

- You may have to use cartridge-operated tools during your career. If you do, you will be given the necessary instructions on the safe and correct methods for using them. They can be dangerous, especially if they are operated by accident or used as toys. This could cause ricochets, which could lead to a serious injury.

Trips and fire hazards

Tools, equipment and materials left lying about, trailing cables and welding hoses, spilled oil and so on cause people to trip, slip or fall. Clutter and debris, oily rags, paper etc. should be cleared away to prevent fire hazards. As an individual you may have no control over the general state of the work place, but you should see that your own work area is kept clear and tidy, as this is the mark of a skilled and conscientious plumbing tradesperson.

Electricity on site

The safe use of electricity on site is covered by the Electricity at Work Regulations 1989, which in turn are covered by the Health and Safety at Work Act (HASAWA) 1974. These Regulations impose specific duties on employers to put into place measures to protect their employees against death or personal injury from the use of electricity at work.

Safety tip

Never be tempted to just 'make do' with whatever tool you may have to hand

Definition

PAT tests – Checking for safety and keeping maintenance records of all portable electrical equipment to ensure it is in safe working order

Employers are required to have specific codes of practice for their employees. These practices must include keeping maintenance records for all portable equipment (PAT tests). The HSE suggests that portable electrical equipment should be tested every 3 months for construction-site applications. Records must show that the equipment is tested regularly by a competent person using suitable test equipment. You must, however, make a visual inspection of a power tool to establish whether it is safe to use EVERY TIME YOU USE IT!

The supply to a work site may be from a generator or from the local public supply. Care should be taken to site a generator so that noise and fumes are reduced to a minimum. Whatever the source of supply, it must be routed to where it is needed on the site. This may involve electric cables being buried underground or, more often, suspended overhead on poles.

Both methods of distribution present hazards. Sites will often have **residual current devices** (RCDs) in place for added protection. These are intended to 'trip out' the electrical supply if there is any current leakage.

You may work on a site where a site distribution system is needed for the temporary supply of electrical power to the various locations around the site. A number of different units are used for this purpose, all of which should comply with the relevant British Standard.

All the units will be clearly marked with details of the output voltage and the 'Danger–electricity' symbol. The installation of this system must conform to the requirements of the current edition of BS 7671. You should never interfere with or alter any installation. Only an authorised person may carry out any alterations.

Electricity on-site safety checklist

✓ Do not use lighting circuits for power tools

✓ Power tools should be double insulated

✓ Never carry a portable electric tool by its cable

✓ Ensure that equipment is not damaged before you plug it in

✓ Always have enough light for the job

✓ Keep lights clean

✓ Check that all cables are correctly insulated and not damaged or frayed

✓ Check that plugs and sockets are clean and in sound condition

✓ Check for current PAT labels

✓ Check that RCD protection is provided where necessary

Lifting and carrying (manual handling)

The manual handling or lifting of objects is the cause of more injuries on worksites than any other factor. Back strains and associated injuries are the main source of lost hours in the building services industries. Manual handling can involve: pushing, pulling, and the lifting and lowering of loads (tools, cylinders boilers, radiators, etc.).

The movement of loads requires careful planning in order to identify potential hazards before they cause injuries. You should follow the safety precautions and codes of practice at all times.

Before we consider the risks and procedures of manual handling, get into the habit of asking yourself whether the load can be moved another way with less risk of personal injury, for example using a sack trolley or a cart.

Before moving a heavy load, do the following basic risk assessment:

1 **The task – does it involve:** <ul><li>stooping?</li><li>twisting?</li><li>excessive lifting or lowering distances?</li><li>excessive carrying distances?</li><li>excessive pushing or pulling distances?</li><li>frequent or prolonged physical effort?</li><li>the sudden risk of the load moving?</li></ul>	2 **The load – is it:** <ul><li>heavy?</li><li>bulky or unwieldy?</li><li>difficult to grasp?</li><li>unstable or with contents that are likely to shift?</li><li>sharp, hot or otherwise potentially damaging?</li></ul>
3 **The working environment – does it have:** <ul><li>space constraints?</li><li>slippery or unstable floors?</li><li>variation in levels?</li><li>poor lighting?</li><li>hot/cold/humid conditions?</li></ul>	4 **The individual – does he or she have:** <ul><li>any restriction on their physical capability?</li><li>the knowledge and training for manual handling?</li></ul>

These are factors that must be taken into account when carrying out manual handling operations. Take a few moments to consider the information in the above lists and the impact each factor could have on the safe movement of tools and materials.

Extreme care must be taken when lifting or moving heavy or awkward objects manually. The load generally accepted as maximum for a fit person to lift is 20 kg. The rules for correctly lifting a load are shown in detail on the next page.

Safety tip

It takes time to become competent. Practise first with light weights until you become proficient.

If possible, use a mechanical means of lifting in preference to manual handling – it reduces the risk of personal injury

FAQ

Why can't I do whatever I like with my own safety helmet?

It is an offence under HASAWA to misuse or interfere with equipment provided for your Health and Safety or the Health and Safety of others.

Testing weight with foot

- Ensure that the path where you want to move the load is clear from obstructions, that any doors you have to pass through are opened and that you have a clear area for placing the load.

- Test the load by gently applying force with your foot – if it feels heavy or difficult, you may need to seek help for a double lift.

Getting hold of load

- If it feels comfortable to move, start from a good base and stand with the feet hip-width apart.

- Grip the load firmly. (Use gloves to avoid injuries if the load has sharp or rough edges.)

- Balance the load, using both hands if possible.

- Take account of the position of the centre of gravity of the load when lifting.

Picking up with legs bent

- Maintain a straight back; bend your knees and let the strong muscles of your legs and thighs do the work.

- Keep your arms straight and close to the body.

- Avoid sudden movements and twisting of the spine.

Moving slowly

- Move slowly and evenly.

- Never obstruct your vision with the load that you are carrying.

Placing carefully

- Place load carefully at the end of the move. Do not drop or slam the load down.

Remember

A safe lifting technique requires thinking before acting, total concentration, correct use of body power and a smooth, rhythmical sequence of activity. Safe handling is a skill.

In cases where team lifting (two or more people) is required, the following points must be remembered:

- team members should ideally be of similar height and build
- all team members must know the lifting sequence
- one member must be nominated to act as coordinator
- good communication when lifting should reduce the risk of accidents happening.

Movement of loads and methods of transport

Making use of various methods of transport can make the movement of loads much simpler and safer. Flat trailers, sack trolleys and fork-lift trucks are some of the moving aids you will come across on-site.

Untrained people must NEVER use equipment such as dumper trucks and similar mechanical devices on construction sites to move equipment; these should only be used by people who are properly trained. It is also vital to ensure that the item of mechanical equipment is safe to use for the task to be carried out.

Remember

Fork-lift trucks and certain other methods of transport can only be operated by a qualified person

Knowledge check 2

1   What type of respiratory protective equipment would normally be used in moderately dusty conditions?

2   What type of gloves should be worn when working with products contaminated by sewage?

3   What is the recommended PAT test period for portable electrical appliances used on construction sites?

4   What is the main safety concern associated with the use of cold chisels?

5   When should a manual handling risk assessment take place – before or after carrying out the work?

Access to work

Access equipment is a term that refers to items such as ladders, roof ladders, trestle scaffolds, independent and putlog scaffolds, and tower scaffolds. Access also includes working in excavations.

Access equipment forms a very important part of site work. It is vital that it is kept in good order. For this reason all access equipment is regularly checked and recorded. We will cover its correct use and limitations.

Ladders

As a plumber you will need to use ladders frequently, either working from the ladder directly or using it to gain access to the place of work or scaffold.

A ladder should be used only for:

- gaining access to a work platform, e.g. a scaffold, or

- relatively short-term working (duration 30 minutes or less) such as repairs to an external waste pipe.

A ladder is not suitable for long-term working; for this type of work another form of access equipment should be used.

Ladders can be manufactured from wood, aluminium or fibre glass. Because ladders are used frequently, their condition tends to be neglected, which can lead to defects. It is advisable to inspect any ladder before use.

You should check the following to ensure the ladder is safe for you to use:

- the **stiles**/strings are not cracked or warped

- the **rungs** are not split or dirty

- tie-rods are not missing and are not damaged

- there is no wood rot

- there are no temporary repairs

- the ladder should not be painted as the paint may be hiding defects.

Short ladders can be carried by one person, on the shoulder in either the horizontal or vertical position. Longer ladders should be carried by two people horizontally on the shoulders, one at either end holding the upper stile. When carrying ladders you should take care in rounding corners or passing between/ under obstacles.

Definition

Stiles – the upright bars of a ladder, into which the **rungs** are set

Cracks on styles

Splits on rungs

Dirt on rungs

Temporary repairs

Wood rot

Damaged tie rods

Warping

Figure 2.5 Ladder defects

There are certain rules for erecting ladders, which must be followed to ensure safe working:

- The ladder should be placed on firm, level ground. Bricks or blocks should not be used to 'pack up' under the stiles to compensate for uneven ground.

- If using extension ladders, these should be erected in the closed position and extended one section at a time. When extended, there must be at least three rungs overlap on each extension.

- If the ladder is placed in an exposed position it should be guarded by barriers.

- The angle of the ladder to the building should be in the proportion of 4 up to 1 out, or 75°.

- The ladder should be secured at the top and, as necessary, at the bottom to prevent unwanted movement. Alternatively the ladder may be 'steadied' by someone holding the stiles, and placing one foot on the bottom rung; this is commonly known as 'footing' the ladder. This person must not, under any circumstances, move away while someone is on the ladder.

- When the ladder provides access to a roof or working platform, the ladder must extend at least 1 m or 5 rungs above the access point.

- Ensure that the ladder is not resting against any fragile surface (e.g. a glass window) or against fittings such as gutters or drainpipes – these could easily give way, resulting in an accident.

- When climbing up ladders you must use both hands to grip the rungs. This will give you better protection if you slip.

- All ladders, stepladders and mobile tower scaffolds should be tested and examined on an annual basis. The results of the tests should be recorded. Ideally the item tested should be marked to show it has been tested. A competent person must carry out this test.

The diagram shows a correctly positioned ladder, securely lashed to the scaffold with sufficient extension past the access point (5 rungs).

4 units

75°

1 unit

Figure 2.6 Angle of ladder to the building

Did you know?

There are three classes of ladder: 1, 2 and 3. A class 1 ladder is intended for industrial use

Safety tip

When accessing scaffolding on a construction site where heavy materials/equipment are moved or there is regular foot access, only a Class 1 ladder may be used

1m

Figure 2.7 Ladder attached to platform

Stepladders

Stepladder

Plumbers use stepladders extensively. The first essential check before using a stepladder is to make sure the ground is level and firm. If it is not, you should not use the stepladder on it. All four legs of a stepladder should rest firmly and squarely on the ground. They will do this provided that the floor or ground is level and that the steps themselves are not worn or damaged.

When using the steps, ensure that your knees remain below the top of the steps. The top of the steps should not be used unless the ladder is constructed as a platform.

On wooden stepladders, check that the hinge is in good condition and that the ropes are of equal length and not frayed. A step ladder should only be used face on to the work, never sideways on, as this increases the risk of the ladder overturning.

Trestles

Some jobs cannot be done safely from a pair of steps. In such cases a working platform known as a 'trestle scaffold' should be used. This consists of two pairs of trestles or 'A' frames spanned by scaffolding boards which provide a simple working platform.

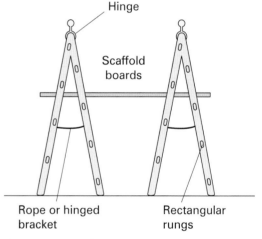

Figure 2.8 Trestles

When erecting trestle scaffolds, the following rules should be observed:

- As with ladders, they should be erected on a firm, level base with the trestles fully opened

- The platform must be at least two boards or 450 mm wide

- The platform should be no higher than two-thirds of the way up the trestle: this ensures that there is at least one-third of the trestle above the working surface

- The scaffold boards must be of equal length and should not overhang the trestle by more than four times their own thickness, for example a 40 mm board must not overhang by more than 160 mm

- The maximum span for boards is 1.3 m for 40 mm thick and 2.5 m for 50 mm thick boards

- If the platform is more than 2 m above the ground, toe boards and guardrails must be fitted and a separate ladder provided for access

- Trestles must not be used where anyone can fall more than 4.5 m.

Trestle scaffold boards

Scaffold boards are made to satisfy the requirements of BS 2482/70 and are the only boards that should be used. The maximum length is usually no greater than 4 m. If a greater length is required, then special staging is used. Scaffold boards should be:

- clean, free from grease and dirt
- straight
- free from decay, damage or any splits
- unpainted.

Scaffolds
Independent and putlog scaffolds

As a plumber you will not be expected to erect either an independent or a putlog scaffold. However, you will have to work from one at some stage, maybe to install guttering, or to access roof weatherings. It is important, therefore, that you are happy that the scaffold has been erected correctly and is safe for you to work from. As a rule of thumb, stand back and look at the scaffold and ask yourself the following questions:

- Does it look safe?
- Are the scaffold tubes plumb and level?
- Are there sufficient braces, guardrails and scaffold boards?
- Is it free from excessive loads such as bricks?
- Is there proper access from a ladder?

Make sure there are ledger-to-ledger braces on each lift for independent scaffolds, and that putlog scaffolds are tied to the building. Look for any gaps between the boards of the platform from which you will be working, and make sure a toe board and guardrail are fitted.

Mobile scaffold towers

Mobile scaffold towers may be constructed of basic scaffold components or may be specially designed 'proprietary' towers made from light alloy tube. The tower is built by slotting the sections together until the required height is reached. Mobile towers are fitted with wheels, static towers are fitted with base plates.

When working with mobile tower scaffolds the following points must be followed:

- The working platform must be provided with suitable edge protection and toe boards. Guard rails should be at least 950 mm high and an intermediate guard rail should be provided so the unprotected gap does not exceed 470 mm.
- When the platform is being used, all four wheels must be locked.
- The platform must never be moved unless it is clear of tools, equipment and people. It should be pushed at the bottom of the base and not at the top.
- The stability of a tower depends upon the ratio of the base width to height. A ratio of base to height of 1:3 gives good stability.

Mobile scaffold tower

- Outriggers can increase stability by effectively increasing the area of the base but, if used, must be fitted diagonally across all four corners of the tower and not on one side only. When outriggers are used they should be clearly marked (e.g. with hazard marking tape) to indicate that a trip hazard is present.

- Access to the working platform of the tower should be by a ladder securely fastened inside the tower, or forming part of the tower structure.

- Towers should only be erected by trained and competent people. There are a number of organisations that provide training for the safe erection, and use, of tower scaffolds against a PASMA competence standard.

- There are specific requirements laid down for the inspection of a mobile tower scaffold, this inspection must be carried out by a person who has been adequately trained.

Using roof ladders

Wheel

Roof ladder

Make sure that the ladder extends at least three rungs above the base of the roof ladder.

Use a ladder stand off to prevent the access ladder bearing onto the plastic gutter

Figure 2.9 Roof ladder

The illustration shows what a roof ladder looks like. It is used by plumbers to gain access to chimneys in order to carry out work such as small repairs to sheet weathering. For a full sheet weathering installation, a scaffold would be needed. Roof ladders are also used to gain access to a roof when installing chimney flue liners for gas fire or gas boiler installations in existing flues. They are made of aluminium, which is extremely light, so that they can easily be manoeuvred up the roof. A roof ladder is positioned by turning it on its wheels, pushing it up the roof, and then turning it through 180 degrees to hook it over the ridge tiles.

The key safety points are:

- the supports under the ladder must all fully rest on the roof surface
- the ladder must be long enough for the roof on which it is being used
- the ladder used to gain access from the ground to the roof must be securely fastened at the top and must be next to the roof ladder
- gaining access to a roof and completing the work using a roof ladder usually requires two people: one to access the roof, the other to assist with providing equipment.

Safety checklist for the roof ladder:

- stiles must be straight and in sound condition
- rungs must be sound
- the ridge hook must be firmly fixed

- wheels must be firmly fixed and running freely
- pressure plates must be sound (these are the parts that rest on the roof surface).

Work in excavations

Work in excavations can be particularly hazardous and therefore a risk assessment must be conducted before the work is commenced to ensure that proper control measures are applied to safely do the work. The following are typical measures:

- An assessment of the ground material type must be undertaken to determine the need for shoring.
- Only properly trained plumbers must work in excavations.
- Any work carried out should be in pairs, one plumber carrying out the work in the trench, the other passing tools and materials.
- Fencing must be erected around trenches over 2 m in depth to prevent persons falling into the trench.
- Material removed from the trench should be safely stored so that it may not fall back into the trench.
- Barriers such as vehicle stop blocks may need to be used where dump trucks are being used in the vicinity of the trench. This is to prevent the vehicle falling into it.

Note on Liquefied Petroleum Gas (LPG)

LPG is heavier than air, and therefore poses a particular danger when working in excavations. If LPG is being used in an excavation, any leakage could result in the formation of pockets of LPG. This is a potentially dangerous situation, as LPG is a highly explosive gas.

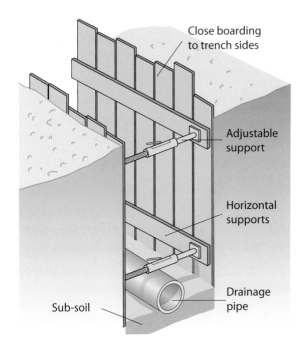

Close boarding to trench sides

Adjustable support

Horizontal supports

Drainage pipe

Sub-soil

Figure 2.10 Shoring of trenches

Safety tip

On no account should you ever move trench supports to locate pipework in excavations: the trench may collapse and the results could be deadly!

Knowledge check 3

1 In what circumstances should a ladder normally be used?

2 What is the preferred angle of a ladder to a building?

3 Can the top step of a step ladder be used as a working platform?

4 Which organisation provides a recognised certification scheme for using mobile tower scaffolds?

5 Over what depth should fencing be erected surrounding an excavation?

Fire safety and emergencies

All of the topics covered within this chapter are designed to keep you safe, but this is a particularly important section. Fire is a constant risk in any industry, but even more so for plumbers, who come into contact with electricity, gas and heating equipment.

Fire safety

Figure 2.11 The Fire Triangle

Fire or burning is the rapid combination of a fuel with oxygen (air) at high temperature. A fire can reach temperatures of up to 1000°C within a few minutes of starting. For a fire to start there are three requirements: a combustible substance (called the fuel), oxygen (usually as air) and a source of heat (spark, friction, match). When these three items come together in the correct combination, a fire occurs.

Fires can spread rapidly. Once established, even a small fire can generate sufficient heat energy to spread and accelerate the fire to surrounding combustible materials. Fire prevention is largely a matter of 'good housekeeping'. The workplace should be kept clean and tidy.

Plumbing is a particularly risky job in terms of potential fire hazards. The use of blow torches and welding equipment – which are often deployed near combustible materials, and sometimes in tight positions or difficult-to-access areas – means you are particularly vulnerable.

Because 'hot working' is such a risky job, your employer is required to provide strict working methods. These procedures constitute basic common sense, but are also a requirement of insurance companies. They usually include:

- providing a fire extinguisher in the immediate working area
- completing work with a blow or welding torch a minimum of an hour before leaving a site.

If you are working close to a combustible material (e.g. timber skirting, joists etc.), always make sure you protect the area around the fitting you are soldering with a heat-resistant mat. You can get these from your local plumbers' merchant. If the pipe is insulated, remove the insulation for about 300 mm on each side of the fitting from the area where you are using the torch.

If you think you may have caught any material with the flame, wet the area and check again after a few minutes to make sure it is all right. If you are working in a customer's home, make sure you pull back carpets or place curtains away from the area you are going to be working in.

Finally, building sites can be dirty paces. Timber shavings and other combustible materials find their way under floors, baths, etc. Make sure the area where you are working is clean before you start as well as after you finish.

Another major cause of fires is electrical faults. All alterations and repairs in electrical installations must only be carried out by a qualified person, and must be to the standards laid down in BS 7671.

You may sometimes work in an occupied building, such as an office block. You must be aware of the building's fire safety procedures, and find out about normal and alternative escape routes. Know where your assembly point is located and report your presence to your supervisor.

If you discover a fire, follow the procedures below:

- raise the alarm immediately
- leave by the nearest exit
- call the fire service.

In addition, windows should be closed to help starve the fire of oxygen, but they should not be locked.

Safety tip

If in doubt stay out!

Classes of fire

Fires are commonly classified into four groups, according to fuel type and how the fires are extinguished:

- Class A – fires involving solid materials, extinguished by water
- Class B – fires involving flammable liquids, extinguished by foam or carbon dioxide
- Class C – fires involving flammable gases, extinguished by dry powder
- Class D – fires involving flammable metals, extinguished by dry powder.

Fire-fighting equipment

If a fire is small, it may be possible to put it out quickly and safely. However, if your efforts to contain a fire prove fruitless and it starts to get out of control, when you finally decide to make your escape you could have difficulty finding your way through the smoke and the fumes. Smoke and fumes are just as lethal as the fire itself.

Fire-fighting equipment, including extinguishers, buckets of sand or water and fire-resistant blankets, should be readily available in buildings. In larger premises, you will find automatic sprinklers, hose reels and hydrant systems. The table below shows each type of fire extinguisher and the use to which they should be put.

Remember

Personal safety must always come before your efforts to contain a fire

Type of extinguisher	Colour code	Main use
Water	Red	Wood, paper or fabrics
Foam	Cream	Petrol, oil, fats and paints
Carbon dioxide	Black	Electrical equipment
Dry powder	Blue	Liquids, gases, electrical equipment

Figure 2.12 Fire extinguishers

FAQ

Do I really have a legal Health and Safety duty?

Yes you do! You *must* take resonable care of your own Health and Safety and that of others affected by you.

The colour coding on extinguishers may not seem immediately obvious. Recent European legislation dictates that all extinguishers must be coloured red whatever substance they contain, but will carry 5 per cent of the colour the extinguisher would have been under the original system: for example, a carbon dioxide extinguisher will be red with a black stripe, triangle or lettering etc. Care should therefore be taken when choosing which extinguisher to use to extinguish a fire that you choose the correct type.

The decision to have all extinguishers coloured red was made because red is recognisable by people with colour blindness. Previously, it was possible for a colour-blind person to use the wrong extinguisher for a specific type of fire. The fact that they are now all the same colour means you have to read the labels to discover their intended use.

Water extinguisher

Powder extinguisher

Fire extinguisher safety checklist

✓ Never use a fire extinguisher unless you have been trained to do so.

✓ Do not use water extinguishers on electrical fires because of the risk of electric shock and explosion.

✓ Do not use water extinguishers on oils and fats as this too can cause an explosion.

✓ Do not handle the nozzle on carbon dioxide extinguishers as this can cause a freezer burn to the hands.

✓ Do not use carbon dioxide extinguishers or halon types in a small room as this could cause suffocation.

✓ Always read the operating instructions on the extinguisher before use.

Accidents

It is recommended that your plumbing apprenticeship training course includes a certificated first-aid course, but in case it does not, here are a few pointers about dealing with accidents.

Carbon dioxide extinguisher

First actions

No matter how many precautions are taken, sooner or later you are likely to witness an accident. Whatever the accident – a cut finger or a fall from a ladder – you must know what to do.

Consider a situation where you are working with another plumber who has just cut her hand. There is a lot of bleeding and your workmate is in distress. What are you going to do? More importantly, *what are you going to do first*? There are a number of things that need to be done after an accident:

1 Seek or administer immediate first aid

2 Get help if necessary, i.e. phone for an ambulance

3 Report the accident to the site supervisor

Which one of these things you carry out first will depend on several things:

- Are you a qualified first aider?
- How severe is the injury?
- Is your supervisor or a qualified first aider immediately available?

Later you will need to:

1 Write down the details in an accident report book

2 Complete a company accident report form.

There are several important things to remember. In this scenario, we are dealing with a cut hand; the patient is bleeding quite heavily so it would be necessary to administer first aid as soon as possible.

However, if the scenario were, for example, an electric-shock casualty, you must first make sure that the area is safe – to ensure that whatever caused the accident is not going to injure you. In this case, you should isolate the supply first and call for an ambulance if necessary; then someone can attend to the casualty and inform the supervisor.

Did you know?

There are many providers of first-aid courses, and many courses are specific to the workplace. You can find plenty of information about them on the Internet. Some providers will run a course at your place of work

Remember

People with back injuries should only be moved by trained personnel

Remember

After an accident – seek first aid, inform your supervisor, complete the accident report form

On the job: Accident report form

You are working with two other plumbers – Alan, a second-year apprentice, and Ishmail, who is the senior tradesperson in charge of the site. Alan was balancing on the top of a stepladder trying to secure some guttering when he overbalanced and fell to the ground. The ground was hard concrete and as a result of this fall Alan has sustained a broken wrist and bleeding to his hand and forehead. Ishmail accompanies Alan in the ambulance. You are left to complete the accident report form.

1 What information do you need to complete the form as well as you can?

2 Complete the accident report form on the next page, using the information contained in the scenario above and making assumptions about the exact time and necessary addresses.

If someone has fallen from a height, do not move them: if they have a serious back injury, your actions could make things worse. If they are conscious, make them as comfortable as possible until help arrives; this may include treating minor wounds. If the person is unconscious, they may be put in the recovery position, but only if you are confident that there is no back injury.

If someone is injured in an accident, the details have to be entered in the accident book. You may have to enter details on a casualty's behalf if they are unable to do so themselves. If the accident involved a piece of faulty equipment, do not tamper with it as it may be subject to an investigation by the Health and Safety Inspectorate.

Accidents that result in three or more days' absence from work must be reported to the Health and Safety Executive. The next page shows a typical example of an accident report form.

Other emergencies

Emergencies other than a fire or an accident could include a bomb threat or chemical spillage. These events are pretty rare and it is unlikely that you will be involved in one of them. However, you still need to be aware of the procedure for dealing with emergencies.

1   Find a telephone in a safe environment well away from the emergency. Make sure you are not going to be trapped while making the call.

2   Dial the emergency service number, which is normally 999. However, this may not always be the case, especially if you are on an internal exchange. You need to familiarise yourself with any special procedures that may apply.

3   Keep calm and listen to what the emergency-service operator has to say.

4   When asked, give your name and the name of the service you require.

5   You will then be connected to the emergency service you require; again, stay calm, listen to what the operator has to say and answer the questions they ask.

6   You will be required to explain the nature of the emergency and where it occurred. Try to give the operator the exact location of the incident, such as the name of the company, its address, etc.

7   When you have completed the call, arrange for someone to meet the emergency services to show them where the incident has occurred.

If you are working in an environment where there is an increased risk of an emergency situation occurring, your employer should provide you with sufficient information on what to do in an emergency. Some of these procedures will be similar to the fire evacuation procedures and should all be about getting outside to a place of safety. There may be other more specific procedures, and it is your duty to familiarise yourself with them and follow them when necessary.

Did you know?

Any act of aggression by a colleague at college or at work that causes a situation of conflict is also considered a dangerous occurrence

Full name of injured person:	
Home address:	Sex: Male/Female
	Age:

Status: Employee
Contractor
Visitor

Date of accident:
Time of accident:
Precise location:

What was the accident and its cause? (You may have to give a detailed written description.)

Name and address of witness if any:

Details of apparent injuries:

Summary:

Figure 2.13 Typical accident report form

Electric shock and first aid

In this section you will be looking at the treatment of an unconscious casualty suffering from an electric shock. We will also be looking at the basic first-aid procedures for the treatment of burns, shock and breaks. You should be doing a purpose-made first-aid course; this section can offer only the basic theory.

Electric shock and its treatment

Electric shock occurs when a person becomes part of the electrical circuit. The severity of the shock will depend on the level of current and the length of time it is in contact with the body.

The lethal level is approximately 50 mA, above which muscles contract, the heart fibrillates and breathing stops (in other words, you die!). A shock current above 50 mA could be fatal unless the person is quickly separated from the supply. Below 50 mA, only an unpleasant tingling sensation may be experienced. However, this may cause someone to fall from a roof or ladder and the resulting fall may lead to a serious injury.

Action in the event of an electric shock

- First of all, check for your own personal safety to ensure that you will not be putting yourself at risk by touching the casualty – they are part of the circuit and they are live too.

- Break the electrical contact to the casualty by switching off the supply, removing the plug, or wrenching the cable free. If this is not possible, break the contact by pushing or pulling the casualty free using a piece of non-conductive material, e.g. a piece of wood. Guide the casualty to the ground, making sure that further injuries are not sustained, e.g. banging their head on the way down.

- Check the casualty's response, to assess their level of consciousness.

- Talk to them and gently shake them to gauge their level of response.

- If the casualty appears unharmed, they may only be shaken and should be advised to rest.

- If there is no movement or any sign of breathing, summon help immediately. If there is someone with you, tell them to get help, i.e. ring 999. If you are on your own with the casualty, you will have to leave them for a moment while you get help.

- As soon as you return to the casualty you need to start cardio-pulmonary resuscitation (CPR).

To assess someone's consciousness level, ask a question or give a command – for example, 'What happened?' or 'Open your eyes'. Speak loudly and clearly, close to the casualty's ear. Carefully shake the casualty's shoulders. A 'slightly conscious' casualty may mumble, groan or make slight movements. A fully unconscious casualty will not respond at all.

Remember

It is not the volts that give you the jolts; it's the amps that put out your lamps!

Remember

CPR is based on the ABC of resuscitation: Airway, Breathing, Circulation. These are the three crucial things your casualty needs to stay alive

Remove obvious obstructions from the mouth

Cardio-pulmonary resuscitation

1 *Open the airway.* An unconscious casualty's airway may be narrowed or blocked, making breathing difficult and noisy or impossible. The main reason for this is that muscular control in the throat is lost, which allows the tongue to sag back and block the throat. Lifting the chin and tilting the head back lifts the tongue away from the entrance to the air passage.

2 *Remove any obvious obstruction from the mouth.* Place two fingers under the point of the casualty's chin and lift the jaw. At the same time, place your other hand on the casualty's forehead and tilt the head well back.

3 *To check for breathing,* put your face close to the casualty's mouth, look and feel for breathing as well as looking for chest movements. Listen for the sound of breathing. Feel for breath on your cheek. Look, listen and feel for 10 seconds before deciding that breathing is absent.

4 *Check for a pulse.* If the heart is beating adequately, it will generate a pulse in the neck (the carotid pulse) – this is where the main carotid arteries pass up to the head. These arteries lie on either side of the larynx (throat), between the Adam's apple and the strap muscle that runs from behind the ear across the neck to the top of the breastbone.

5 To check for the pulse ensure the head is tilted back and feel for the Adam's apple with two fingers. Slide your fingers back into the gap between the Adam's apple and the strap muscle, and feel for the carotid pulse. Feel for 10 seconds before deciding that the pulse is absent.

Mouth-to-mouth ventilation is given with the casualty lying flat on their back.

1 Open the airway by tilting the head and lifting the chin.

2 Remove any obvious obstruction, including broken or displaced dentures, from the mouth.

3 Close the casualty's nose by pinching it with your finger and thumb.

4 Take a full breath, and place your lips around the casualty's mouth, making a good seal.

5 Blow into the casualty's mouth until you see the chest rise. Blow for about 2–5 seconds for full inflation.

6 Remove your lips and allow the chest to fall fully.

7 Deliver subsequent breaths in the same manner.

In this situation, you need to deliver two complete breaths.

Fifteen chest compressions are now given:

Chest compressions

1 With the casualty still lying flat on their back on a firm surface, kneel beside them, and find one of the lowest ribs using your index and middle fingers.

2 Slide your fingers upward to the point in the middle where the rib margins meet the breastbone.

3 Place your middle finger over this point and your index finger on the breastbone above.

4 Place the heel of your hand on the breastbone, and slide it down until it reaches your index finger. This is the point at which you will apply pressure.

5 Place the heel of your first hand on top of the other hand, and interlock fingers.

6 Leaning well over the casualty, with your arms straight, press down vertically on the breastbone to depress it approximately 4–5 cm, then release the pressure without removing your hands.

7 Repeat the compressions, aiming for a rate of approximately 15 compressions per 15 seconds (4 cycles per minute).

Return to the head and give two more ventilations, then a 15 further compressions. Continue to give two ventilations to every 30 compressions until help arrives. Do not interrupt CPR to make pulse checks unless there is any sign of a returning circulation. If a pulse is confirmed, check breathing, and if it is still absent continue with ventilation. Check the pulse after every 10 breaths, and be prepared to re-start chest compressions if it disappears. If the casualty starts to breathe unaided place him or her in the recovery position. Re-check breathing and pulse every two minutes.

The recovery position

Any unconscious casualty should be placed in the recovery position, so long as they have no obvious back injury. This position prevents the tongue from blocking the throat, and, because the head is slightly lower than the rest of the body, it allows liquids to drain from the mouth, reducing the risk of the casualty inhaling stomach contents should they vomit.

The recovery position

The head, neck and back are kept in a straight line, while the bent limbs keep the body propped in a secure and comfortable position. If you must leave an unconscious casualty unattended, they can be left safely in the recovery position while you get help.

The technique for turning is described here and assumes that the casualty is found lying on their back from the start.

1 Kneeling beside the casualty, open the airway by tilting the head and lifting the chin. Straighten the legs. Place the arm nearest you out at right angles to the body, elbow bent and with the palm uppermost.

2 Bring the arm furthest from you across the chest, and hold the hand, palm outwards, against the casualty's nearer cheek.

3 With your other hand, grasp the outside of the knee furthest from you and pull the knee up, keeping the foot flat on the ground.

4 Keeping the casualty's hand pressed against their cheek, pull the knee towards you to roll the casualty towards you and on to their side.

5 Tilt the head back to make sure the airway remains open. Adjust the hand under the cheek if necessary so that the head stays in this tilted position and blood flow to the hand is not restricted.

6 Adjust the upper leg, if necessary, so that both the hip and the knee are bent at right angles.

A casualty who has received an electric shock, or who has been unconscious for any reason, should be sent to hospital in an ambulance.

Treatment of burns

The treatment of burns is essentially based on the severity of the burn.

Severe burns and scalds

The priority is to cool the injury. The longer the burning goes unchecked, the more severely the casualty will be injured. Resuscitate the casualty only when cooling is underway. All severe burns carry the danger of shock. Key actions are:

- If possible lay the casualty down, protecting the burned area from contact with the ground.

- Douse the burn with copious amounts of cold liquid. Thorough cooling may take 10 minutes or more, but this must not delay the casualty being taken to hospital

- While cooling the burns, check airway, breathing, and the pulse, and be prepared to resuscitate if necessary.

- Gently remove any jewellery, watches, belts, shoes, or smouldering clothing from the injured area, before it starts to swell. Carefully remove burned clothing unless it is sticking to the burn.

- Cover the injury with a sterile burns sheet or other suitable non-fluffy material, to protect from infection. A clean plastic bag or kitchen film may be used. Burns to the face should be cooled with water, not covered.

- Ensure that the emergency service is on its way. While waiting, treat the casualty for shock. Monitor and record breathing and the pulse, and resuscitate, if necessary.

Do not:

- overcool the casualty – this may dangerously lower the body temperature

- remove anything sticking to the burn – this may cause further damage and cause infection

- touch or interfere with the injured area

- burst blisters

- apply lotions, ointment, or fat to the injury.

Minor burns and scalds

Minor burns and scalds are usually the result of domestic accidents. Prompt first aid will generally enable them to heal naturally and well, but the advice of a medical practitioner should be sought if there is doubt as to the severity of the injury. Key actions include:

- Cool the injured part with copious amounts of cold water for about 10 minutes to stop the burning and relieve the pain. If water is unavailable, any cold, harmless liquid, such as milk or canned drinks, will suffice.

- Gently remove any jewellery, watches, or constricting clothing from the injured area before it starts to swell.

- Cover the injury with a sterile dressing, or any clean, non-fluffy material to protect from infection. A clean plastic bag or kitchen film may be used.

Do not:

- use adhesive dressings
- break blisters, or interfere with the injured area
- apply lotions, ointments, creams, or fats to the injured area.

Treatment for shock

Shock can develop when the heart pump fails to work properly, causing a reduction in the pressure of the circulating blood. The most common cause of this type of shock is a heart attack. Shock can develop as a result of a reduction in the volume of fluid circulating around the body. The most common examples of this are external or internal bleeding, or the loss of other bodily fluids through severe diarrhoea, vomiting or burns. The blood supply is diverted from the surface to the core of the body. The main symptoms and signs of shock relate to such redistribution of the circulation. Key actions are:

- Treat any cause of shock which can be remedied (such as external bleeding).
- Lay the casualty down, keeping the head low.
- Raise and support the casualty's legs (be careful if suspecting a fracture).
- Loosen tight clothing, braces, straps or belts, in order to reduce constriction at the neck, chest and waist.
- Insulate the casualty from cold, both above and below. Contact the emergency services.
- Check and record breathing, pulse and level of response. Be prepared to resuscitate the casualty if necessary.

Do not:

- let the casualty move unnecessarily, eat, drink, or smoke
- leave the casualty unattended – reassure the casualty constantly.

Treatment for broken bones

If a broken bone is suspected, get expert help. Do not move the casualty unless they are in a position which exposes them to immediate danger.

Treatment for bleeding

Minor cuts, scratches and grazes

- Wash and dry your own hands.
- Cover any cuts on your own hands and put on disposable gloves.
- Clean the cut, if dirty, under running water. Pat dry with a sterile dressing or clean lint-free material. If possible, raise affected area above the heart.
- Cover the cut temporarily while you clean the surrounding skin with soap and water and pat the surrounding skin dry. Cover the cut completely with a sterile dressing or plaster.

Severe bleeding

- Put on disposable gloves.

- Apply direct pressure to the wound with a pad (for example, a clean cloth) or fingers until a sterile dressing is available.

- Raise and support the injured limb. Take particular care if you suspect a bone has been broken.

- Lay the casualty down to treat for shock.

- Bandage the pad or dressing firmly to control bleeding, but not so tightly that it stops the circulation to fingers or toes. If bleeding seeps through the first bandage, cover with a second bandage. If bleeding continues to seep through bandage, remove it and re-apply.

- Treat for shock.

- Dial 999 for an ambulance.

Treatment for eye injuries

All eye injuries are potentially serious. The casualty will be experiencing intense pain in the affected eye with spasms of the eyelids.

Before attempting to treat, wash your hands. If there is something in the eye, irrigate the eye with clean, cool water or sterile fluid from a sealed container to remove loose material. Do not attempt to remove anything that is embedded. If chemicals are involved, flush the open eye with water or sterile fluid for at least 10-15 minutes. Apply an eye pad and send the casualty to the hospital.

First-aid kits

Employers should provide first-aid kits. The HSE suggests that a basic first-aid kit should contain the following items:

- twenty adhesive dressings (assorted sizes)

- two sterile eye pads

- six triangular bandages

- six unmedicated wound dressings, 10 cm × 8 cm

- two unmedicated wound dressings, 13 cm × 9 cm

- six safety pins

- three extra sterile unmedicated wound dressings 28 cm × 17.5 cm.

The kit should only contain basic/first-aid items. Additional items such as headache tablets should never be stored in first-aid kits.

In circumstances where plumbers work in domestic properties, it is usual practice for a first-aid kit to be provided with every company vehicle.

First-aid kit

FAQ

What is dermatitis?

Dermatitis is an inflammation of the skin normally caused by the hands coming into contact with irritating substances. Skin can become sore, cracked and easily infected. Protect your hands by wearing appropriate gloves, washing your hands with approved cleaners and using barrier creams.

Knowledge check 4

1 What is involved in a Class C fire?

2 What should a red fire extinguisher be used on?

3 What type of accidents should be recorded in an accident book?

4 What first aid treatment should be offered by a first aider to a casualty who is not breathing?

5 How should foreign bodies in the eye be removed by the first aider?

Control of Substances Hazardous to Health

COSHH

COSHH stands for the Control of Substances Hazardous to Health. These Regulations were first introduced in 1988 and provide a legal framework for the control of substances hazardous to health in all types of business, including factories, farms, offices, shops – and plumbing!

The Regulations require that employers make an assessment of all work that is liable to expose any employee to hazardous solids, liquids, dusts, fumes, vapours, gases or micro-organisms. Any risks to health must be evaluated and a decision must be taken on the action to remove or reduce those risks. For example, if you were replacing waste pipework in a college laboratory, you would expect protective clothing to be provided to prevent injury to yourself while carrying out your work.

What is a substance hazardous to health?

Substances that are 'hazardous to health' include substances labelled as dangerous (i.e. very toxic, toxic, harmful, irritant or corrosive). They also include micro-organisms and substantial quantities of dust and indeed any material, mixture or compound used at work, or arising from work activities, that can harm a person's health. There is a wide variety of substances that fall under COSHH that are used in plumbing, including items ranging from fluxes to working with irritants such as loft insulation and working with human waste such as urine: all require specific actions to enable plumbers to work safely with them.

TOXIC

HARMFUL

CORROSIVE

IRRITANT

OXIDISING

EXTREMELY FLAMMABLE

Figure 2.14 Categories of hazardous substance

What do COSHH regulations require?

- The risk to health arising from working with a substance must be assessed.

- The necessary measures to prevent exposure or control the risk must be identified.

- There must be control measures to ensure that equipment is properly maintained and procedures observed.

- Employees must be informed, instructed and where necessary receive the relevant training about the risks and the precautions that need to be followed.

Prevention will usually involve the use of protective clothing and equipment – respiratory equipment masks, dust masks, goggles, gloves and so on.

Chemical safety

Chemicals are substances that form the basis of many common items. However, they can be harmful to the people working with them if they come into contact with the skin by accident. When not contained or handled properly, chemicals can be:

- inhaled as a dust or gas

- swallowed in small doses over a long period

- absorbed through skin or clothing

- touched by or spilled on unprotected skin.

Some chemicals can cause:

- injury to eyes, skin, organs – from fires and burns etc.

- silent illnesses – effects are not noticeable immediately after exposure, but can cause medical problems after many months or years of exposure (e.g. asbestosis)

- allergy to the skin as a rash, coughing and breathing problems

- death – some poisonous chemicals can kill outright.

Why do chemical accidents happen?

- Hurrying, overconfidence, fooling around or not adhering to instructions can lead to accidents.

- Spills and leaks can be dangerous if not wiped up.

- Vapours may build up where there is no proper ventilation.

- The exposure of some chemicals to heat or sunlight can cause explosion, fire and poisonous reactions.

- Contact between a chemical and the wrong material can cause harmful reactions.

- Neglect or failure to dispose safely of certain old chemicals is dangerous, as chemical changes can happen over time.

There are four main types of chemicals: toxic agents, corrosives, flammables and reactives. You need to know what they are and how to guard against their hazards.

Safety tip

More haste, less speed! Try not to rush, that's when accidents happen

Toxic agents

Poisons such as hydrogen sulphide and cyanides can cause injury, disease and death. To protect yourself:

- close containers tightly when not in use
- be sure that the work area is well-ventilated
- wear personal protective equipment
- wash hands often
- carry cigarettes in a protective packet
- safely dispose of contaminated clothing
- keep any antidotes handy.

Corrosives

These are irritants such as acids and alkaline, which are especially dangerous to the skin, eyes and respiratory tract. To protect yourself:

- wear personal protective equipment – goggles, breathing devices, protective gloves
- make sure ventilation is good
- run for the nearest water if corrosives come into contact with you; use a safety shower if one is available
- if your eyes are affected, flush with water for 20 minutes and get medical aid. Many larger first-aid kits contain eye-irrigation kits.

Flammables

These are liquids and gases that burn readily such as methanol, ethanol, ether, LPG and petrol.

To protect yourself:

- make sure no flames, sparks, or cigarette lighters are allowed near flammables
- keep only a small quantity of flammables in the work area
- store and dispose of flammables safely.

In an emergency:

- evacuate the area, if possible
- turn off all flames and sparking equipment
- clean up flammables and ventilate the area thoroughly
- call the emergency services.

Remember

Burning requires three things: the right amount of flammable fuel, oxygen, and a spark or other source of ignition

Reactives

These are substances that can explode, such as nitrogen compounds. To protect yourself:

- Know your chemicals before working with them: read about them and test them for stability.

- Handle reactives with great care – for example certain proprietory cleaners mixed with bleach can give off hydrogen peroxide.

- At the first sign of trouble, close the doors and evacuate the room through doors that do not lead to the danger area.

When using chemicals of any sort, you must understand the dangers involved. This information will usually be communicated to you using a risk assessment – a typical example produced by the British Plumbing Employers Council (BPEC) is shown on the following page. The packaging of hazardous substances usually provides essential safety information – you must follow all recommended safety rules and procedures and you must know what to do in an emergency.

The example shows an assessment sheet for using sulphuric acid – a particularly nasty product. You should follow all the guidance detailed on the sheet to use the substance safely, paying particular attention to the control measures you must use.

Asbestos dust safety

Asbestos is a particularly nasty substance and you must take great care if you encounter it on site. Asbestos is the cause of a sizeable number of deaths in the plumbing and heating industry – *so be warned!*

Asbestos is a mineral found in many rock formations. When separated from rock it becomes a fluffy, fibrous material that has many uses. About two-thirds of all the asbestos produced is used in the construction industry. It is used in cement production, roofing, plastics, insulation, floor and ceiling tiling, and fire-resistant boards for doors and partitions.

Asbestos becomes a health hazard if the fibres are inhaled through the nose or mouth; they cannot be absorbed through your skin. Some of the fine rod-like fibres may work their way into the lung tissue and remain embedded for life. This will become a constant source of irritation and can cause chronic illness.

Breathing in asbestos fibres can lead to you developing one of three fatal diseases:

- asbestosis – a scarring of the lung leading to shortness of breath

- lung cancer

- mesothelioma – a cancer of the lining around the lungs and stomach.

SULPHURIC ACID (H_2SO_4) LOCATED: DRAINAGE WASTE SOIL-PIPE DISCHARGES MAIN SEWER SYSTEMS	CORROSIVE	HAZARDS FUMES CONTACT
HAZARDOUS OPERATIONS FUMES – INHALATION OF FUMES WHERE INSUFFICIENT VENTILATION CONTACT – INGESTION VIA EXPOSED SKIN OR INTO THE EYE		
HEALTH RISKS SULPHURIC ACID IS A STRONGLY CORROSIVE LIQUID WHICH WILL CAUSE SEVERE BURNS TO SKIN AND COULD ON ENTRY TO THE EYE CAUSE PERMANENT DAMAGE. RESPIRATORY IRRITANT CAUSING SHORTNESS OF BREATH AND AS A RESULT OF LONG- TERM CONTACT, DELAYED PULMONARY OEDEMA		

SPILLAGES COVER WITH SODA ASH AND LIFT. RINSE WITH COPIOUS AMOUNTS OF WATER	FIRST AID SEEK MEDICAL ADVICE IF INGESTED FROM SKIN CONTACT OR ENTRY INTO THE EYE. WASH WITH PLENTY OF WATER
FIRE N/A	STORAGE NEVER ADD WATER TO SULPHURIC ACID. STORE IN A SECURE LOCATION

RISK FACTORS DURATION	PROCESS	OUTDOORS
0–1 MINUTE	UNBLOCKING DRAINS	CRITICAL
1–10 MINUTES	UNBLOCKING DRAINS	CRITICAL
10 + MINUTES	UNBLOCKING DRAINS	CRITICAL

CONTROL MEASURES
CRITICAL RISKS   EYE PROTECTORS, IMPERVIOUS GLOVES, IMPERVIOUS OVERALLS AND CARTRIDGE-TYPE RESPIRATOR CAPABLE OF REMOVING FUMES

Figure 2.15 Risk assessment sheet for working with sulphuric acid

There are three main types of asbestos:

- chrysotile – white (accounts for about 90 per cent of asbestos in use)
- amosite – brown
- crocidolite – blue.

They cannot be identified by colour alone; laboratory analysis is required.

There is no cure for asbestos-related diseases.

It is a legal requirement to know the hazards of your job; any asbestos on site should have been identified before you start work.

Where restricted work is permissable, you may come across asbestos in the following:

- asbestos cement flue pipes
- asbestos gutter systems
- asbestos soil pipes
- artex ceilings containing asbestos.

If so, you need to work to the following guidelines:

- Where possible, keep asbestos materials damp when working on them.
- Never use power tools on asbestos materials as they create dust; use hand tools instead.
- Wear and maintain any personal protective equipment provided – and use it properly.
- Practice good housekeeping – only use special vacuums and dust-collecting equipment.
- Keep asbestos waste in suitably sealed containers and ensure that they are labelled to show that they contain asbestos.
- Report any hazardous occurrences, e.g. unusually high dust levels, to your supervisor.
- Your employer should provide all the necessary information, instruction and training if you are going to work on an asbestos cement-based product. The procedures that you follow should be based on the fact sheet series provided by the HSE at www.hse.gov.uk/asbestos

The removal of asbestos, or working in an area of asbestos other than those in the examples listed above, must only be carried out by specialist contractors. Consider your health at all times.

Safe use of LPG, and lead working

LPG

LPG stands for Liquefied Petroleum Gas and is any concentration of commercial propane or butane used in heating processes. Even small quantities of LPG, when mixed with air, create an explosive mixture. If 1 litre of LPG is boiled or evaporated, it becomes 250 litres of gas (LPG is a gas when its temperature rises above –42°C). This is enough to make an explosive mixture in a large shed, room, store or office.

LPG is widely used in construction and building work as a fuel for burners, heaters and gas torches. The liquid, which comes in cylinders and containers, is highly flammable and needs careful handling and storage.

Whether in use or stored, cylinders must be kept upright in open-air secure compounds. When not in use cylinder valves should be closed and protective dust caps should be in place. When handling cylinders do not drop them or allow them to come into heavy contact with other cylinders.

When using a cylinder with an appliance, including a blowtorch, ensure it is connected properly, in accordance with the instructions you have been given. The amount of draw-off from LPG cylinders MUST be limited because, if this rate is too high, the control valve may freeze; this could potentially lead to leakage.

It is essential to make sure the gas does not leak. LPG is heavier than air; if it leaks, it will not disperse into the air, but will sink to the lowest point and form an explosive concentration, which could be ignited by a spark. Leakages are especially dangerous in trenches and excavations because the gas cannot flow out of these areas.

If you are to assemble blowtorch equipment or if you suspect a leak from gas heating equipment, the equipment should be checked using a soap solution. If there is a leak, the soap solution will bubble and identify the leak source for you to repair.

Transportation of LPG

The transportation of LPG is covered by the Carriage of Dangerous Goods by Road Regulations. When using LPG, everyone should understand the procedures to be adopted in case of an emergency. The appropriate fire extinguishers (dry powder) should always be available. When LPG is being transported in an enclosed van, only small quantities should be carried, and two dry-powder fire extinguishers must be available at all times. (It is also preferable that the vehicle is ventilated and there must be an LPG hazard warning sign displayed on the vehicle.) If significant quantities are being transported, an open vehicle should be used, and the driver will need to undertake special training in order to receive a TREMCARD – a requirement for moving significant quantities of LPG.

Lead safety

The vast majority of pipework and tube that you come into contact with will be copper, steel or plastic, but during your time as a plumber you will probably come across lead. Most dwellings are now free of lead pipework, but lead is still used on roof weatherings, and plumbers are still considered as lead-work specialists. Therefore, it is necessary for you to be aware of the hazards involved when working with lead. Extended unprotected exposure to lead working can lead to chronic illness.

The Control of Lead at Work Regulations set out measures similar to those concerning safe work with asbestos. Working with lead takes two forms:

- handling tasks – controlled by simple protective methods
- heating-process tasks – which require much more stringent control methods.

Handling lead materials can effectively be controlled by blocking the likelihood of lead entering the body through the skin by the use of PPE such as barrier cream, gloves, eye protection and overalls.

When lead is heated, as it would be during lead-welding processes, poisonous fumes are given off. It is therefore necessary to take precautions to guard against breathing these fumes in. This is usually achieved through using respiratory protective equipment or, preferably, local extraction ventilation systems.

Harmful lead particles can be inhaled, ingested (swallowed) or absorbed through the skin. To ensure you are kept safe from the adverse effects of lead, your employer should provide areas to eat, drink and smoke well away from the worksite.

And remember: wash your hands, arms and face after a lead-working session. Lead safety is revisited in Chapter 12, *Sheet lead working*.

Safety tip

You should never smoke while working with sheet lead, as the lead can be ingested via the cigarette.

Knowledge check 5

1   A substance that leads to extreme scratching when brought into contact with the skin is known as?

2   What is a substance said to be when it readily ignites when exposed to an ignition source?

3   Who should remove white asbestos insulation material?

4   At what point in a property will leaking LPG gas collect?

5   What is the main safety risk associated with using sheet lead?

Additional information sources

Sources of additional information on the HSE regulations covered in this chapter include:

Construction regulations – www.hse.gov.uk/construction
Asbestos – www.hse.gov.uk/asbestos
Hazardous substances – www.hse.gov.uk/coshh
Electrical safety – www.hse.gov.uk/electricity
Work at height – www.hse.gov.uk/falls
PPE at work – www.hse.gov.uk/pubns/indg174.pdf
Work with lead – www.hse.gov.uk/pubns/indg305.pdf
Manual handling – www.hse.gov.uk/pubns/indg143.pdf
Safety in excavations – www.hse.gov.uk/pubns/cis08.pdf
RIDDOR – www.hse.gov.uk/riddor
First aid – www.hse.gov.uk/firstaid

chapter 3

Effective working relationships

OVERVIEW

As well as excellent technical skills – and you should always aim to be a top-class plumber – you need different skills when you have to deal with people. These are known as 'personal skills'.

Plumbing is part of the construction industry. In this chapter, we will look at the construction industry and some of the people who work in it, so that you will know what they do.

In this chapter, we will look at:

- **The construction industry**
  - company structure

- **The construction team**
  - non-craft jobs
  - craft trades
  - other occupations
  - how plumbing relates to the construction industry
- **Communication with customers/co-contractors**
  - communication and job responsibilities of plumbing staff
  - the communication process
  - effective communication in the workplace

The construction industry

During your time as a plumber you are likely to meet a wide range of people. It is important that you know how to conduct yourself in all sorts of situations and that you can communicate clearly with other people.

Ninety-five per cent of employers in the construction industry employ between 1 and 13 people, 4 per cent 14–299, and 1 per cent 300+.

Small companies will tend to concentrate on one-off house-building, extensions, maintenance and repair work. They may also be subcontracted by larger construction companies, which usually means providing a specific service.

Larger contractors will develop construction projects from start to finish. These may be private or public housing developments or major construction projects such as the new Wembley Stadium. To do this type of work, companies need to employ a range of managers, office-based staff, designers etc. as well as the staff who will run the job on site.

Larger contractors are also involved in facilities management. This is the ongoing management of all aspects of running a large building (e.g. a hospital) or an estate of houses (local authority, housing association or private). We will look at what a facilities manager does later.

Many of the larger contractors carry out work on behalf of insurance companies. At one time, individual householders arranged for a contractor to effect repairs; now, most insurers place this work directly in the hands of the contractor, who handles everything from receiving the emergency call or claim to doing the job.

Company structure

You will appreciate by now that the construction industry is diverse in terms of both company size and the range of work undertaken. Whatever the size of company, it has to have a 'legal status'. This means it will be one of the following:

- sole trader
- partnership
- limited company (private)
- public limited company (plc).

Sole trader

One person will own the business and look after everything: quoting for work, arranging the jobs, organising materials, doing the job and then sorting out all the paperwork, invoices, tax and so on.

A sole trader

Advantages	Disadvantages
Owner entitled to all profits	Owner liable for all business's debts
Minimal paperwork	Owner risks personal assets if business fails
Do not need to register with Companies House	No job security, pension, annual leave or sick leave benefits
Freedom to make own decisions	Owner entirely responsible for business

Figure 3.1 Advantages and disadvantages of being a sole trader

Did you know?

Sole traders account for the largest number of companies in the construction industry

Partnerships

This is also a popular format for small construction and plumbing businesses. Often a partnership develops from trades people working together and deciding to go into business; a joiner and bricklayer setting up a partnership is a typical example, although partners do not have to have separate trades, and often two, or even more, plumbers set up partnerships.

Construction involves collaboration between trades

Advantages	Disadvantages
Partners bring additional expertise to the business	Partners are liable for all the business's debts
Usually all partners will invest some money in the business	Partners risk personal assets if business fails
Do not need to register with Companies House	If one partner wants to leave they have to be bought out
Responsibility for business is shared	Profits are shared

Figure 3.2 Advantages and disadvantages of partnerships

Limited companies (private)

Most businesses, if they begin to expand, tend to become limited companies. This means the company is registered with Companies House and has a legal identity. Its owners have shares in the company.

The owners of private limited companies are known as directors. The annual accounts produced by the organisation must be lodged with Companies House as part of the registration process, together with ongoing details of who the directors of the organisation currently are. The organisation will usually be required to hold at least one formal meeting annually to approve the annual accounts prior to their submission to Companies House. In order to assist with the smooth running of the business, and the interaction with Companies House, a company secretary is normally appointed to deal with the paperwork.

Advantages	Disadvantages
The owners are not personally liable for business debts, only for an amount up to their investment in the company	Limited companies have to comply with more regulations
There are tax advantages	Companies have to register with Companies House
Money can be raised for investment by selling shares	Lots of paperwork and accurate accounts and records must be kept

Figure 3.3 Advantages and disadvantages of private limited companies

Public limited companies (plcs)

These tend to be the largest companies. They are allowed to trade their shares on the stock market provided they have a share capital of at least £50,000. Some of the major contractors in the construction industry are plcs.

The structure of sole trader and partnership businesses is relatively simple. Limited companies and plcs usually have a board of directors, and, depending on their size, will have a chairman, a managing director (MD) or chief executive officer (CEO), who are usually responsible for overseeing the management of the business and planning the business strategy. Companies' structures vary but the illustration on the next page shows what the structure might look like for a company employing around 80–100 craft operatives.

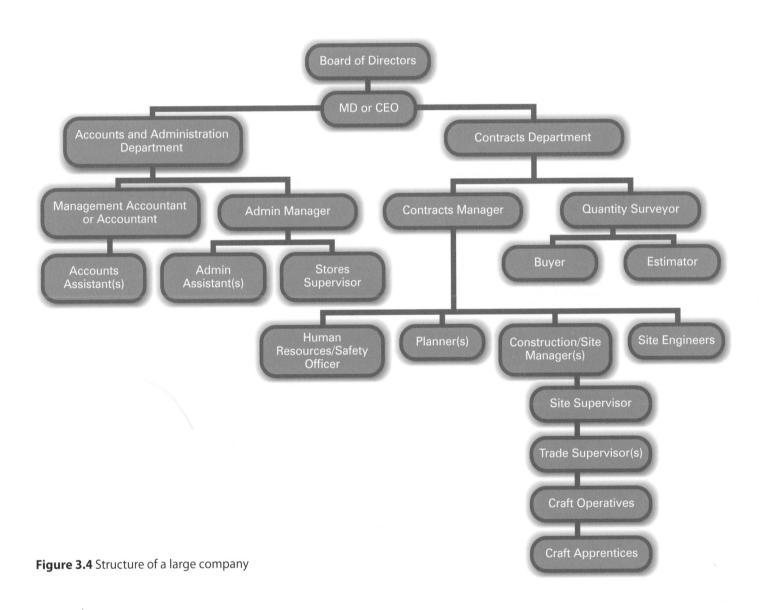

Figure 3.4 Structure of a large company

The construction team

The following outlines the key jobs of the main members of the construction team that you may come into contact with on a construction site.

Job	Main responsibilities	Other aspects
Contracts manager	• Responsible for running several contracts • Works closely with construction management team • Link between the other sections of the business and the MD/CEO • Makes sure the job is running to cost and programme	• Office-based • Some site visits
Construction manager (Site manager, site agent or building manager)	• Responsible for running construction site or a section of a large project • Develops strategy for the project • Plans ahead to solve problems before they happen • Makes sure site and construction processes are carried out safely • Communicates with clients to report progress and seek further information • Motivates workforce	• On larger contracts, the construction manager could have the support of supervisory staff
Site supervisor (Foreman)	• Responsible for the trade supervisor and trade operatives • Supervises the day-to-day running of the job	• Reports to the construction manager
Trade supervisor (Charge hand)	• Supervises operatives from a specific trade (e.g. bricklayers)	• Reports to the site supervisor • This role is usually only needed on very large jobs
Planner	• Ensures project completed on time and within budget • Reschedules projects as necessary • Works closely with estimators to establish working methods and costs • Plans the most effective use of time, people, plant and equipment • Schedules events in a logical sequence • Visits sites to monitor progress	• Works with construction manager

Figure 3.5 Non-craft jobs in construction

Job	Main responsibilities	Other aspects
Site engineer	• Ensures the technical aspects of the construction projects are correct • Sets out the site so that things are in the right place • Interprets original plans, documents and drawings • Liaises with workforce and subcontractors on practical matters • Checks quality • Refers queries to the relevant people • Provides 'as built' details	• Key role in ensuring things are built correctly and to the right quality • May supervise parts of the construction
Quantity surveyor – contractor (Commercial manager, Cost consultant)	• Advises on and monitors the costs of a project • Allocates cost effectively work to specialist subcontractors • Manages costs • Negotiates with client's quantity surveyor on payments and final account • Arranges payments to subcontractors	
Buyer (Procurement officer)	• Identifies suppliers of materials • Obtains quotations from suppliers • Purchases all the construction materials needed for a job • Negotiates on prices and delivery • Resolves quality or delivery problems	• Liaises with other members of the construction team
Estimator	• Calculates how much a project will cost including plant, materials and labour • Identifies the most cost-effective construction methods • Establishes costs for labour, plant, equipment and materials • Calculates cash flows and margins • Seeks clarification on contract issues affecting costs	• This information is used as the basis of the tender the contractor submits to a client prior to getting the contract
Accounts department	• Invoices for work carried out • Pays staff, suppliers and subcontractors • Produces financial reports and budget forecasts • Deals with Inland Revenue and Customs and Excise (VAT)	• Works with Construction manager
Administration department	• Works closely with accounts to keep financial records • Deals with customers and other enquiries • Records and files work records, time sheets etc	• Most systems are electronic, but companies often keep hard-copy records

Figure 3.5 Non-craft jobs in construction *(continued overleaf)*

Job	Main responsibilities	Other aspects
Human resources department	• Oversees employees' training and development needs • Assists with the recruitment and selection of new staff • Deals with industrial relations matters	• Some businesses will also employ a human resources officer, who looks after health and safety matters. (Otherwise, health and safety would be the responsibility of the Construction or Contracts manager)

Figure 3.5 Non-craft jobs in construction *(continued)*

Any construction job is likely to involve several craft trades. The following table outlines the major trades and the tasks they are generally responsible for.

Craft or trade	Main aspects of job	Typical job details
Carpenter and joiner	Positions and fixes timber materials and components from roofs and floors to doors, kitchens and stairs.	Work on site is usually divided into two phases or 'fixes'. The **first fix** is usually completed before the building has been made watertight and includes fixing floor joists, boards and sheets, stud partitions, and roof trusses and timbers. The **second fix** takes place when the building is watertight. Staircases, doors, kitchen units, architraves and skirtings are common second-fix items.
Roof slater and tiler	Creates a waterproof covering for a building by applying individual slates and tiles to a basic framework.	Roofing felt is laid over the roof timbers and tacked down. Timber battens are then fixed horizontally at centres to suit the roof dimensions and type of slate or tile. The roof is loaded with slates or tiles and then they are laid. Tiles have to be cut to fit at valleys, hips and gable ends. Ridge tiles are bedded on mortar. On re-roofing or maintenance work, tiles and slates are removed and timber checked and replaced where necessary. Reclaimed tiles and slates must be checked and sorted before reuse.
Built-up felt roofer	Built-up felt roofing is used mainly on flat roofs and sometimes on sloping roofs and vertical surfaces.	It is called 'built-up' because it involves putting layers of felt on top of each other using bitumen to form a waterproof surface.

Figure 3.6 Main tasks of construction trades

Craft or trade	Main aspects of job	Typical job details
Bricklayer	Uses bricks and blocks to construct interior and exterior building walls as well as tunnel linings, archways and ornamental brickwork.	Working on new buildings, extensions, maintenance and restoration of older buildings. Building foundations, bringing brickwork and blockwork up to damp-proof course level. Working at height from trestles, hop-ups and scaffolds. Construction of drainage and concrete work.
Construction general operative (labourer)	Provides varied support on site in a number of areas.	Skills include concreting, form working, steel fixing, kerb laying and drainage work. Some operatives specialise in areas such as spraying or repairing concrete.
Plasterer	Applies plaster to interior walls so that decoration can be applied.	Renders external walls and sometimes lays floor screeds. Some specialise in fibrous plasterwork. Others work in maintenance, conservation and restoration of existing buildings.
Painter and decorator	Applies paint, wall coverings and other materials to the inside and outside of buildings.	Work can include maintenance, conservation and restoration of existing buildings as well as on new-build, commercial and industrial sites. Wall coverings need to be accurately measured, cut and hung using the correct adhesive. Some decorators help clients with interior design.
Wall and floor tiler	Fixes tiles on exterior and interior walls, as well as on other surfaces such as floors and swimming pools.	On maintenance and restoration work, most surfaces need to be repaired. Tiles are normally laid on adhesive, cut along the edges and around obstacles, and finally grouted up. Setting out the tiles is critical when intricate and complex patterns are being produced. Tiles come in a variety of shapes, sizes, textures and colours and materials (e.g. ceramic, stone, terracotta).
Floor layer	Prepares and levels floors for new and old buildings. The most common materials are carpet, cork, plastic and timber.	Prepares floor measurements and plans laying to avoid waste. Different techniques are used for different materials. These include gripper fixings, glues, adhesives and secret nailing.
Plant operator	Uses cranes and other plant for moving and transporting construction materials.	Cranes can be mobile, track-mounted or tower cranes. Transporting plant includes excavators, specialised earth-moving equipment, fork-lift trucks and power-access equipment.

Figure 3.6 Main tasks of construction trades *(continued overleaf)*

Craft or trade	Main aspects of job	Typical job details
Scaffolder	Erects scaffolding and working platforms for construction workers, fixes edge protection. Scaffolders need to be competent persons who have been properly certificated to do their work.	Scaffold needs a firm foundation and careful planning. It can be made from traditional tubes and fittings or may be a purpose-made system. Working platforms are made using scaffold boards. Access is usually provided with ladders. Hoists, special loading platforms, safety nets and guard rails are all features of scaffolding erection. Most scaffolds need to be lifted and modified as construction work progresses.

Figure 3.6 Main tasks of construction trades *(continued)*

Did you know?

Roofers use a range of specialist tools to cut slates and tiles

Did you know?

There are approximately 8,000 floor layers in the UK. They usually work in pairs for specialist firms or are self-employed

Other occupations

There are a number of other occupations that play an important role in the overall construction process.

Architect

Architects plan and design buildings. The range of their work varies widely and can include the design and procurement (buying) of new buildings, alteration and refurbishment of existing buildings and conservation work. An architect's work includes:

- meeting and negotiating with clients
- creating design solutions
- preparing detailed drawings and specifications
- obtaining planning permission and preparing legal documents
- choosing building materials
- planning and sometimes managing the building process
- liaising with the construction team
- inspecting work on site
- advising the client on their choice of a contractor.

Project manager/clerk of works

The project manager or clerk of works takes overall responsibility for the planning, management, coordination and financial control of a construction project. They work for architects, clients such as local authorities, or as a consultant.

A project manager will ensure the client's requirements are met and that the project is completed on time and within budget. Depending on the project, their responsibilities can start at the design stage and continue through to completion and handover to the client.

A project manager's/clerk of work's job includes:

- representing the client's interests
- providing independent advice on the management of projects
- organising the various professional people working on the project
- making sure that all the aims of the project are met
- ensuring quality standards are met
- keeping track of progress
- accounting, costing and billing.

Structural engineer

Structural engineers are involved in the structural design of buildings and structures such as bridges and viaducts. The primary role of the structural engineer is to ensure that these structures function safely. They can also be involved in the assessment of existing structures, perhaps for insurance claims, to advise on repair work or to analyse the viability of alterations and adaptations.

Building surveyor

Building surveyors are involved in the maintenance, alteration, repair, refurbishment and restoration of existing buildings. A building surveyor's work includes:

- organising and carrying out structural surveys
- legal work including negotiating with local authorities
- preparing plans and specifications
- advising people on building matters such as conservation and insulation.

Building-control officer

Working for local authorities, building-control officers ensure that buildings conform to regulations on public health, safety, conservation and access for the disabled. The job involves the inspection of plans and of work-in-progress at various stages relating to Building Regulations. A building-control officer's work includes:

- checking plans and keeping records of how each project is progressing
- carrying out inspections of foundations, drainage and other major building elements
- issuing a completion certificate when projects are finished
- carrying out surveys of potentially dangerous buildings
- meeting with architects and engineers at the design stage
- using technical knowledge on site.

Facilities manager

A facilities manager ensures that buildings continue to function once they are occupied. He or she is responsible for maintaining the building and carrying out any changes needed to ensure the building continues to fulfil the needs of the organisations using it. A facilities manager's work includes:

- planning how the inside of the building should be organised for the people occupying it

- managing renovation works

- managing routine maintenance

- managing the installation and maintenance of computer and office equipment

- managing the building's security

- managing the cleaning and general upkeep of the building

- negotiating with contractors and service suppliers.

Building-services engineer

Building services include: water, heating, lighting, electrical, gas, communications and other mechanical services such as lifts and escalators. Building-services engineering involves designing, installing and maintaining these services in domestic, public, commercial and industrial buildings. A building-services engineer's work includes:

- designing the services, mostly using computer-aided design packages

- planning, installing, maintaining and repairing services

- making detailed calculations and drawings.

Most building-services engineers work for manufacturers, large construction companies, engineering consultants, large mechanical and electrical contracting companies, architects' practices or local authorities.

Their role often involves working with other professionals as part of a team on the design of buildings, for example with architects, structural engineers and contractors.

When working for a consultant, this job is mainly office-based at the design stage. Once construction starts, there will be site visits to liaise with the contractors installing the services. When working for a contractor, the building services engineer may oversee the job and even manage the workforce and is therefore likely to be site-based. When working for a services supplier, the role will require being involved in design, manufacture and installation and may involve spending a lot of time travelling between the office and various sites.

How plumbing relates to the construction industry

Plumbing businesses often work for construction companies on a subcontracting basis. This means that the main contractor is a construction company, and that they contract directly with the customer or client. The main contractor then sublets a part of the contract, for example the plumbing work, to a plumbing contractor (subcontractor).

The plumbing subcontractor deals directly with the construction company and not the client. The tender price, contract details and payment of work are between the plumber and the construction company.

Plumbing is not the only part of the contract that is subcontracted out. Other areas include:

- electrical
- heating and ventilating – domestic, industrial and commercial
- refrigeration and air-conditioning
- service and maintenance
- ductwork
- gas – domestic, industrial and commercial.

Revision

Many companies in the plumbing industry often seek specialist support and assistance through involvement, or membership of, the various industry trade bodies, for example JTL, the Association of Plumbing and Heating Contractors (APHC), the JIB for PMES, etc. It is important that you have a full understanding of the role of these various industry bodies in ensuring that effective working relationships are developed and maintained in the plumbing industry. At this point in the chapter we recommend that you familiarise yourself again with the key roles of the main industry organisations by reading Chapter 1 pages 4 to 7.

Knowledge check 1

1   Approximately what percentage of plumbing businesses employ between one and four staff?

2   What type of business does not trade its shares on the stock market but is required to register with Companies House?

3   Which member of the construction team is responsible for the purchase of materials for a construction contract?

4   Which construction trade weatherproofs the flat roofs of buildings with building felt?

5   Which member of the construction team could provide detail on whether a load bearing building structure is capable of taking the weight of a cold-water storage cistern?

Communication with customers and co-contractors

Communication and job responsibilities of plumbing staff

A great deal of plumbing work involves communication and problem solving. As your career progresses you well constantly need to understand the level of responsibility that you have in your company. This will lay down how you deal with communication and problem solving issues. How companies deal with this can vary. Here are some typical examples:

- Apprentice plumber – works directly to a qualified member of staff who provides the necessary level of work instruction and supervision to undertake the work. A plumbing apprentice would not usually be directly involved in communications with customers and co-contractors related to work problems/issues and would usually pass these on his/her supervisor to handle.

- Level 2 (or equivalent) qualified plumber – usually does not take full responsibility for the job. Normally he/she works under minimal supervision and generally responds to queries for information from customers and co-contractors. Requests for additional work, or major changes, are usually forwarded to the supervisor for action.

- Level 3 (or equivalent) qualified plumber – usually takes full responsibility for the job including dealing with queries from customers and co-contractors. A Level 3 plumber may be given the responsibility for dealing with requests for additional work, including the confirmation of the work, pricing information and customer complaints.

Remember

It is good practice to maintain regular communication with your employer on the extent of your job responsibilities as your career progresses

So essentially an apprentice or a Level 2 qualified plumber will more than likely have to forward requests for work, or complaint issues, to other staff in the business to deal with.

The communication process

Communication is the passing and receiving of information between one or more people, which achieves understanding and gets the required result – hopefully. Poor communication can lead to dispute and disagreement on a job so your ability to effectively communicate with customers and fellow workers is likely to be as important to employment in the industry as your technical skills in doing the job. Communication and the passing of information usually takes place either:

* verbally – by mouth
* in written form
* visually (graphic, body language)
* by a combination of one or more of the above.

Whatever the method, remember:

* the person sending the communication must know what they want
* the message must be delivered to the person receiving it in the appropriate style and in a form that can be understood
* the message must get the desired response from the receiver
* the sender must know that the message has been acted upon.

So, communication is a two-way process between sender and receiver(s). Taking this further, there are a number of essentials of communication.

Plumber in discussion with customer

Clarity	The communication should be presented in simple, direct language that the receiver is able to understand.
Correct	The communication should contain accurate information.
Concise/ complete	Keep the communication short without being curt (rudely brief), but make sure it includes all the relevant information, and enough of it, so the receiver can respond with the information you are after.
Constructive	The communication should enable the receiver to respond in a way to achieve their purpose.

Figure 3.7 Essentials of communication

The message should also be in the correct style for its intended purpose and for the person who is receiving it. This can be:

- formal – a written record is produced
- informal – communication by word of mouth.

There are many ways in which information may be communicated, including:

- letters
- memos
- reports
- faxes
- text messages
- email
- conversations face-to-face
- meetings

- telephones
- signs
- charts
- graphs
- video conferencing
- video
- DVD.

Communication can therefore be:

- verbal
- written

- graphics.

Many jobs will require a combination of communication methods to be used, for example pricing for a job with a customer may include:

- arranging a site visit by phone
- meeting with the customer to discuss his/her needs
- preparing a written list of information and tasks whilst on the job
- obtaining a price for the material for the job by email from the merchants
- producing a written quotation for the job
- and the list goes on…

The important point is that we use the right type of communication for a particular purpose.

Verbal communication

This is an example of informal communication, i.e. there is no written record of what has taken place.

Advantages

- More personal contact between sender and receiver.
- Tone and body language can be used to help pass the information on, and to establish whether the receiver has in fact understood it.
- Feedback is usually instant.

Remember

Certain types of information may be considered private and confidential, for example profits made on a job, the purchase price of materials. You therefore need to be careful in identifying the types of information that can be passed on as part of your job

Disadvantages

- No formal record unless the conversation is recorded – this can be a real problem if a dispute arises as there is no record of what has been discussed or agreed.

- Not much good if you need to send a lot of detailed or technical information.

- The receiver can be easily distracted.

Verbal communication tends to be used to pass on relatively 'low-risk' information, such as advice and guidance to a customer on how a component operates, arrangements for protecting the customer's furniture, confirmation of an appointment on-site, etc.

Written communication

Advantages

- Provides a permanent record for future reference, particularly useful with:
  - legal documents
  - contracts
  - agreements including estimates, quotations and job orders
  - key points from meetings.

- Is a more reliable method for reporting complex or technical information.

Disadvantages

- Producing written material takes time.

- Feedback can be delayed, or not given at all.

- Chance of misunderstandings, as a face-to-face explanation will not be provided.

- Lacks the personal touch.

- Can be used for negative reasons, delaying tactics.

Written communication tends to be used for 'higher-risk' activities such as those which may have an element of financial risk, for example agreement to the price for doing a job and the range of tasks to be carried out.

Graphics

These are generally used to support other formal or written documents to assist with understanding of the subject.

Advantages

- Useful for explaining complex technical details.

- Useful where it's difficult for people to read.

Disadvantages

- No control over interpretation and feedback unless you're there to explain.

- No personal contact if graphics are forwarded onto the receiver.

Use of information technology

Information technology enables written, or visual, communication to be forwarded quickly by the use of email. Email is basically a 'written or graphic' electronic message sent down the phone line.

Documents of all sizes can be forwarded as attachments to the e-mail and then downloaded by the receiver either in hard copy, or read on-screen and stored on their system. Information technology is improving constantly, and mobile telephones have text facilities and internet connections. Some phones have photo messaging, or software that will enable video messages to be forwarded – all increasing the methods of effective communication.

Effective communication in the workplace

Effectively dealing with customers or co-contractors on-site is essential to the smooth running of a job. Here are a few pointers to effective communication:

Definition

Toolbox talk – a briefing given by management/ supervisory staff on subjects such as safety, customer care, etc.

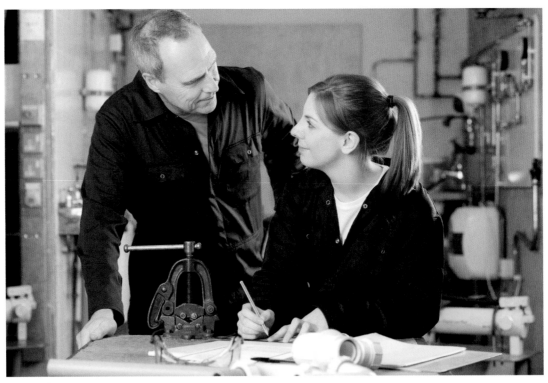

Plumber in discussion with fellow worker

- Always be polite – many problems on-site arise from a failure to communicate properly, in a proper manner, or with incorrect or wrong information.

- Always deal with a query if it is raised – the vast majority of complaints raised by customers in the plumbing industry relate to failure to provide information or failure to respond to a query while the work is being carried out.

- If a query is raised, remember to take all the details. Only deal with queries that are regarded as your normal responsibilities and, if the query needs to be passed on to someone else for action, arrange for it to be passed on and try and set a timescale for the response.

- A good plumbing company will see good communication skills as a major part of their staff recruitment requirements. A company will want to ensure that effective communication takes place between its staff and customers/co-contractors, as well as effective communication between its management/supervisory staff and its plumbers through activities like **toolbox talks** designed to deal with better ways of carrying out the work.

Resolving disputes

It would be a perfect world if the job always went according to plan. Part of the work of a plumber will involve communication and dealing with dispute situations or solving customer complaints. How effectively we deal with these issues can have a major impact on any plumbing business. The extent that you will be involved in resolving disputes really depends on the extent of the job holder's responsibility in the business (mentioned earlier in this chapter). Here are a few key points:

- Customer complaints should be actioned as soon as possible – if left they have a tendency to 'spiral out of control'. Ensuring that the right person gets the information to deal with the complaint is essential if you don't have the authority to deal with it yourself.

- Plumbers can sometimes see quite serious problems between work colleagues on-site. If this occurs, most businesses usually require their operatives to report the matter – again it's an item of good practice. Disputes such as this need urgent resolution as they have a tendency to affect the work carried out, including potential costs to the company and may leave a bad impression of the company on its customers and co-contractors.

- Most companies encourage staff to raise personnel issues such as pay and conditions with them direct rather than through 'chatting with work colleagues'. Typically this will affect the work carried out if not properly addressed.

- Formal disputes between staff and management in larger companies may be dealt with via the trade union if key personnel involved in the dispute are members. Dispute resolution may mean formal meetings between management and affected personnel. These meetings are usually well-structured with a formal agenda and purpose for the meeting. In smaller companies it is usually the boss sitting down and discussing the points with the affected operatives directly.

- In the event that disputes between the trade union and management of large companies cannot be resolved then the matter is usually handed over to independent arbitration – this is where a third party reviews the case and recommends solutions to overcome the problem.

Did you know?

Arbitration – guidance on resolving employment disputes and the arbitration process is provided by the Advising, Conciliation and Arbitration Service (ACAS)

Did you know?

Plumbing companies sometimes arrange social events outside work in order to assist with team-building at work and assist with the smooth running of jobs

Knowledge check 2

1   What level of plumbing qualification is normally required to deal with a wide range of site communications such as job pricing or customer complaints?

2   An email is an example of which type of communication method?

3   Provide one example of where informal communication may be appropriate.

4   How should a plumbing apprentice deal with a request from a customer for a price for carrying out a job?

5   What action should be taken by a plumber who suspects that an initiation ceremony for an apprentice has taken place on-site?

Additional information sources

Sources of additional information on topics in this chapter include:

SummitSkills (various reports outlining the structure of the building services industry) – www.summitskills.org.uk
Business Link (providing guidance on the legal structure of companies) – www.businesslink.gov.uk
ACAS (providing guidance on personnel requirements in companies) – www.acas.co.uk

chapter 4

Key plumbing principles

OVERVIEW

This chapter covers the science of plumbing and acts as an introduction to the basic materials, theories and concepts that you will encounter and work with on a daily basis. It includes:

- **Plumbing materials**
  - standards and regulations
  - materials and their properties

- **Plumbing science**
  - units and measurement
  - properties of water and heat
  - capillary action and siphonage
  - principles of electricity

Plumbing materials

A plumber's job involves the installation, maintenance and service of the following systems:

- hot and cold water
- central heating
- above-ground drainage
- sheet lead weathering
- rainwater
- below-ground drainage (connection only).

The plumber will work with a range of materials used for system pipework, fittings and components, including:

- copper
- steel
- lead
- brass
- plastics
- ceramics.

It is important that you have a good understanding of the **properties** of these materials and their suitability for the type of work for which they will be used.

Standards and Regulations in the use of materials

Materials used in everyday plumbing work are required to meet minimum standards of performance. It is also important that there is standardisation for the sizes and dimensions of fittings and components. Imagine what it would be like if you could buy a range of 15 mm fittings that all had a slightly different internal diameter from the external diameter of a 15 mm pipe!

British Standards

You may be familiar with the BSI kitemark. This symbol can be used only on materials and equipment that meet the standards of the British Standards Institution (BSI), the organisation for standards in the UK.

Figure 4.1 The BSI kitemark

The BSI ensures standards of quality and also sets standard dimensions for such items as pipes and fittings.

British Standards all start with the letters BS followed by the number of the standard. For example BS 6700 is one of the main standards for the plumbing industry. It's a specification for design, installation, testing and maintenance of services supplying water for domestic use within buildings and their **curtilages**.

We will refer to BS 6700 often throughout this book.

Definition

Properties – physical and working attributes of materials

Did you know?

The BSI is an independent organisation and was set up in 1901 under a Royal Charter

Definition

Curtilage – an area attached to a dwelling house and forming one enclosure within it

Codes of Practice

Codes of Practice (CoPs) make recommendations related to good practice. While they are not legal documents, CoPs are widely used by specifiers such as clients and architects.

The Water Supply (Water Fittings) Regulations 1999

These Regulations apply only to England and Wales, and have replaced Water Bye-laws. The Regulations are national Regulations made by the Government's Department of the Environment, Food and Rural Affairs (DEFRA), while the Bye-laws that they replace were made locally and applied in that area, for example, Yorkshire Water, Thames Water, Northwest Water etc.

These Regulations will be discussed in more detail later. They are mentioned here because they cover materials and substances in contact with water.

European Standards

These start with the letters EN followed by the standard number, in the same way as British Standards. Where a product is certified to an EN standard it means the manufacturer will have taken the product through a series of tests that are regularly checked under EC Quality Control Schemes.

International Standards

International Standards start with ISO followed by the number, for example ISO 9000, which refers to a standard for quality.

Water Regulations Advisory Scheme (WRAS)

Formally known as the Water Bye-laws Scheme, the WRAS has been carrying out fitting testing for many years and will continue to advise on Water Regulations in the future. As part of their work they produce a *Fittings and Material Directory*, which lists all approved fittings and is an important guide for all who aim to comply with or enforce the Water Regulations. Products approved by WRAS carry the symbols shown in the illustration.

Materials and their properties

The properties of materials relate to things such as how strong they are, how well they conduct heat or electricity or how flexible they are. We will examine the different properties materials have before looking at common plumbing materials in more detail.

Materials are classified according to a variety of properties and characteristics. Properties can be measured by the way materials react to a variety of influences.

Figure 4.2 WRAS approved product symbol

Figure 4.3 The properties of materials

Definition

Thermal conductivity – how well or poorly a material will conduct heat

Electrical conductivity – how well or poorly a material will conduct electricity

Hardness

Hardness is a measure of a material's resistance to permanent or plastic deformation by scratching or indentation. It is an important property in materials that have to resist wear or abrasion – moving parts in machinery, for example – and frequently needs to be considered along with the strength of materials. Hardness is measured on a scale of 1 to 10 based on the hardness of 10 naturally occurring minerals.

Strength

The strength of a material is the extent to which it can withstand an applied force or load (stress) without breaking. The load is expressed in terms of force per unit area (newtons per square metre, N/m^2), and can be in the form of:

- compression force, as applied to the piers of a bridge or a roof support

- tensile or stretching force, as applied to a guitar string, tow rope or crane cable

- shear force as applied by scissors or when materials are torn (see figure 4.5).

Materials are therefore described as having compressive, tensile or shear strength.

1	Talc
2	Gypsum
3	Calcite
4	Fluorite
5	Apatite
6	Feldspar
7	Quartz
8	Topaz
9	Corundum
10	Diamond

Figure 4.4 Mohs' scale

Did you know?

The hardness scale is called Mohs' scale. It is named after the German mineralogist Friedrich Mohs (1773–1839), who developed the scale in 1812

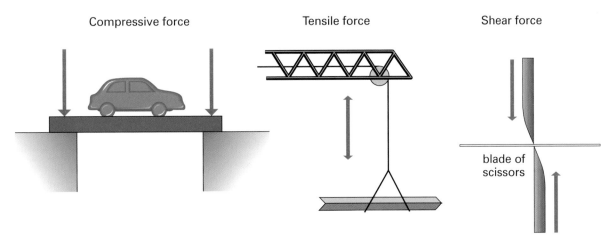

Figure 4.5 Compressive, tensile and shear forces

Materials that can withstand a high compression loading include cast iron, stone and brick, hence the common use of these materials for building purposes. However, they are brittle and will break if subjected to high tension. If a building is to be designed to resist tensile strain – in an earthquake-prone area, for example – steel, which has high tensile strength, would be a more suitable building material.

Elasticity and deformation

Almost all materials will stretch to some extent when a tensile force is applied to them. The increase in length on loading, compared to the original length of the material, is known as strain. As loading continues, a point is reached when the material will no longer return to its original shape and size on removal of the load, and permanent **deformation** occurs; the material is said to have exceeded its elastic limit or yield stress, and is suffering plastic deformation – it has been stretched irreversibly. Eventually, at maximum stress, the material reaches its breaking point – its ultimate tensile strength – and failure or fracture rapidly follows. This sequence is illustrated for four different materials below.

Find out

What can we learn about the materials from the stress/strain diagram?

Figure 4.6 The effects of stress and strain on materials

Elasticity	The ability of a material to resume its normal shape after being stretched or squeezed
Plasticity	The exact opposite of elasticity – a material that is not able to return to its original shape when deformed
Ductility	The ability of a material to withstand distortion without fracture, such as metal that can be drawn out into a fine wire (plastic deformation under tension)
Durability	A material's ability to resist wear and tear
Fusibility	The melting point of a material, i.e. when a solid changes to a liquid
Malleability	The ability of a metal to be worked without fracture
Temper	The degree of hardness in a metal
Tenacity	A material's ability to resist being pulled apart
Thermal Expansion	The amount a material expands when heated

Figure 4.7 Properties of metals

Copper and plastic pipes

- Mild steel has little elasticity but has the highest yield stress of all the samples; it is fairly ductile, i.e. it has a large range over which it can sustain plastic deformation, and it has the highest ultimate tensile strength.

- Cast iron is brittle – it has the least elasticity of the four samples, and no ability to sustain plastic deformation, although its tensile strength is higher than that of concrete.

- Copper has little elasticity but is the most ductile of the four samples. It has an ultimate tensile strength less than half that of mild steel.

- Concrete has little elasticity, and the lowest tensile strength of the four samples.

Pipework materials

There is no perfect pipework material that is suitable for all applications; different materials perform better in relation to different factors and conditions such as pressure, type of water, cost, bending and jointing method, corrosion resistance, expansion and appearance. There are two basic types of pipework material:

- metal
- plastic.

Metal

Metals rarely occur in their pure form. More often they occur as ores, which are **compounds** of the metal and have unwanted impurities. To produce the required metal, a process of smelting is necessary. Metals commonly used in the manufacturing industry include iron, copper, lead, tin, zinc and aluminium.

The most common method of producing metals is by removing the oxygen from the ore by a process known as reduction.

The industrial production of iron and steel

Iron ore (haematite – iron oxide) is loaded into a blast furnace along with coke and limestone.

Hot air is blasted into the base of the furnace and carbon from the coke reacts with oxygen from the air to form carbon monoxide.

Carbon monoxide reacts with oxygen from the haematite (iron oxide) to form carbon dioxide and iron.

Limestone combines with impurities in the ore (mainly silicates) to form slag.

The molten iron is tapped from the base of the furnace and solidifies into billets known as 'pigs' – hence the term 'pig iron'.

At this point the iron is impure. To form steel, which is an alloy of iron and carbon, it is necessary to reheat the iron to drive off the impurities, and then to add up to 1.5 per cent of carbon. Other metals can give the steel particular properties – the addition of chromium will produce stainless steel, for example. Alloys can be produced either by mixing different metals or by mixing metals with non-metallic elements, such as carbon.

Did you know?

Most metals that occur in their pure form tend to be unreactive – and valuable, such as gold and silver

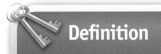

Definition

Compound – a mixture of two or more elements

Metals commonly used in the plumbing industry

Alloys

An **alloy** is a type of metal made from two or more other metals.

Some commonly used alloys		
Brass	Copper and zinc	Used for electrical contacts and corrosion-resistant fixings (screws, bolts etc.) and pipe fittings
Bronze	Copper and tin	Used for decorative or artistic purposes and corrosion-resistant pumps
Solder	Lead and tin, tin and copper	Used for electrical connections Used as a jointing material
Duralumin	Aluminium, magnesium, copper and manganese	Used in aircraft production
Gunmetal	Copper, tin and zinc	Used for underground corrosion-resistant fittings

Figure 4.8 Alloys

Copper

Copper tube has been used as a material for pipework for over 100 years. It is a malleable and ductile material which you will use frequently throughout your plumbing career. There are four main types of copper tube used in the plumbing industry:

- R250 half hard (also called Table X) is the copper tube most commonly used above ground for most plumbing and heating installations. It is fairly rigid and will usually need to be bent using a bending machine

- R290 hard (also called Table Z) is a more rigid copper tube. Its increased hardness means that the walls of the pipe don't need to be as thick, so the internal diameter of the tube or bore can be wider than that of the R250 type of tube; the tube cannot easily be bent

- R220 soft coils (also called Table W) is a copper tube used for micro-bore pipework, typically on central heating systems

- R220 soft coils (also called Table Y) is a softer copper tube which is most commonly used underground for the supply of water.

The outside diameter is the same for each type of pipe, but there are differences in the internal bore due to the variations in the pipe wall thickness. R250 half hard lengths and R220 soft coils are available with a plastic coating. This type of tube can be used in potentially corrosive environments as the plastic coating protects the copper from corroding.

Steel

1 Low carbon steel

Low carbon steel (LCS) or mild steel is an alloy made from iron and carbon. It is frequently used in the plumbing and heating industry and is manufactured to BS 1387. The tube comes in three grades of weight: light, medium and heavy. As with copper tube, the outer diameter is similar, but the internal bore and wall thickness varies.

Grade	Wall thickness	Bore	Colour code
Light LCS tube	Thin walls	Larger bore	brown
Medium LCS tube	Medium walls	Medium bore	blue
Heavy LCS tube	Thick walls	Smaller bore	red

Figure 4.9 Grades of LCS or mild steel

Light LCS is usually used for conduit. As a plumber, you will come across it on occasion, but you will work far more frequently with medium and heavy LCS. Medium and heavy LCS tubes are used for water-supply services, as they are capable of sustaining the pressures involved. When LCS tube is used for domestic water supplies, it must be galvanised.

2 Stainless steel

Stainless steel is the most recently developed pipe material used for water services. It is a complex alloy made up of a number of elements, as shown in the table.

The tube has a shiny appearance due to the chromium and nickel content and is protected from corrosion by a microscopic layer of chromium oxide, which quickly forms around the metal and prevents further oxidisation. This tube is produced with bores of 6 mm to 35 mm and has an average wall thickness of 0.7 mm. The outside diameters are similar to those of R250 copper tubes.

Stainless steel is commonly used where exposed pipework and sanitary appliances are needed, as it is a very strong metal (much stronger than copper) and is easy to clean. Stainless steel is also commonly used for:

- sink units
- urinal units and supply pipework
- commercial kitchen discharge pipework.

Element	%
Chromium	18
Nickel	10
Manganese	1.25
Silicon	0.6
Carbon	0.08
Iron	70 (approx)
Sulphur	trace
Phosphorus	trace

Figure 4.10 Composition of stainless steel

Lead

Lead is a very heavy, valuable metal which requires specialist handling. It is one of the oldest known metals and is highly ductile, malleable and corrosion-resistant. During your career, you will come into contact with lead in sheet form, which is used for weatherings on buildings. It was used in the past for mains, sanitary and rainwater pipework, but this practice stopped in 1986 with the Model Water Bye-laws, due to the possibility of lead poisoning. It has now been superseded by the use of such materials as plastics.

Did you know?

The main use of lead in plumbing today is the weathering of buildings

Cast iron

Cast iron is an alloy of iron and is approximately 3 per cent carbon. It is very heavy but quite brittle, although it can stand years of wear and tear. It has been used in the plumbing industry for many years for above- and below-ground sanitary pipework. You will probably come into contact with it on older properties and new industrial/commercial properties.

Plastics commonly used in the plumbing industry

Plastics (polymers) are products of the oil industry.

Ethene, a product of crude oil, is a building block of plastics. It is made up of carbon, hydrogen and oxygen atoms. Molecules of ethene (monomers) can link together into long chains (polymers) to make polythene (poly + ethene) when they are heated under pressure with a **catalyst**. If the ethene monomer is modified by the replacement of one of the hydrogen atoms with another atom or molecule, further monomers result, producing other plastics. This process is called polymerisation.

Definition

Catalyst – a substance that increases the rate of a chemical reaction while itself remaining unchanged

There are two main categories of plastics used in the plumbing industry:

- thermosetting plastics
- thermoplastics.

Thermosetting plastics are generally used for mouldings. They soften when first heated which enables them to be moulded, then they set hard and their shape is fixed; it cannot be altered by further heating. WC cisterns can be made of thermosetting plastic.

Thermoplastics can be resoftened by heating them. Most of the pipework materials you will come into contact with fall into this category. The different types of thermoplastics share many of the same characteristics:

- strong resistance to acids and alkalis
- low specific heat (that is, they do not absorb as much heat as metallic materials)
- poor conductors of heat
- affected by sunlight (this leads to the plastic becoming brittle, also called degradation).

Material	Max usage Temp °C	Main plumbing industry purpose
Polythene – low density	80	Flexible pipe material used to channel chemical waste
Polythene – high density	104	More rigid, again used for chemical or laboratory waste
Polypropylene	120	Tough plastic with a relatively high melting temperature, can be used to channel boiling water for short periods of time
Polyvinyl Chloride (PVC)	40–65	One of the most common pipework materials, used for discharge and drainage pipework
Unplasticised Polyvinyl Chloride (UPVC)	65	More rigid than PVC, used for cold water supply pipework
Acrylonitrile Butadiene Styrene	90	Able to withstand higher temperatures than PVC, used for small-diameter waste, discharge and overflow pipework

Figure 4.11 Types of thermoplastic

Other materials relevant to the plumbing industry

Ceramics include those products that are made by baking or firing mixtures of clay, sand and other minerals: bricks, tiles, earthenware, pottery and china. The kiln firing process fuses the individual ingredients of the product into a tough matrix. The main constituent of all these products is the element silicon (Si) – clay is aluminium silicate; sand is silicon dioxide (silica). Ceramics also include those products made by 'curing' mixtures of sand, gravel and water with a setting agent (usually cement) to form concrete, or mortar, using a sand, water and cement mixture.

Types of glass – pyrex, lead crystal and normal

Glass is also produced by the melting together of minerals. The basic ingredients are sand (silicon dioxide), calcium carbonate ($CaCO_3$) and sodium carbonate ($NaCO_3$). The resulting mixture of calcium and sodium silicates cools to form glass. Again, additives can change the character of the product: boron will produce heat-resistant 'Pyrex'– type glass, and lead will produce hard 'crystal' glass.

Knowledge check 1

1   What is the key quality standards sign used by the British Standards Institution known as?

2   What is the ability of a material to be stretched then to resume its normal shape known as?

3   What colour code is medium duty low carbon steel tube?

4   What is the most common type of copper tube used in plumbing systems?

5   What type of plastics when formed or moulded can be re-softened by heating?

Plumbing science

Units of measurement

In the UK there are two principal systems of measurement: metric and imperial. The units of measurement you will come across in the plumbing trade will usually be metric (metres, kilograms etc.), although you may hear reference to imperial measurements such as feet, inches, pounds and ounces.

The standard international measurement system, commonly known as SI units, predominantly uses metric measurements. The table overleaf gives an indication of the basic SI units and appropriate metric and imperial equivalents.

Did you know?

SI stands for 'Système Internationale d'Unités'

Attribute	SI unit	Abbreviation	Conversion	Imperial unit(s)	Imperial abbreviation	Conversion
Length	metre centimetre	m cm		inch, feet, mile	in, ft .	1 in = 2.54 cm 1 ft = 0.3048 m
Mass	kilogram	kg		ounce, pound	oz., lb .	1 oz = 28.35 g 1 lb = 0.4536 kg
Time	second	s	60 s = 1 min 60 mins = 1 hour			
Electric current	ampere	A				
Temperature	kelvin	K	degrees Centigrade (°C) K = °C + 273.15	degrees Fahrenheit	°F	$°C = \frac{5}{9}(°F - 32)$
Angle	radian	rad	1 rad = 57 degrees (°) 1 degree = 60 mins (°) 1 min = 60 seconds (°)			

Area	square metre	m^2	1 hectare = 10,000 m^2	square inches, acre		
Volume	cubic metre	m^3	cc = cm^3	cubic inches		
Capacity	litre	l	1 ml = 1 cc or cm^3	pint, gallon	1 pint = 0.5663 l	
Speed	metres per second	m/s		miles per hour feet per second	mph fps	
Acceleration	metres per second per second	m/s^2 or ms^2				
Force	newton	N		pounds per square inch	lb/in^2	

Figure 4.12 SI Units

FAQ

Why do I need to know all this stuff about science?

This all affects the way in which plumbing systems are chosen, installed and work, so it is essential to have a good awareness of the key science principles.

Mass and weight

In its simplest terms, mass is the amount of matter in an object and is measured in grams (or for larger weights, kilograms). Under normal circumstances and as long as it remains intact, an object should always maintain the same mass. For example, a nail will maintain the same mass whether it is:

- on a workbench
- on the moon.

But will it weigh the same?

The weight of an object is the force exerted by its mass as a result of acceleration due to **gravity**. On earth all objects are being accelerated towards the centre of the planet due to the earth's gravitational pull. The 'pull' exerted by gravity on the mass of an object is known as its weight, which is measured in newtons. A newton is a unit of measurement equivalent to 1 metre per second (m/s) per 1 kg of mass.

Therefore: weight (in newtons) = mass × acceleration due to gravity

To give an example of the difference between mass and weight we need to go to the moon:

On earth

The gravitational pull of the earth is 9.8 m/s^2

Therefore an object with a mass of 1 kg on earth would weigh 9.8 newtons.

On the moon

The gravitational pull of the moon is approximately 1.633 m/s^2

Therefore an object with a mass of 1 kg on the moon would weigh 1.633 newtons.

The mass of the object does not change whether it is on the earth or the moon, but the weight of an object changes considerably because of the reduced gravitational pull of the moon.

Density

Density of solids

Solid materials, which have the same size and shape, can frequently have a completely different mass. This relative lightness or heaviness is referred to as **density**. In practical terms the density of an object or material is a measure of its mass (grams) compared to its volume (cm^3) and can be worked out using the following formula:

$$\text{density} = \frac{\text{mass}}{\text{volume}}$$

The densities of all common materials you'll come into contact with during your plumbing career are known. Lists of these comparative densities can be found in a variety of reference sources.

Density of liquids

Like solids, liquids and gases also have differing densities depending on the number of molecules that are present within a particular volume of the substance. As a plumber it will be important to understand the density of water and that this changes with the water's temperature. Water is less dense when it is heated:

- 1 m³ of water at 4°C has a mass of 1,000 kg
- 1 m³ of water at 82°C has a mass of 967 kg.

This is because heat energy excites the molecules so that they move further apart and the water becomes less dense.

Relative density

Relative density (also occasionally known as specific gravity) is an effective way of measuring the density of a substance or object by comparing its weight per volume to an equal volume of water. Water has a relative density (specific gravity) of 1.0. For example:

- 1 m³ of water has a mass of 1,000 kg
- 1 m³ of mild steel has a mass of 7,700 kg.

The mild steel is 7.7 times heavier than water and therefore has a relative density of 7.7.

Density of gases

In the same way that water is said to have a relative density of 1.0 to enable comparisons of relative density for solids and liquids, air has a relative density of 1.0 to enable comparisons between gases which are classified by whether they are lighter or heavier than air. Think about helium-filled balloons for instance – do you think helium is lighter or heavier than air?

Pressure

Pressure is defined as force applied per unit area and is measured in newtons per square metre (N/m^2), a unit also known as a Pascal (Pa).

You will probably come across other terms used to identify pressure such as the 'bar' or 'pounds per square inch' (lbs/in^2). These can be expressed as:

- 1 bar = 100,000 N/m^2
- 1 lbs/in^2 = 6,894 N/m^2

Pressure is therefore a measurement of a concentration of force. The effect of a concentration of pressure can be seen if water flowing through a pipe is forced through a smaller gap by reducing the diameter of the pipe. (Think about a hosepipe, and how to maximise the force of the jet of water.)

Figure 4.13 The effects of pressure in solids and liquids

Pressure can also be lowered by 'spreading' the applied force over a wider area. For example, rescue teams will often spread themselves over fragile roofs or on thin ice to minimise the chance of the surface giving way.

As a plumber you will need to have a basic understanding of the effects pressure has on the pipes and fittings you will install. The internal pressure in a pipe or vessel will be affected by what is being transported (water or gas), and must be considered when deciding which material and which size of pipe or vessel should be used.

Pressure in liquids

The pressure in a liquid increases with depth, so in a plumbing system water pressure is higher at the lowest points of the system and lower at the highest points of the system.

Water pressure is measured using a number of different units –

- metres head (m)
- the pascal (Pa), also the Newton, per metre squared (N/m^2)
- bar pressure (bar).

> 1 metre head = approx. 10,000 Pa (10kPa) = approx. 0.1 bar

An understanding of the **pressure head** created in systems is important in determining component sizes, for example pipe sizes and confirming that components will be able to withstand the pressure of water created within them.

Intensity of pressure and total pressure

Intensity of pressure is the force created (kPa) by the weight of a given mass of water acting on a unit area (m^2).

Total pressure is the intensity of pressure multiplied by the area acted on.

Example 1:

Calculate the intensity of pressure and total pressure acting on the base of the lower cistern.

$$\text{Intensity of pressure} = \text{head} \times 9.81 \text{ kPa}$$
$$= 3 \times 9.81$$
$$= 29.43 \text{ kPa}/m^2 \text{ or approximately 0.3 bar pressure.}$$

An alternative method of calculating this is to multiply the head $\times$ 0.1 bar (0.1 bar = 1m head) = 3×0.1 bar = 0.3 bar pressure.

$$\text{Total pressure} = \text{intensity of pressure} \times \text{area of base}$$
$$= 29.43 \times (2 \times 1.5)$$
$$= 88.29 \text{ kPa}$$

Example 2:

If a tap is sited 5 metres below a plumbing cistern feeding it, the pressure created at the tap will be

5.0 metres head $\times$ 0.1 bar pressure = 0.5 bar

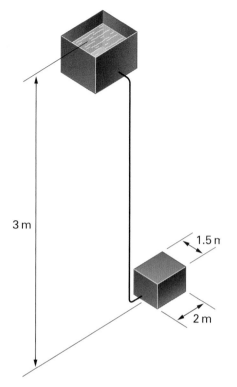

Figure 4.14 Pressure calculation

3 m

1.5 m

2 m

Atmospheric pressure

The pressure exerted by the weight of the earth's atmosphere pressing down on the ground varies depending on height above sea level. The pressure at the top of Mount Everest is not as high as the pressure in the bottom of a valley below sea level (such as the Great Rift Valley in Africa). The pressure at sea level is 101,325 N/m^2 (approximately 1 bar).

Plumbers must be aware of the effects of atmospheric pressure to ensure that they avoid creating 'negative' pressure or vacuums within pipework systems. Negative pressure can damage components.

The siphon

The siphon uses atmospheric pressure to operate. A quantity of water is forced through a short leg of pipe (column A) and over the crown of the siphon. The weight of water in column B longer leg (under positive pressure) is greater than in column A. The water continues to flow owing to the difference in the weight of water with positive (atmospheric pressure in column B) and negative (suction pressure) in column A.

Figure 4.15 The principle of the siphon

The principle of the siphon is used to good effect in plumbing applications. Examples include:

- siphonic WC pan
- WC flushing siphon
- siphoning the contents of a hot water storage cylinder using a hose pipe.

Properties of water

Water is a chemical compound of two gases: hydrogen and oxygen (H_2O). It is formed when hydrogen gas is burned.

One of the most important properties of water is its solvent power. It can dissolve numerous gases and solids to form **solutions**. The purest natural water is rainwater collected in the open countryside. It contains dissolved gases such as nitrogen, oxygen and carbon dioxide, but this does not affect its potability (suitability for drinking).

Water may be classified as having varying degrees of **hardness** or softness.

Hard water

Water is classified as hard if it is difficult to obtain a lather with soap. Hard water is created when it falls on ground containing calcium carbonates or sulphates (chalk, limestone and gypsum), which it has dissolved and taken into solution.

Soft water

Water is said to be soft when it is easy to produce a lather with soap. This is because of the absence of dissolved salts such as calcium carbonates and calcium sulphates.

Soft water can cause corrosion in plumbing components because it is relatively acidic

Water hardness

Hard water is undesirable in domestic installations as it can produce limescale in pipework, heating equipment and sanitary appliances. This can lead to high maintenance costs. Hard water also requires the use of much more soap and detergent for washing purposes, as the 'hardness' makes it far more difficult to produce a lather.

Water hardness can be described as temporary or permanent.

- Permanent hardness is a result of ions of nitrates and sulphates. It makes it difficult to form a lather and *cannot* be removed by boiling.

- Temporary hardness is a result of the amount of carbonate ions in the water. Temporary hardness *can* be removed from the water by boiling, which results in the carbonate being precipitated out as limescale. This hard scale accumulates inside boilers and circulating pipes, restricting the flow of water, reducing the efficiency of appliances and components and ultimately causing damage and system failure.

pH value and corrosion

The term **pH value** refers to the level of acidity or alkalinity of a substance. As a plumber, you will need to be particularly aware of the potential effects the acidity or alkalinity of water can have upon materials, appliances and components.

Both acids and alkalis can cause **corrosion** and thereby damage plumbing materials; metals are at particular risk.

FAQ

Why is it more difficult to obtain a good lather in different parts of the country?

This is because the hardness of water varies in different parts of the country. The harder the water, the more difficult it is to obtain a lather. Is the water where you live soft or hard?

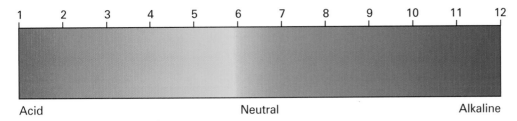

All water has a 'pH value'. Rainwater is naturally slightly acidic, due to small amounts of carbon dioxide and sulphur dioxide in the atmosphere being dissolved into it, forming very weak carbonic and sulphuric acids. The pH value of groundwater is affected by the different rock types it passes though. For instance, water with dissolved carbonate from chalk or limestone is alkali.

1	2	3	4	5	6	7	8	9	10	11	12

Acid   Neutral   Alkaline

Figure 4.16 The pH scale of acidity

The main causes of **corrosion** are:

- the effects of air
- the effects of water
- the direct effects of acids, alkalis and chemicals, e.g. from environmental sources
- electrolytic action.

Atmospheric corrosion

Pure air and pure water have little corrosive effect, but together in the form of moist air (oxygen + water vapour) they can attack **ferrous** metals such as steel and iron very quickly to form iron oxide or rust. The corrosive effects of rusting can completely destroy metal.

Various other gases (carbon dioxide, sulphur dioxide, sulphur trioxide), which are present in our atmosphere, also increase the corrosive effect air can have on particular metals, especially iron, steel and zinc. These gases tend to be more abundant in industrial areas as they are often waste products from various industrial processes.

Coastal areas also suffer from increased atmospheric corrosion due to the amount of sodium chloride (salt) from the sea which becomes dissolved into the local atmosphere.

Non-ferrous metals, such as copper, aluminium and lead, have significant protection against atmospheric corrosion. Protective barriers (usually sulphates) form on these metals to prevent further corrosion. This protection is also known as **patina**.

Corrosion by water

Ferrous metals are again particularly vulnerable to the effects of corrosion caused by water. These are commonly seen in central heating systems as black ferrous oxide and red rust build-up in radiators. A by-product of this process is hydrogen gas, which accumulates in the radiator, leading to the need to be 'bled'. (See Chapter 9.)

Definition

Ferrous – metals that contain iron

Figure 4.17 Corrosion effects in steel panel radiators

In certain areas of highly acidic water, copper may become slightly discoloured. This won't affect the quality or safety of the drinking water. However, in areas where lead pipework is still in use, if the water is very soft (acidic) there is a risk of it dissolving minute quantities of lead, thus contaminating the water – with potentially toxic effects, especially for children.

Corrosive effects of building materials and underground conditions

Some types of wood (such as oak) have a corrosive effect on lead, and latex cement and foamed concrete will adversely affect copper. Certain types of soil can damage underground pipework. Heavy clay soils may contain sulphates which can corrode lead, steel and copper. Ground containing ash and cinders is also very corrosive as they are strongly alkaline; if pipes are to be laid in such ground they should be wrapped in protective material.

Electrolytic action and corrosion

Electrolytic action describes a flow of electrically charged ions from an **anode** to a **cathode** through a medium known as the **electrolyte** (usually water) as shown in Figure 4.18.

Electrolytic corrosion takes place when the process of electrolysis leads to the destruction of the anode. The length of time it takes for the anode to be destroyed will depend on:

- the properties of the water that acts as the electrolyte: if the water is hot or acidic the rate of corrosion will be increased
- the position of the metals that make up the anode and the cathode in the electromotive series.

Did you know?

A simple battery uses electrolysis to produce electricity

The electromotive series

The list below shows the common elements used in the plumbing industry; the order in which they appear indicates their electromotive properties.

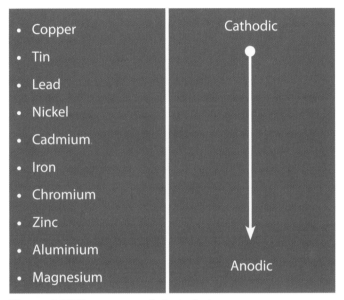

- Copper
- Tin
- Lead
- Nickel
- Cadmium
- Iron
- Chromium
- Zinc
- Aluminium
- Magnesium

Cathodic

Anodic

Figure 4.18 The process of electrolysis

Ammeter

A

Cathode

Anode

Electrolyte

The elements higher up in the list will destroy those lower down through the process of electrolytic corrosion. The further apart in the list the materials appear, the faster the corrosion will take place. For example, copper will destroy magnesium at a faster rate than lead will destroy chromium.

As a plumber you will need to be aware of the potential for electrolytic corrosion when two very dissimilar metals such as a galvanised tube and a copper fitting are in direct metallic contact. If these metallic elements are then surrounded by water (of a certain type) or damp ground, a basic electrical cell is effectively created and electrolytic corrosion can take place.

Find out

Try to think of some practical examples of electrolytic corrosion

Soil acts as an electrolyte

Anode – corroding

Low carbon steel pipe

Corrosion current

Return path of current

Cast iron pipe

Cathode protected

Figure 4.19 Example of possible electrolytic corrosion in pipework systems

Behaviour of water at different temperatures

Matter, such as water, can exist in what is known as three different states:

Solid (ice) ⟶ Liquid ⟶ Gas (steam)

In moving from one state to another a **change of state** must take place, during which the water is either cooled or heated.

Figure 4.20 Change of state diagram – atmospheric pressure

Referring to the diagram showing the behaviour of water at atmospheric pressure, in its natural state between 0°C to 100°C water is in the form of a **liquid.** When heated during its liquid state the water will increase in volume by up to 4 per cent (it will expand). Using the information previously covered on density, for the same mass of water when heated an increase in volume will take place, the result will be a decrease in density, and i.e. it will become lighter. Water is at its maximum density at a temperature of 4°C.

The **freezing point** of water is 0°C and at temperatures of 0°C and below water changes state to ice.

On cooling below 0°C the water turns to ice and expands. This expansion or increase in volume in an enclosed space can result in components rupturing, for example a burst pipe.

The **boiling point** of water is 100°C. At temperatures of 100°C and above water under atmospheric conditions changes state to steam.

When changing to steam a rapid increase in volume takes place (up to 1,600 times). Its original volume when in liquid form can have explosive effects if the water is stored in an enclosed space.

Did you know?

Hot water must be very effectively controlled to below 100°C in order to avoid potentially serious explosions.

Water stored at above atmospheric pressure

It is also worth noting that the pressure at which the water is stored is also linked to water temperature and water volume. So if water is stored in an enclosed space, such as a storage cylinder (a constant volume), at above atmospheric pressure, when water is heated, the temperature at which it boils will rise above 100°C.

As an example the boiling point of water at 1 bar pressure is approximately 120°C. So why is water stored at above atmospheric pressure and above temperatures of 100°C so dangerous if it is not boiling?

Quite simply if someone were to open a tap, or a storage cylinder ruptured, then the water pressure would rapidly reduce to atmospheric pressure, almost instantly causing the water to boil. This would result in a change of state from water to steam, causing a rapid increase in volume of the gas (up to 1,600 times its original volume) which the vessel or system more than likely could not withstand so would rupture, like a bomb in fact!

Properties of heat

Difference between heat and temperature

The main difference between heat and temperature is that heat is recognised as a unit of energy, measured in **joules** (J).

- Temperature is the degree of hotness of a substance.
- Heat is the amount of heat energy (J) that is contained within a substance.

For example, imagine an intensely heated short length of wire and a bucket of hot water:

The wire has a temperature of 350°C

The water has a temperature of 70°C

The wire is far hotter, but actually contains less heat energy.

Measuring temperature

The SI unit of temperature measurement is the degree kelvin (°K), but the unit you'll deal with most frequently is degrees Celsius (or centigrade), written as °C.

Temperature is measured using thermometers. There are many types, but the most common depend upon the expansion of either a liquid or a **bi-metallic strip**.

The two most common liquids used in thermometers are alcohol and mercury. These liquids are used because they expand at a uniform rate when exposed to heat.

Bi-metallic strips work on the principle of thermal expansion and contraction (covered in more detail later) and the fact that some metals expand and contract at a faster rate than others. Bi-metallic strips are frequently used in thermostats, where they will bend (see Figure 4.21) when a particular temperature is reached. This process will break an electrical circuit and turn off the heating.

Did you know?

Examples of the bi-metallic strip may be found in many thermostats, e.g. those used in ovens

Figure 4.21 A bi-metallic strip. The brass expands more than the iron on heating, thus bending the strip.

Brass

Iron

Before heating

After heating

Simple temperature measuring devices

Simple thermometer

In its simplest form a thermometer is a glass tube containing a substance such as alcohol or mercury that expands or contracts at a constant rate in response to temperature changes. A metal bulb is sited at one end of the thermometer for taking the temperature or immersing in water. This type of thermometer can be used for checking the water temperature at taps or outlets.

Clip-on thermometer

The clip-on, clamp-on or often called pipe thermometer, is used for taking surface temperatures of pipes. The thermometer uses the bi-metallic strip principle as its method of operation. This type of thermometer can be used when commissioning central heating systems where two clip-on thermometers are used to measure the temperature of both the flow and return pipework in order to establish the correct water flow through each.

Simple thermometer

Differential digital thermometer

This device comes with a range of attachments that can be used to measure simple water temperatures by immersing a probe in the water stream, or by connecting a device known as a **thermistor** to pipe surfaces. When commissioning central heating systems two thermistors are attached by means of a strap to the flow and return pipework and the digital thermometer can measure the temperature of each pipe and the difference in temperature between each pipe.

Differential digital thermometer

Definition

Thermistor – a device that changes electrical resistance in response to a temperature change

Specific heat capacity

To size various plumbing components, such as boilers and radiators, plumbers need to be able to understand the concept of heat. Heat is different from temperature. Heat is a measure of the amount of energy in a substance. The standard unit of measurement of heat is the joule.

In order to work out the amount of heat required to heat a substance we need to be able to measure the amount of heat required over time or the **power** required. This is a measure of the energy divided by the time taken to heat the substance and is expressed in in kW/h.

Material	kJ/kg °C
Water	4.186
Aluminium	0.887
Cast iron	0.554
Zinc	0.397
Lead	0.125
Copper	0.385
Mercury	0.125

Figure 4.22 Specific heat values

Find out

What do you think happens when materials are cooled down?

$$1 \text{ kW/hr} = \frac{1000 \text{ joules}}{1 \text{ second}} \times 3{,}600 \text{ seconds (number of seconds in one minute)}$$

In order to be able to undertake plumbing calculations involving heat we usually need to be able to work out the amount of heat required to raise a quantity of a substance such as water from one particular temperature to another. To do this we need to know the substance's **specific heat capacity**.

The specific heat capacity of a substance is the amount of heat required to raise the temperature 1kg of a substance by 1°C. The specific heat capacity of water is 4.186 kJ/kg/°C.

Example:

Calculate the heat energy and power required to raise 200 litres of water from 10°C to 60°C (assume 1 litre of water to roughly weighs 1kg).

> Heat energy = 200 litres × 4.186 kJ/kg/°C × (60°C – 10°C)
> = 41,860 kJ

Power required to heat the water in 1 hour (assuming no energy is lost)

$$= \frac{41{,}860}{3{,}600} = 11.63 \text{kW}$$

The power calculation is essential in determining factors such as the amount of energy required to re-heat a hot-water storage cylinder within a specific period of time. For example had the water required to be reheated in a 30 minute time period in our example power calculation, then the power required would be double as the re-heating period has been halved (from 3,600 seconds to 1,800 seconds).

$$= \frac{41{,}860}{1{,}800} = 23.26 \text{Kw}$$

Thermal expansion and contraction

Most materials will expand when heated. This is because all substances are made up of molecules (groups of atoms) which move about more vigorously when heated. This in turn leads to the molecules moving further apart from each other – resulting in the materials taking up more volume.

As the material cools, the molecules slow down and move closer together; thus the material gets smaller or contracts. The amount that the material expands in length when heated can be calculated using the following formula:

length (m) × temperature rise (°C) × coefficient of linear expansion

This table shows the coefficient values for some of the most common materials used in the plumbing industry.

Material	Coefficient °C
Plastic	0.00018
Zinc	0.000029
Lead	0.000029
Aluminium	0.000026
Tin	0.000021
Copper	0.000016
Cast iron	0.000011
Mild steel	0.000011
Invar	0.0000009

Figure 4.23 Coefficients of expansion

Example:

Find the amount a 6 m-long plastic discharge stack will expand due to a temperature rise of 19°C.

6 × 19 × 0.00018 = 0.02052 m or 20.52 mm

We need to take this expansion and contraction into account in plumbing systems that are constantly being subjected to heating and cooling processes – if we don't, the system or component may break down, causing, for example, leakage.

Heat transfer

As a plumber you will need a good understanding of the methods of heat transfer because you will be dealing with the effects of this process on a daily basis. There are three methods of heat transfer:

- conduction
- convection
- radiation.

Conduction

Conduction is the transfer of heat energy through a material. It takes place as a result of the increased vibration of molecules, which occurs when materials are heated. The vibrations from the heated material are passed on to adjoining material,

Safety tip

Plastics tend to expand the most and we quite often need to leave an expansion gap in the pipework system when working with plastics to prevent the material from failing which then heats up in turn. Some materials are better at conducting heat than others. For example:

- metals tend to be good conductors of heat
- wood is a poor conductor of heat.

Gases and liquids also conduct heat, but poorly. In the plumbing industry, conduction is principally between solids. Of the metals commonly used by plumbers, copper has a higher conductivity than steel, iron or lead. Wooden, ceramic and plastic materials, which are poorer conductors of heat, are known as **thermal insulators**.

Find out

How good are the following materials as heat conductors: plastic, rubber, ceramic, wool, carbon?

Convection

Convection is the transfer of heat by means of the movement of a locally heated **fluid** substance (usually air or water). As a fluid is heated, the process causes expansion, which in turn causes a lowering of density. The less dense warm fluid begins to rise,

Figure 4.24 Gravity circulation taking place owing to the effects of convection

and is replaced by cooler, denser fluid from below. Eventually, convection currents are set up which allow for a continuous flow of heat upwards from the source. Examples of systems that use convection currents for heat transfer are:

- convector heaters, which warm the air at one place in a room; the resulting convection currents transport the heat around the room

- domestic hot water systems, which depend on convection currents to transfer heat from an immersion heater (similar to the 'element' in an electric kettle) to the rest of the water in the hot tank (cylinder).

It is easy to demonstrate the 'updraught' part of a convection current by hanging a piece of light material above a convector heater. The movement of the hanging material will clearly show the presence of rising currents of warm air. Figure 4.24 shows a practical example of the use of convection currents in domestic hot-water systems.

Radiation

Radiation is the transfer of heat from a hot body to a cooler one without the presence of a material medium (other than air), by means of 'heat' waves. Heat radiation can be felt as the 'glow' from a fire or as the heat from the sun.

Heat radiation is better absorbed by some materials than others; in these instances colour can be an important factor. Dull, matt surfaces will absorb radiated heat more efficiently than shiny, polished surfaces.

Convected heat 85%

Radiated heat 15%

Heat production from panel radiator

Capillary action

Capillary action is the process by which a liquid is drawn or hauled up through a small gap between the surfaces of two materials. The relevance of this phenomenon is especially important for plumbers, as capillary action can affect the way in which water can get into buildings.

Forces of attraction

Surface tension

Surface tension describes the way in which water molecules 'cling' together to form what is effectively a very thin 'skin'. This can be demonstrated by filling a glass beaker right to the top and examining the top of the glass. The water will appear above the upper limit of the glass – why doesn't it spill down the side?

The answer is because of the **cohesion** (water molecules 'sticking together') as a result of surface tension.

Figure 4.25 Surface tension and cohesion   Figure 4.26 Adhesion

Adhesion

Adhesion is the force of attraction between water molecules and the side of the vessel the water is contained in. This leads to the slightly curved 'skin' that appears when water is held in a vessel as shown in the illustration. The correct name for this skin is a meniscus. These processes of adhesion and cohesion cause capillary action.

Practical examples of capillary action

Plumbers must consider the possible affect of capillary action when planning lead roofing work. Water can find its way into a building's lapped roof if materials are close together. The diagram shows how this occurs and how it can be remedied.

Water enters at A and rises up behind splash lap, entering building at B.

Splash lap

Undercloak

The introduction of an anti-capillarity groove at C prevents water entering the building.

Figure 4.27 Avoiding capillary action in fixing sheet leadwork

Capillary action can occur in 'S bend' drainage traps, which are found under sinks. If a length of waste material, such as a piece of dishcloth, becomes lodged in the 'S bend', capillary action could take place. This can lead to the loss of seal at the bottom of the trap, allowing bad smells from the drainage pipe to filter back into the home.

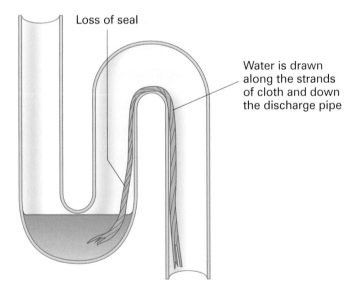

Loss of seal

Water is drawn along the strands of cloth and down the discharge pipe

Figure 4.28 Trap seal lost due to capillary action

Water flow

The rate of water flow through pipework is affected by friction. This is an important consideration when designing pipework installations. To get an idea of how friction works, compare the speed of a car driving freely through a tunnel with the speed of a car that is in contact with the tunnel walls. As you can imagine, the result of rubbing on the wall will cause a considerable reduction in speed.

Pipe walls have the same effect on water flow, i.e. the water is 'rubbing' on the pipe. This friction is further increased if the inside surface of the pipe wall is rough. This effect is known as frictional resistance or frictional loss.

Comparison of flow in smooth and rough pipework

- Water flowing through a 10 m length of 25 mm diameter pipe would flow at approximately 22 litres/minute for copper and plastic pipes, which have a smooth internal surface.

- Water flowing through a 10 m length of 25 mm diameter pipe would flow at approximately 18 litres/minute for galvanised LCS pipe, which has a rougher inner surface.

- The use of knuckle bends, or elbows, provides a greater resistance to the flow of water than the use of manufactured machine bends. The use of certain types of valves such as stop valves creates resistance to the flow of water in a pipeline due to the valve including reduced dimensions/minor obstructions within its design.

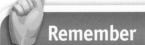

Remember

Frictional resistance is also increased by the number of fittings in a pipework run

Knowledge check 2

1   If 1m³ lead weighs 11,300 kg, what is the lead's relative density in comparison to water?

2   What is the intensity of pressure at a tap if the head of water above it is 6 m?

3   Water with a pH of 6 could be described as what?

4   What type of corrosion occurs when there is direct contact between dissimilar metals in a damp environment?

5   What is the transfer of heat from a hot body to a cooler one known as?

Principles of electricity

Try to imagine a world without electricity. Everything from computers to stereos, washing machines to lighting and heating relies on it.

So what precisely is it? You can't see it with the naked eye; you cannot smell it or hear it. But if you misuse or touch it, it can hurt or even kill. Because of its possible dangers, a deeper understanding of the subject is required.

What are electrons?

Molecules and atoms

Every known substance is made up of molecules. A molecule is a very tiny part of matter which can only be seen using special microscopes. Molecules are always in a state of rapid motion, and the ease with which they move around determines the form of the substance they make up:

- When molecules are densely packed together, their movement is restricted and the substance formed by them is called a **solid.**

- When the molecules are less densely bound together, and where they can move more freely, the substance is called a **liquid.**

- A substance that allows the molecules almost unrestricted movement is known as a **gas**.

These three conditions, solid, liquid and gas, are sometimes known as the three states of matter.

Molecules are made up of **atoms**, and we need to take an imaginary look inside the atom to begin to understand what electricity is. Atoms are the smallest parts of matter that can be subdivided, and they are not solid. They have a **nucleus** at their centre, made up of very tiny particles, known as **protons** and **neutrons**. Protons are said to contain a positive charge (+) and neutrons are electrically neutral (that is, they have no charge). We can think of neutrons as the 'glue' that holds the nucleus together. Around the nucleus orbits a third type of particle – **electrons** – which contain a negative charge (–).

You may remember from your school science studies that like charges repel each other (+ and + or – and –) and unlike charges attract each other (+ and –)

All atoms contain equal numbers of protons and electrons, and, in this unaltered state, the matter is said to be electrically neutral (in other words, no electricity is flowing). In some cases, it is possible to add or remove electrons, leaving the atom with a positive (+) or negative (–) charge.

Now to the part that assists the definition of electricity. Consider our own solar system, where the planets (including earth) rotate around, or 'orbit', the sun.

In each tiny atom, there is the nucleus containing protons and neutrons (the 'sun') and electrons (planets), which orbit the nucleus. Sometimes these electrons 'break free' and flow to a neighbouring atom. It is these wandering, or free, electrons, moving through the material structure, that give rise to electricity.

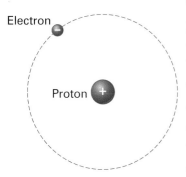

Figure 4.29 A hydrogen atom

What are conductors?

In some materials, it is very easy to get the electrons to move; these materials are called **conductors**. In other materials, it is very hard to get the electrons to move; these materials are called **insulators**. Examples of good conductors of electrons are copper and aluminium (gold is probably the best conductor, but it is too expensive to use in everyday installations). Typical insulators are wood, plastic and rubber.

Measuring electricity

For practical purposes, three things need to be present to create an electrical circuit:

* current
* voltage
* resistance.

Current

If electricity is the movement of electrons through a conductor, we need to be able to measure *how much* electricity will flow in any given circuit in a given time and to control how much electricity is flowing to create a practical electric circuit. A single electron, because it is only a tiny part of an atom, is much too small a quantity to have any practical use or to usefully measure. Instead, we 'group' millions and millions of electrons together into useful amounts and then measure the groups. These 'groups of electrons' are known as **coulombs**, and a coulomb is an extremely large number of electrons.

As plumbers, we can use water analogies to help us to understand more about electricity. If we consider the electron as a single drop of water, ask yourself whether it is practical to measure water flow in terms of drops of water flowing in a second. It is more practical to measure *water current flow* in litres per second.

Similarly, an electrician would consider the flow of electricity through a conductor in terms of coulombs per second, which brings us nicely to one of the main properties of an electric circuit: the electric **current** flow, measured in **amperes** (usually abbreviated to amps and given the symbol I).

Voltage

The next essential ingredient is **electro-motive force** measured in terms of the number of joules of work required to push one coulomb of electrons along the circuit. It is measured in joules/coulomb, more commonly refered to as the **volt**. When a coulomb of electrons leaves a battery or generator it has a potential energy, but as it travels round the circuit this energy is used up. The amount of energy used up by one coulomb in its passage between two points in a circuit is known as the **potential difference**, measured in volts.

To continue with our water comparison, voltage can be thought of as the pressure that pushes the water (electricity) around the system. By applying a voltage to the end of a conductor, we provide an electrical pressure that causes a current to flow. The voltage may be supplied from a battery or a mains supply. A current that is produced

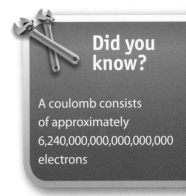

Did you know?

A coulomb consists of approximately 6,240,000,000,000,000,000 electrons

Definition

Ampere – a flow of one coulomb in one second. In other words, the quantity or amount of electricity that flows every second

Remember

We measure the amount of current flowing in amps using an instrument called an ammeter, which is connected into a circuit in series (more about series circuits later)

by connecting a battery to a circuit is called **direct current** (DC). The electricity that we use in our homes and which is produced by power stations is called **alternating current** (AC). There's more on the differences between them later.

Resistance

The third item we need to consider to understand electricity properly is **resistance**, given the symbol R. In our water comparison think about water being supplied to a bath via a cistern using gravity. The head of water in the cistern will provide the pressure (voltage in electrical terms). With the tap open, the water will flow into the bath at a fixed rate per second (current in electrical terms). In order to flow from the cistern to the bath the water passes through a pipe of a certain diameter. If we increase the size of the pipe (and tap), the natural resistance to the flow of water will decrease and the bath will fill more quickly. If we decrease the size of the pipe, the resistance to the flow will increase and the bath will take longer to fill. This resistance to current flow in an electrical circuit is measured in units called ohms (sometimes denoted by the Greek letter omega, Ω) and given the symbol R. In plumbing terms, the bigger the pipe the lower the resistance to water flow; in electrical terms, the bigger the conductor the lower the resistance to current flow.

Gravity drives the water through this system, but for this to happen there must be a difference in level between the supply (the tank) and the load (the bath). This gives the equivalant of potential difference. The tap performs the same function as an electric switch.

Figure 4.30 Analogy of a water system as an electrical circuit

One very important point to note here is that ammeters and voltmeters are connected into live electrical circuits to measure current and voltage, and they pick up the electricity that they need to operate from the live circuit.

The three major components that make up an electric circuit – voltage, current and resistance – are interrelated, and, if we know any two of the quantities, we can calculate the third by using a basic rule known as 'Ohm's Law'.

Ohm's Law states the relationship between the electrical quantities as: voltage equals current multiplied by resistance. This can be expressed by the equation:

$$V = I \times R$$

where V = voltage, I = current and R = resistance. It can also be shown in a simple form, often called the 'Ohm's Law Triangle'.

Remember

We use an instrument known as a voltmeter to measure the voltage of a circuit. Voltmeters are connected into a circuit in parallel (more later)

Remember

The resistance of an electric circuit is measured with an instrument known as an ohmmeter. The ohmmeter is connected across the component whose resistance we want to measure

Safety tip

The ohmmeter has its own internal power supply from a battery, and an ohmmeter must NEVER be connected into a live circuit

Using the Ohm's Law Triangle

If you know two of the quantities, you can find the third by covering up the quantity that you are looking for with your finger and calculating the other two.

Suppose you want to determine the voltage, V, of a circuit and you know the current, I, and resistance, R.

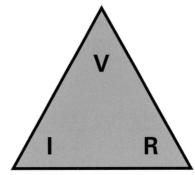

Figure 4.31 Ohm's Law Triangle

Covering the V with your finger shows you that you need to multiply the current I by the resistance, R, to find voltage, V.

To find the current I, cover I and you will see that you now need to divide voltage, V, by resistance, R, as V is shown above R.

For example:

A circuit has a resistance of 120 ohms and a current of 2 amps is flowing. What would be the voltage applied to the circuit?

$$V = I \times R = 2 \times 120 = 240 \text{ volts}$$

How an electric circuit works

If we were to bend a conductor, such as a copper wire, into a loop and connect a battery across the two ends, we would have a complete electric circuit and a current would flow. Unfortunately this would not create a practical circuit that we could use. A single length of copper wire has a very low resistance on its own (depending on the length and cross-sectional area it may have only a few millionths of an ohm of resistance). If we were to connect this across a normal domestic voltage supply of 230 volts, using Ohm's Law we could show that, as $I = \frac{V}{R}$ and R is very low, thousands of amps of current will flow, causing serious damage to the circuit. The flow of electric current causes a corresponding heating effect, and the higher the current the higher the heating effect, causing a serious fire risk.

Even if we could have the circuit described above it would still not be practical because there is no load to limit the current, no way to switch the current off and no protection for the circuit from the high current flowing due to the very low cable resistance.

For all practical electric circuits, therefore, we must have:

- conductors through which current can flow (usually copper wire)
- a source of supply (such as a battery) connected across the ends to provide a potential difference (voltage) and to make a complete circuit
- a load such as a lamp or a device that needs electric current to make it work but which will also act to limit the amount of current flowing
- a device such as a fuse or miniature circuit breaker to protect the circuit should too much current flow
- a switch for opening and closing the circuit to switch the current on and off.

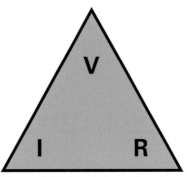

Did you know?

Ohm's Law is named after a German physicist, George Ohm, who defined electrical resistance in 1827

Figure 4.32 shows a practical electric circuit containing a battery as a supply source, a lamp as a load, a switch to open and close the circuit and a fuse to protect the circuit from excess current flow.

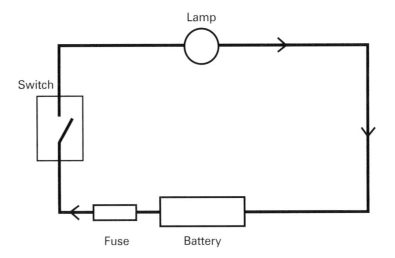

Figure 4.32 A simple electrical circuit

Fuses and circuit breakers

Why are fuses needed?

Fusing is a safety measure which aims to prevent a high electrical current passing through wires that are not designed to carry such large charges. Why is this important? Because when a current flows, it causes a heating effect and a temperature rise in the wire; if a current that is too high for a wire is passed through it, the overheating that results presents a serious risk of fire.

How do fuses help?

The various different types of fuse that exist all contain fuse wire which will melt or 'blow' if an electric current above a specified level is passed through the wiring. Fuses come in different sizes to protect against different levels of current.

Rewirable fuses

Cartridge fuses

In modern electrical installations you will also come across miniature circuit breakers (mcbs). An mcb is a device that will trip a switch to break the electrical circuit if excessively high current is detected. When electricity flows, it also produces a magnetic field around the conductors as well as the heating effect previously mentioned. The mcb uses a combination of the heating and magnetic effects to break the circuit under fault conditions. Mcbs are more accurate and expensive than fuses but are re-settable and are now found in all electrical consumer units within newer domestic properties.

You will also come across residual current devices (RCDs). These are very sensitive devices providing a high degree of protection to high-risk parts of electrical systems, such as plug-socket outlets and electric showers. An RCD measures the difference between the current in the electrical conductors in the system, e.g. live and neutral, and measures changes in the electrical current. If a small change occurs, the system is automatically disconnected. A typical operating current for an RCD in the event of a fault would be 30 mA (that is, 30 one thousandth parts of one amp.) This may seem very low, but it has been calculated that the average person can 'feel' electricity at current levels of around 1 mA.

Miniature circuit breaker (mcb)

Fuse rating

Ensuring that the appropriate size of fuse is used – called the fuse rating – can be worked out using the simple formula:

$$amps = \frac{watts}{volts}$$

Most items of electrical equipment – lamps, televisions, washing machines, hi-fi systems, vacuum cleaners – are rated in watts, which is the amount of electrical power that the item consumes. Frequently in domestic environments fuses are overrated: i.e. a fuse with a too high rating is used for safety reasons. For example, if a lamp contains a 100-watt light bulb, the fuse rating would be calculated as follows:

$100 \div 230$ (the voltage of domestic mains supply) = 0.434 amps

Manufacturers of fuses do not produce fuses rated at 0.434 amps, and the next practical size up would be used, in this case a 3-amp fuse. In this example, the bulb may be used in a stand-alone table lamp fitted with a cord and 13-amp plug top, which should be fitted with a 3 A fuse. Plugs are normally supplied with a 13 A fuse, but this is unnecessarily high in this example and provides insufficient protection for the user. Technically, this set up would be classed as unsafe.

If the lamp were being used as one of a number connected to a fixed lighting circuit, the entire circuit would usually be fitted with a 6 A fuse or circuit breaker.

There is potential for confusion here, because 1 amp of current is more than sufficient to kill at 230 volts (the average that a person can 'feel' is 1 mA).

Remember

The purpose of the fuse is mainly to protect the wiring; an earth is provided to protect the user. We will learn more about earthing later

Circuits

There are two basic types of electrical circuit:

- series
- parallel.

It is also possible to connect components into a circuit using a combination of these two ways. These are called 'series/parallel' circuits, and they will be considered later when we start to look at control circuits for central heating and boiler installations etc.

Series circuits

If we take a number of different resistors (items of electrical equipment such as lamps etc.) and connect them together end to end and then connect the free ends to a battery, we will find that the current will have only one route to take around the circuit. This type of connection is known as a series circuit. Most switches and controls such as thermostats are usually connected into circuits in series, for example in those used for central heating systems.

Figure 4.33 The series circuit

The main features of a series circuit are:

- the total resistance of the circuit is found by simply adding together all of the resistances within the circuit:

  R = R1 + R2 + R3

- the total current flowing (I) can be found by dividing the supply voltage by the resistance:

$$I = \frac{V}{R}$$

(you will recognise this as Ohm's Law)

- the current flowing will have the same value at any point in the circuit

- the potential difference or voltage across each resistor is proportional to its resistance. We know from our study of electrons that we use voltage to push electrons through a resistor. How much voltage we use depends on the size of the resistor. The bigger the resistor the more voltage we use for a given current (revise Ohm's Law).

Therefore the voltage across resistor R1 = I (remember that the current is the same at all points of a series circuit) × value of R1.

Considering the circuit shown in the illustration:

V1 = I × R1   V2 = I × R2   V3 = I × R3

- the supply voltage of the circuit (V) is equal to the amount of the individual voltages across each resistor, as shown by the formula:

V = V1 + V2 + V3

Calculations with series circuits

Example:

Two resistors of 6.2 ohms and 3.8 ohms are connected in series with a 12V battery, as shown in the diagram. Calculate:

(a) total resistance of the circuit   (b) total current flowing

(c) the potential difference across each resistor.

(a) Total resistance

For series circuits the total resistance is the sum of the individual resistances.

R = R1 + R2 = 6.2 + 3.8 = 10 Ω

(b) *Total current*

Using Ohm's Law:

$$\text{Current} = \frac{\text{Voltage}}{\text{Resistance}} = \frac{12}{10} = 1.2 \text{ Amps}$$

Figure 4.34 Total resistance

Current flow

(c) *Potential difference across each resistor*

V = I × R

Therefore: Across R1: V1 = I × R1 = 1.2 × 6.2 = 7.44 volts

Across R2: V2 = I × R2 = 1.2 × 3.8 = 4.56 volts

Parallel circuit

An alternative way to connect components into a circuit is via a parallel circuit. These are commonly used for lighting and power circuits, as one important feature of this type of connection is that the same voltage appears across each of the resistances in the circuit. This is important in lighting circuits where we want all the lamps to burn with the same brilliance and not to burn dimly because the available voltage is shared between them.

Figure 4.35 The parallel circuit

In a parallel circuit a number of resistors are connected together so that there are two or more routes for the current to flow as shown in the illustration.

In a parallel circuit the current 'splits up' and divides itself among the various branches of the circuit. However, while the current divides within a parallel circuit, the voltage across each branch remains the same. This means that we can take components out of the circuit and the remaining ones will still work. In a series circuit, if we removed a single component the circuit would be broken and would not work.

It is this feature that makes parallel circuits popular, and indeed all lighting and power circuits in domestic premises are connected in parallel; only the switches and certain other control devices are connected in series.

The main features of a parallel circuit are:

- the total current (I) is found by adding together the current flowing through each of the branches

$$I = I1 + I2 + I3$$

- the same voltage will occur across each branch of the circuit

$$V = V1 + V2 + V3$$

the total resistance can be found by using the formula:

$$\frac{1}{R} = \frac{1}{R1} + \frac{1}{R2} + \frac{1}{R3}$$

Some calculations with parallel circuits

Using the example shown above, calculate the total current flowing through the circuit.

$$\frac{1}{R} = \frac{1}{R1} + \frac{1}{R2} + \frac{1}{R3}$$

Therefore:

$$\frac{1}{R} = \frac{1}{16} + \frac{1}{24} + \frac{1}{48}$$

And therefore:

$$\frac{1}{R} = \frac{3 + 2 + 1}{48}$$

Giving us:
$$\frac{1}{R} = \frac{6}{48}$$

Turn this upside down to find R:
$$R = \frac{48}{6} = 8 \text{ ohms}$$

Now, using Ohm's Law:
$$I = \frac{V}{R}$$

We see that:
$$I = \frac{240}{8} = 30 \text{ amps}$$

Direct and alternating current

The illustration shows the flow of electrons in both AC and DC circuits.

Figure 4.36 Flow of electrons in AC and DC circuits

Direct current (DC)

In a direct current electrical circuit, the electron flow is in the same direction all the time. One example would be from the anode to the cathode of a battery around a simple circuit.

Direct current from batteries is produced by a chemical reaction where plates containing dissimilar metals are placed in a solution known as an 'electrolyte'. When the battery is charged and a load is connected across it, a current flows. Batteries and direct current supplies are generally only used where very small amounts of electricity are required – for example in torches, light current back-up power supplies and battery-operated power tools.

As plumbers you will probably only use direct current when using small batteries or perhaps in heating or boiler control circuits where the direct current has been produced from an alternating current source using a transformer to reduce the voltage and a rectifier (which is a solid state electronic device) to convert the alternating current to direct current.

Alternating current (AC)

Most of the work that plumbers do and the installations that plumbing systems are connected into will be supplied with alternating current. The supply in a domestic building will almost always be single phase and supplied at 230 volt and a frequency of 50 hertz.

We said earlier that electricity was invisible to the naked eye and we could not see it or smell it. This is true, but there is an instrument known as an oscilloscope with a screen like a small television which allows us to look at electricity.

If we look at direct current we see a straight line where the electricity flows in one direction only from one pole of the battery to the other, as shown in the diagram.

If we look at alternating current, however, we see a shape that is called a 'sine' wave .

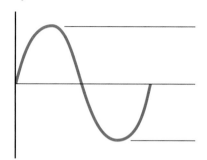

Figure 4.37 The alternating current sine wave

The sine wave represents one complete 'cycle' of alternating e.m.f. which would be 'induced' into a coil of copper wire when it is rotated within a magnetic field. During the top half of the cycle, the current will flow in one direction, and during the bottom half of the cycle, the current will flow the other way (hence 'alternating'). In the UK, the electricity supply has 50 of these single cycles produced every second. This is known as the frequency of the supply, and the unit used for frequency is the Hertz, (abbreviated to Hz). The frequency is therefore stated as 50 Hz.

Whereas direct current is caused as a result of a chemical reaction in a battery, alternating current is produced as a result of **electromagnetism**:

- As well as the heating effect already mentioned, all electrical currents produce a magnetic force. This is the basic principle that underpins the creation of almost all the electricity used today, and is known as electromagnetism.

- The application of this fact was in principle demonstrated by Michael Faraday in the 1830s. He discovered that electricity could be generated by moving a magnet in and out of or around a coil of wire wound on to a soft iron core.

- The same effect is present if we move a coil of wire on an iron core within a fixed magnetic field.

Knowledge check 3

1   Which electrical term describes the electrical pressure that causes the current to flow in a conductor?

2   Using Ohm's Law, if a 240 V circuit has a current rating of 3 A, what is the resistance in the circuit?

3   What are the two main types of electrical circuit?

4   What is the type of electric current that is produced in the form of a sine wave described as?

5   If a central heating system supplied with electricity at 230 V has a load of 600 watts, what is the approximate fuse rating required?

Additional information sources

Sources of additional information on topics covered in this chapter include:

- Sentinel (manufacturer of corrosion treatment products) – www.sentinel-solutions.net

- Fernox (manufacturer of corrosion treatment products) – www.fernox.com

chapter 5

Common plumbing processes

OVERVIEW

This chapter is one of the most important in the book. It provides you with the knowledge to support the basic skills you will need as a competent plumber:

- **Tube bending and marking out**
    - bending methods – copper pipe
    - bending methods – low carbon steel pipe

- **Pipework materials and jointing processes**
    - copper tube and fittings
    - methods of jointing copper tube
    - LCS and fittings
    - methods of jointing LCS pipe
    - plastic tube and fittings
    - methods of jointing plastic pipes
    - jointing different materials
- **Hand and power tools**
- **Fixing devices, pipe supports and brackets**
- **Associated trade skills**
    - lifting floor surfaces
    - drilling a hole in a timber floor joist
    - cutting holes in the building fabric

- **Generic systems knowledge**
    - pipework and plumbing symbols
    - measuring jobs and estimating materials
    - documentation
    - work programmes

Tube bending and marking out

Bending, measuring and marking out are essential basic skills for any plumber. We will concentrate here on bending copper pipe – which can be done by hand or machine – and low carbon steel pipe, which is usually bent by hydraulic machines.

There are also steel pre-formed 90° brackets for tighter bends, into which the pipe can be clipped.

Bending pipe, rather than using ready-made fittings, has the following advantages:

- it produces larger-radius bends than elbow fittings (larger-radius bends have less frictional resistance)

- using bends costs less than using fittings

- long sections of pipework can be prefabricated before installation, saving time.

Bending methods – copper pipe

The pipe grade suitable for bending is R250 grade X. R290 grade Z is tempered, and has thin walls and cannot be bent. R250/220 grade Y is softer, with thick walls, and is supplied in coils; it is not usually used internally in dwellings. There is also R220 grade W, which is used for micro-bore installations.

Copper pipe can be bent by:

- hand

- machine.

Bending by hand

This is a method of bending pipe when carrying out maintenance and repair work. You may be working in a loft space, for example, where it would be much quicker to fabricate the pipe in situ. The best way of doing this is by using a **bending spring**.

Bending springs can be used externally or internally. In either case you pull the bend against your knee to get the desired angle. When using an internal spring there are a few things to remember.

- Don't try and bend a piece of pipe that is too short. You would have to apply a good deal of physical pressure, which could lead to injury, and kink the pipe, or both.

- Don't make the radius of the bend too tight, or you could find that you can't remove the spring.

Using an external spring

- Always pull the bend slightly further than the required angle, and then pull it back to the required angle. This will help to release the spring.

- The spring has a 'hook' at the end. This can be twisted clockwise and will tighten the spring coil inwardly, again making it easier to remove the spring.

Using an internal spring

Overpull for a 90° bend

Spring

Bar used to tighten spring to make removal easier

Figure 5.1 Pipe bending with internal spring

Spring pipe-bending can be set out with the same accuracy as machine bending. All bends appear to gain length when bent. This is because, before the pipe is bent, the length A to B to C is longer than the actual bend A to C. It is possible to work out accurately the length of pipe to be bent in relation to fixed points by finding out the length of the pipe actually occupied by the bend.

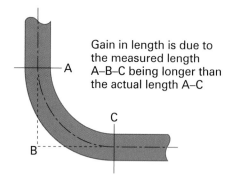

Gain in length is due to the measured length A–B–C being longer than the actual length A–C

Figure 5.2 Apparent gain in length when tubing is bent

This is done by deciding the centre-line radius of the bend. This is usually four times the diameter of the pipe (4D).

Next, you apply the following formula:

$$\frac{\text{Radius} \times 2 \times 3.14}{4}$$

For example, for a 15 mm diameter pipe:

radius of bend = 4D or 4 ×15 = 60 mm

Now, to find the length of bend, use the formula:

$$\frac{\text{Radius} \times 2 \times 3.14}{4}$$

Substituting for the radius:

$$\frac{60 \times 2 \times 3.14}{4}$$

this gives 94.26, rounded to 95 mm. This is the length of pipe required to form the bend. The illustration shows how this is applied.

Figure 5.3 Finding the length of pipe required to form a bend

Now that you have calculated the length of pipe taken up by the bend, you can now add this to the pipe lengths to give an overall pipe length:

500 mm + 400 mm + 95 mm = 995 mm total pipe length.

Bending by machine

This is the most common method used for bending copper tube. Bending machines can be either hand held (**hand bender**) or free-standing (**stand bender**), and they work on the principle of leverage.

Small mini-bore (8/10 mm) bender

Hand bender

Stand bender

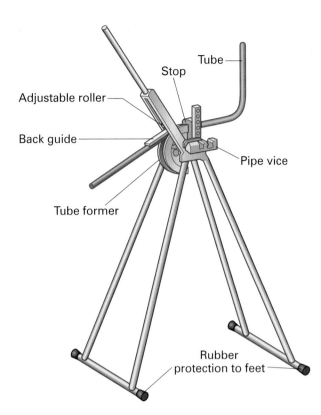

Figure 5.4 Stand bender

The small hand-held bender is used for pipe sizes of 15 mm and 22 mm, and is light and portable. The free-standing bender can handle pipes up to 42 mm and uses a range of sizes for the back guide and former. It is important that the machines are set up properly. If the roller is adjusted so that it is too loose, this will cause rippling on the inside of the radius of the pipe. If it is too tight, it will reduce the pipe diameter at the bend. This is called **throating**.

Types of machine bend

There are three main types of bend:

- 90° or square
- offset
- passover.

90° or square bend

You may be given a measurement from a site drawing, but it is more likely that you will take a site measurement yourself. Measurements should be taken from a fixed point to the back of the bend.

Figure 5.5 Taking the measurement

A pencil mark is made from the fixed point (end of tube) to the required length.

Figure 5.6 Marking and tubing

The tube is set up in the machine with the mark squared off from the outside of the former.

Figure 5.7 Positioning the tube in the bender

Then, the back guide is positioned, the roller correctly adjusted and the bend pulled. Return bends can be made using the same technique, only now the back of the first bend becomes the fixed point.

Offset

The first set is made to the desired angle. The first angle is not critical, but will usually depend on the profile of the obstacle you want to offset the pipe around.

Now you have made the first set, the tube is reversed and returned to the machine.

Figure 5.8 The first set

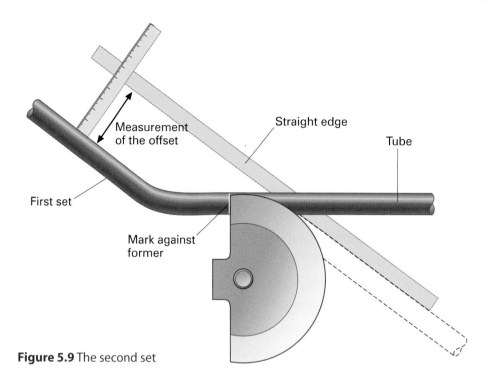

Figure 5.9 The second set

A straight edge is placed against the former, parallel to the tube, and the measurement for the offset is taken from the inside of the tube to the inside edge of the straight edge.

Once the tube has been adjusted against the stop of the machine, it is a good idea to mark the edge against the former, just in case of any movement.

Note the dotted line on the illustration showing the finished position of the second set. This shows that the measurements are taken from the inside to the back of the tube, so in effect it is a centre-to-centre line measurement.

Passover bend

These are used to clear other obstacles, such as another pipe, and could be a passover offset or a crank passover bend.

Figure 5.10 Passover offset

Figure 5.11 Crank passover bend

The measurements for a passover bend are taken in the same way as for an ordinary offset. The angle of the first pull will be governed by the size of the obstacle it has to 'pass over'.

Figure 5.12 Measurement of crank passover

A straight edge is placed across the bend at the distance of the obstacle and the pipe marked. This will be the back of the finished offset.

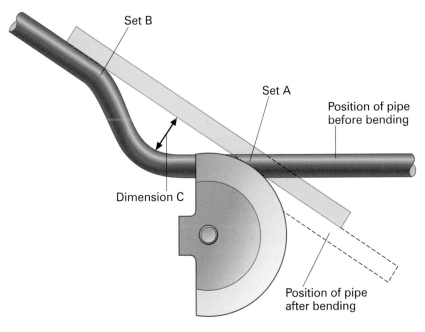

Figure 5.13 The final set in a crank passover bend

The pipe is then returned to the machine, and when the first mark lines up with the former, it can be pulled to the position shown; the pipe is then turned around in the machine, the second mark lined up in the former and the pipe pulled to complete the passover.

Bending methods – low carbon steel pipe

LCS pipes are usually bent using a hydraulic pipe bender.

Only medium- and heavy-grade tubes are used in plumbing systems; they are manufactured to BS 1387, and are supplied in black-painted or galvanised coatings. Galvanised pipe should not be bent, as it this will cause it to lose its coating. It is unlikely, however, that you will come across this material other than on maintenance work. You are more likely to work on medium-grade tube.

Bending with a hydraulic machine

Hydraulic machines are needed to bend low carbon steel tubes, owing to the strength of the material and the thickness of the pipe. For this reason, the pipe does not need to be fully supported with a back guard, unlike copper pipe. Hydraulic bending machines are used to form most bends, including 90° and offsets.

The hydraulic mechanism is usually oil-based, and as liquids are incompressible, once under pressure they can exert a considerable force on the pipework.

Hydraulic bending machine

Forming a 90° bend

Six steps to forming a 90° bend:

1 Mark a line on the pipe at a distance from the fixed point, where the centre line of the finished bend is required.

2 From this measurement you deduct the nominal bore of the pipe. This is because there is a gain in length of one pipe diameter when bends are made.

3 Make sure the correct size former is in the machine.

4 Put the pipe in the bending machine and line up the mark with the centre of the former.

5 The machine can then be worked to apply pressure and bend the pipe to 90°.

6 Due to the elasticity of the metal, you need to take it to approximately another 5° over 90° to allow it to 'spring back'.

Five steps to forming an offset

It is a good idea to make a template from steel wire to help you achieve the required offset profile. The method of marking out the offset is similar to that of copper:

90° bend to be produced

Diameter of pipe

Former

Figure 5.14 Forming the 90° bend

1 Mark off the required measurement for the first set onto the pipe.

2 Place the pipe in the machine, but do not make any deduction. The measurement X mm is from the fixed end of the pipe to the centre of the set.

3 Pull the first set to the required angle.

4 Take the pipe from the machine and place a straight edge against the back of the tube. The measurement of the offset is marked at point A.

5 Replace the tube in the machine and line the mark up with the centre of the former. The second set is pulled and checked against the wire template. Again, allow a 5° overpull for the spring back.

Figure 5.15 Locating the pipe for the first bend

Figure 5.16 Setting out for the second set

Remember

During the bending operation, the pipe can become wedged into the former. Do not hammer the former or the pipe to remove it. You can place a timber block on the pipe and give it a sharp tap with a hammer. Alternatively, remove the pipe and former. Place a wooden block on the floor and strike the end of the pipe on the block. Try to hold the end of the former so that it doesn't fall to the ground

Faults and defects

The main problems encountered with low carbon steel pipe bending are usually down to incorrect marking out, for example not making the deduction for the nominal bore, so the bend does not fit when it is pulled. In the case of offsets, if the pipework is not set up level in the machine, then the resulting offset will be screwed or twisted.

FAQ

How can I make copper tube easier to bend by hand?

By annealing the copper tube first. This is a process of heating the copper tube cherry red and then cooling in water/allowing to cool. The tube will be in a much softer state for bending with a spring.

Knowledge check 1

1   What are the two most common methods of bending copper tube?

2   What are the two types of spring that can be used to bend copper tube?

3   The pipe in a bending machine is bent around what component?

4   If a roller is loosely adjusted in a stand bending machine what may the bend suffer from?

5   What type of bending machine is used to bend low carbon steel pipe?

Pipework materials and jointing processes

Pipes, fittings and jointing materials acceptable for Water Regulation purposes are listed in the Water Fittings and Materials Directory. BS 6700 also states the minimum requirements for pipe joints and fittings. You will cover this in much more detail at Level 3, but you should remember that the following factors must be taken into account when selecting materials for use in plumbing systems:

- affect on water quality.
- vibration, stress or settlement.
- internal water pressure.
- internal and external corrosion.
- compatibility of different materials.
- ageing, fatigue, durability and other mechanical factors.
- **permeation**.

In domestic plumbing installations the main materials that you will work with are copper, low carbon steel and plastic. To a lesser extent, and usually only on maintenance work, you will also come across cast iron and stainless steel.

Copper tube and fittings

Copper tube is available in four grades (under a new BS Standard the terminology used for copper grades has changed slightly, although the grades shown below are still widely used):

Grade X (R250 half hard lengths)

Grade X is widely used for domestic installations. It is classified as half hard, and has pipe diameters ranging from 12 mm to 54 mm. Its pipe diameter is always specified as the external measurement, and the tubes are normally available in 6 m lengths, although most merchants will also supply in 1 m, 2 m and 3 m lengths. Grade X tube should not be used underground. Grade X is also available in chromium plate. It is used where pipework is exposed to the eye and an attractive finish is required.

Definition

Permeation – the possibility of microscopic particles of oxygen entering the water supply through the external wall of the pipe

Remember

You may find it useful to refer back to the section on plumbing materials in Chapter 4 when reading this

Grade Z (R290 hard lengths)

Grade Z is not as popular as grade Y, one of the main reasons being that it is unbendable. This is because its wall is 0.2 mm thinner than grade X, so the pipe is hardened during the manufacturing process to make it stronger. Grade Z is also available in 1 m, 2 m, 3 m and 6 m lengths, and in pipe diameters of 12 mm to 54 mm. Like grade X, grade Z is available in chromium plate.

Probably the most likely place where you would find chrome-plated pipe is in exposed pipe runs to instantaneous showers in a bath or shower room, particularly where the appliance is a new addition to an existing suite.

Find out

Can you think where you might see chrome plated pipework in a domestic dwelling?

Grade Y (R220 soft coils)

Grade Y is used for external underground installations. Its wall thickness is 1 mm, making it the thickest grade, and it is classified as fully annealed, which means it is soft. The size range is 15 mm and 22 mm, and it is supplied in 25 m coils. It is easy to form into wide-radius bends by hand, without affecting the bore of the tube, so the use of bending equipment is not required. Grade Y can be supplied in either a blue (water service) or yellow (gas) plastic coating.

Grade W (R220 soft coils micro-bore)

Grade W is used for micro-bore heating and, like grade Y, is fully annealed. The sizes used for micro-bore are usually 4 mm, 5 mm, 6 mm, 8 mm, 10 mm and 12 mm. The pipe is supplied in coils ranging from 10 m to 30 m. Again, like grade Y, the pipe lengths can be formed into wide-radius bends or offsets without the use of bending equipment. For short-radius bends, a number of purpose-made hand formers are on the market. Grade W can also be supplied with yellow (gas) or white (heating) plastic coating or with air channels in the plastic coating to improve thermal insulation. This is used where pipework is installed internally in solid floors.

Compression joints

Methods of jointing copper tube

There are three main methods of jointing:

- compression joints
- soldered capillary joints
- push-fit joints.

Soldered capillary joints

Compression joints

Compression joints are of two types:

- manipulative – type B
- non-manipulative – type A.

Plastic push-fit joints

Manipulative joint

The illustration shows a typical manipulative fitting detail; these are sometimes referred to as type B.

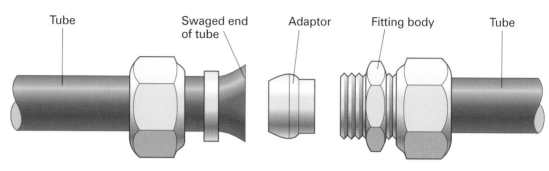

Figure 5.17 Manipulative compression fitting

Figure 5.18 Stages of assembly of a manipulative compression fitting

The jointing process is as follows:

1   Cut the tube to the required length, using either tube cutters or a hacksaw.

2   De-burr the inside of the pipe (if cut with tube cutters) or the inside and outside of the pipe (if cut with a hacksaw).

3   Place the nut and compression ring onto the pipe.

4   Using a **swaging tool** or drift, flare out the end of the tube so that it fits the angle of the adaptor.

5   This process is applied to each pipe end being used in the fitting (2 for a straight coupling or elbow, 3 for a tee).

6   The fitting can now be assembled. The angled side of the adaptor fits in to the flared end of the tube, the other side in to the body of the fitting. The joint should be finger tight, with all the fitting components fully engaged. The fitting can then be completely tightened using adjustable grips or spanners.

Manipulative joint being coupled

This type of copper compression fitting tends to be rarely used owing to the increased amount of time required to form the joint. However, compression fittings used on underground cold-water supply pipework must always be of the manipulative type.

Non-manipulative joint

Backnut   Compression ring   Fitting body   Tube

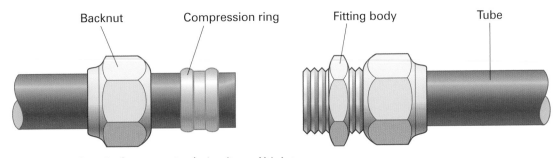

Figure 5.19 Detail of non-manipulative (type A) joint

Sometimes referred to as type A, this is similar to the manipulative fitting. The main difference is that the pipe end is not flared to receive an adaptor; instead, a **compression ring** or **olive** is used to provide a watertight seal on the pipe end. The cutting and deburring process is as before. Again, the joint is hand-tightened first, making sure the pipe is fully pushed into the body of the fitting, before completely tightening with adjustable grips or spanners.

Remember

This type of fitting is suitable for use on grade X, Z and W pipes. Their use is not allowed on underground services

Non-manipulative joint being coupled

Brass fittings

A range of brass fitting designs, compatible to pipe sizes, is available. For example:

- straight adaptor
  - with parallel male thread
  - with parallel female thread
  - with taper-made thread
- straight couplers
  - equal compression and reducers
  - imperial to metric
  - with drain valve
  - with air release valve
  - compression to solder-ring capillary
- tap connectors
- stop ends

- tees
  - equal
  - reducing
- elbows
  - equal
  - with drain tap
  - with air release valve
  - compression to parallel female thread
  - compression to parallel male thread
- wall-plate elbows
- bent and straight tank connectors.

Soldered capillary joints

Soldered joints can be classified as soft-soldered and hard-soldered. Hard-soldered joints, for example using silver and silver alloys including copper, require heating to a much higher temperature than soft-soldered joints. They are rarely used on the majority of domestic plumbing installations and will not be discussed here.

Soft-soldered joints are made using two types of fittings:

- integral solder ring
- end feed.

Figure 5.20 Soldered capillary joints

The difference between the two jointing methods is that the solder ring has solder contained in a raised ring within the body of the fitting (integral). When the fitting is heated, the solder melts (at between 180 and 230°C) and the solder is drawn into the fitting by capillary action. The same principle is used for the end feed, except that the solder is end-fed separately from a spool of wire solder.

The jointing process is fairly straightforward:

1   The tube should be cut square and de-burred.

2   Because the inside of the fitting and the outside of the tube are oxidised, the solder will not adhere to the joint correctly; these surfaces should be cleaned to a 'bright shiny light orange' colour (Step 1).

3   Flux is then applied to the surface of the tube and fitting using a brush – not your fingers! (Step 2).

4   The joint is assembled and heat is applied to the body of the fitting (Step 3).

5   If using solder ring, wait until you see a complete film of solder around the end of each fitting. If using an end feed, apply solder from a wire spool and feed it in until the joint is complete.

Remember

Do not use too much solder: it will leave an unsightly appearance when it cools

Only use lead-free solder on hot- and cold-water supplies – a mixture of tin and copper. The spool should be clearly marked *lead free*

Step 1 Clean and de-burr tube

Step 2 Apply flux using brush

Step 3 Apply heat to body of fitting

6   When the fitting is cool, clean off any excess flux with a cloth. The system should also be flushed to remove flux residue from inside the pipework, as this could corrode the inside of the pipes (Step 4).

The table below provides a summary of what type of fitting can be used on the various grades of copper tube.

Did you know?

There are a number of fluxes on the market that are heat activated. A cleaning action takes place during the heating process, so you don't have to pre-clean the pipe end and fitting. When using these fluxes you should make sure they are non-acidic, non-toxic and WRAS-approved

Jointing method	Grade X	Grade Y	Grade Z	Grade W
Capillary fitting	✓	✓	✓	✓
Type A Compression fitting	✓	✗	✓	✓
Type B Compression fitting	✓*	✓	✗	✗

*Not underground

Figure 5.21 Correct fittings for different grades of copper tube

Capillary fittings

As with compression fittings, there is a vast range of soldered capillary fittings (the same design patterns apply to solder ring and end feed):

- adaptors
  - capillary to male and female thread
- slip couplings
- straight couplings
- tank connectors
- reducing fittings
- metric to imperial converters
- stop ends
- tap connectors
  - straight
  - bent

- return bends
- cylinder unions
- elbows
  - male to female
  - female to female
  - capillary to **BSPT**, male or female
- wall connectors (or back-plate elbows)
- tees
  - equal
  - reducing
- offsets

Push-fit joints

There are a number of types of push-fit joints available for use on hot- and cold-water supplies.

Push-fit joints are made from plastic or metal. In the illustration, a grab ring is used to lock the pipe in place, and a neoprene 'O' ring makes it watertight. The fittings are bulky, so they do not look attractive where they are exposed.

Figure 5.22 Typical push-fit joint

Low carbon steel pipe (LCS) and fittings

Often referred to as mild steel, low carbon steel pipe (LCS) is supplied in three grades:

- light – colour-coded brown
- medium – colour-coded blue
- heavy – colour-coded red.

It is supplied painted in black or with a galvanized coating. Black-painted LCS must only be used on wet heating systems, oil or gas supply pipework, and not on hot and cold supplies.

Generally speaking, light-grade tube is not used for plumbing pipework. You are most likely to work with medium-grade pipes, and occasionally heavy-grade.

Medium and heavy grades are available in approximately 6 m lengths, ranging from 6 mm to 150 mm in diameter, specified as nominal bore. Nominal bore means that the figure is not the actual bore of the pipe, as this will vary depending on the thickness of the pipe wall, which in turn will be determined by the grade.

Methods of jointing LCS pipe

For smaller installations, there are two main jointing methods:

- threaded joints
- compression joints.

Threaded joints

Jointing LCS pipe can be done by cutting threads into the end of the LCS pipe to give a British Standard Pipe Thread (BSPT), then jointing them together with a range of female threaded fittings made from steel or malleable iron. The threads are cut using **stocks and dies** or a threading machine.

The jointing process is as follows:

1 The LCS pipe is cut to length using heavy-duty pipe cutters while the pipe is held securely in a vice. The LCS pipe can also be cut using a large frame hacksaw.

Pipe vice

Pipe cutter

2 Remove any burrs from the inside and outside of the pipe

3 You need to **chamfer** the end of the pipe that is being threaded to provide a leading edge for the dies to catch the pipe wall.

Definition

Stocks and dies – the stocks are the body and handle of the tool, the dies are the actual cutter

Definition

Chamfering – the process of smoothing the cut edge by slightly bevelling or angling it

Adjustment graduations

Cam plate

Socket-threaded handle

Locking handle

Socket-threaded handle

Chaser die set (one of four)

Ratchet control

Guide-adjusting handle

Figure 5.23 Chaser die stock

4 Cutting compound should be applied to the end of the pipe before starting the thread-cutting operation. A thread should be cut that is approximately $1\frac{1}{2}$ to 2 threads longer than the length of the inside of the fitting.

5 The excess cutting compound is wiped off.

6 Threaded pipe sealant, which includes hemp and paste, PTFE sealing tape or gas thread sealing tape, can then be applied to the thread. The fitting is then screwed in place.

Assembling steel pipework using a vice

7 The pipe and fitting are joined together using adjustable pipe grips or alternatively using a short length of pipe as a lever.

Because of the way in which screwed joints are made and installed (rotated on the pipe), it is sometimes difficult to remove or assemble lengths of pipework. Where this is the case, a union connector, which allows the pipework joint to be 'broken', should be used.

Using a pipe-threading machine

Pipe-threading machines provide a quicker and easier method of forming threads for LCS pipes. 'All in one' machines have a pipe cutter, de-burring reamer, and stock head and dies.

Threaded pipe fittings for LCS

These fittings can be made of steel or malleable iron. Steel fittings can withstand higher pressure but are more expensive than malleable iron. They are manufactured to BSEN 10241 for steel and BS 1256 for malleable iron.

Malleable cast-iron fittings are adequate for smaller installations, and as with copper tube fittings there is a wide range available, including:

- bends
  - male to female
  - female to female

- elbows
  - male to female
  - female to female
  - female to female union
  - female to female reducers

- tees
  - female to female to female equal
  - female to female to female reducing and increasing
  - female to female to female pitcher equal

- crosses
  - female to female to female to female equal

- sockets
  - female to female equal

- unions
  - female to female equal

- bushes

- nipples.

Compression joints

There are a number of manufacturers' designs for compression joints; a typical one is shown in figure 5.25.

The fitting is designed to allow steel pipes to be formed without threading. Made of malleable iron, compression couplings use locking rings and seals which are tightened onto the pipe. They can be used on gas supplies and water, and although they are more expensive than threaded joints, they do save time on installation.

Figure 5.24 LCS union connector

Electric pipe-threading machine

Stainless steel backing washer

Rubber compression ring

Figure 5.25 LCS compression coupling

Plastic tube and fittings

Plastics fall into two main categories:

1 products of the polymerisation of ethers:
   - polythene
   - polyethylene
   - polypropylene.

2 UPVC and ABS.

The first group are used mainly for service-supply pipework (coded blue for underground services), hot water and hot-water heating.

The second group are used mainly for waste distribution pipework and cold-water installations, although the latter will not be discussed here.

The following table summarises the typical jointing methods for plastic pipes.

Types of plastic	Mechanical joints	Solvent welding	Fusion welding	Push-fit 'O' ring*
Polythene	✓	✗	✓	✗
Polyethylene	✓	✗	✗	✗
Polypropylene	✓	✗	✓	✓
UPVC	✓	✓	✗	✓
ABS	✓	✓	✗	✓

*Discharge pipes and overflows only

Figure 5.26 Typical jointing methods for plastic pipes

Methods of jointing plastic pipe

Fusion welding: polythene and polypropylene

This process requires specialist equipment and is used mainly for gas and water mains. It will not be discussed here.

Mechanical jointing

This applies to the jointing of:

- polythene pipework
- UPVC and ABS.

Polythene pipe – compression joint

Used for underground services, this pipe is identified by blue colour coding. It is also available, coded black, for internal use on cold-water services, although for domestic situations it is mainly used underground. The joints are made using gunmetal or brass or plastic fittings.

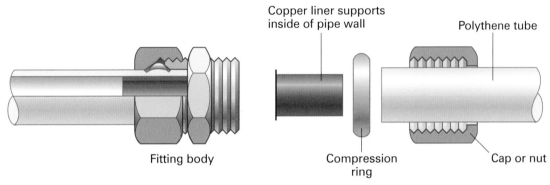

Figure 5.27 Typical polythene pipe fitting

The pipe is cut to length and de-burred. The nut and compression ring are slid onto the pipe, and the copper liner inserted into it. This stops the plastic pipe from being squashed when the nut is tightened onto the body of the fitting and the compression ring applies pressure. Make sure the pipe is fully inserted in the fitting, and hand tighten. Complete the tightening process using adjustable grips or spanners.

UPVC and ABS – compression joint

These joints are restricted to use on waste pipes. Traps are a typical example, using a rubber 'O' ring or rubber plastic washer to make the seal.

UPVC and ABS – solvent-welded jointing

This method is used to joint both UPVC or ABS, using solvent cement. The cement temporarily dissolves the surface of the pipe and fitting, causing the two surfaces to fuse together. The joint sets initially within 5–10 minutes but will take 12–24 hours before it is fully set.

Its use is primarily for joints on soil or waste pipes.

Applying solvent to joint

UPVC and ABS – push-fit joints

These are used mostly on UPVC or ABS soil and waste applications. The pipe is cut to length, making sure it is square and de-burred. The outside edge of the pipe is chamfered to give it a leading edge to make it easier to push into the fitting. This also prevents the 'O' ring being dislodged.

Lubricant is applied to the pipe and fitting, and the pipe is pushed home. Make sure the pipe is withdrawn from the fitting to a length of about 10–15 mm to allow for expansion.

Push-fit connectors

These are also known as **flexible push-fit plumbing systems** for hot and cold water and central heating supplies. A number of manufacturers produce pipe and fittings for these systems, and the fittings can be used on either plastic or copper.

Plastic pipe is supplied in diameters ranging from 10 mm to 22 mm and in 3 m and 6 m lengths, or in coils of 25 m, 50 m and 100 m.

Barrier pipes (which prevent **permeation**) must be used for vented and sealed central heating systems.

They are manufactured from a cross-linked polythene pipe to BS 7291.

Figure 5.28 Push-fit joint used for low-pressure plastic pipework

Figure 5.29 Typical de-mountable fitting for polybutylene pipe

A wide range of fittings is available that are similar to their copper equivalents. In a typical system a mix of copper and **polybutylene** will be used: copper for exposed pipework

and polybutylene for hidden pipework. The jointing and installation processes are fundamentally different from copper, but generally take less time to complete.

The following shows the cutting and assembly sequence of the HEP20 flexible plumbing system.

Stage 1

Use only HEP20 specialist pipe cutter to cut the pipe. Never use a hacksaw. Wherever possible, cut the pipe at the 'V' marks provided. To ensure a clean, square cut, rotate the pipe whilst maintaining pressure on the cutter until the pipe is severed.

Stage 2

Ensure the pipe end is clean, and free from burrs and surface damage. Insert a HEP20 Support Sleeve into the pipe end.

Stage 3

Push the pipe firmly into the pre-lubricated fitting. A secure joint has been made when the end of the retaining cap has reached the next 'V' mark on the pipe. Never knock the fitting onto the pipe, or slacken the retaining cap prior to pipe insertion as these actions will not ease jointing.

Stage 4

Tug back on the pipe to ensure the grab wedge engages correctly. Do not undo the retaining cap after pipe insertion.

Lead to MDPE

Lead to copper

Lead to end feed

Figure 5.30 Fittings to joint different materials

Jointing different materials

Fittings are available that are designed to join copper to plastic; usually they are the plastic push-fit variety. Fittings for jointing copper to LCS are also available.

Although lead pipework is no longer allowed for new installations, it may be necessary to join into an existing supply, either to extend a system where a long run of lead pipework cannot be replaced, or where the joint is at the end of an underground service pipe.

There are several types of fitting on the market for jointing plastic to lead for below-ground use, and for jointing lead to copper. Only use WRAS-approved products – joint wiping lead to copper is no longer permitted under the Water Regulations.

Stainless steel

Stainless steel is not used on domestic pipework installations. You may, however, come across it on systems installed during the copper shortage of the late 1960s, when stainless steel was used as a replacement.

Stainless steel was jointed using compression fittings, and as the outside diameter was that of copper pipe, it would be logical to connect to copper using a compression fitting if carrying out maintenance work.

Cast iron

You are unlikely to install this on new domestic installations, except on historic or listed buildings, but you could come across it during maintenance work, as it was once used extensively on gutters, rainwater pipes and soil pipes. It is still widely used on industrial/commercial buildings due to its resistance to mechanical damage.

Knowledge check 2

1 What type of copper tube is most likely to be used for underground cold-water supply pipework?

2 What is the nut on a non-manipulative compression fitting tightened onto?

3 What type of solder should be used on soft soldered capillary fittings in hot and cold pipework?

4 What type of jointing material is used to make a watertight seal to low carbon steel central heating pipework?

5 Polythene underground cold-water service pipework to a property is normally jointed using what type of fittings?

Hand and power tools

As a plumber, you are required to measure, mark out, cut, fabricate, joint and fix a range of materials. In most cases, this involves the use of tools. Good-quality tools of a well-known brand are expensive but are a good investment.

Once you have put together a toolkit, keep the tools clean and well maintained. This should ensure a long life for them so that you do not have to keep buying replacements.

Hand tools

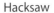

Tool safety and maintenance checklist

✓ Make sure your tools are cleaned regularly

✓ Lubricate the working parts of tools

✓ Once cleaned, lightly coat the tools with an oil spray to prevent rusting

✓ Don't overdo the oiling

✓ Always use the right tool for the job. Screwdrivers are not chisels!

✓ Keep file or rasp teeth clean using a wire brush

✓ Never use tools if they have split handles

✓ Always replace used, worn or defective hand tools

Did you know?

Cheap tools are not always a bargain. They are more likely to be easily damaged or wear out, simply because they are not up to the job

Hacksaws

- Make sure the teeth are pointing away from the forward cut
- Don't use hacksaws with defective or worn teeth
- When using a large hacksaw, make sure the blade is tightened correctly
- Use the right blade type for the job – 32 teeth per inch for light-gauge pipe such as copper and plastics, and 24 teeth per inch for heavier-gauge LCS.

Hacksaw

Pipe cutters

- Replace damaged or blunt cutter wheels
- Use the correct blade for the material being cut
- Make sure that the wheel and the rollers are lubricated and move freely
- Use a pipe cutter or de-burring tool to de-burr the inside of pipework.

Plastic pipe cutters

Pipe cutter

Wood chisels

- Never use chisels with split handles
- Keep chisels sharp using a grinder or whetstone
- Always keep the plastic guard on the chisel
- Make sure handles aren't loose.

Bevelled wood chisel

Cold chisels

- Keep the cutting edge sharp using a grinder
- Keep the striking end of the chisel free from the 'mushrooming effect', again using a grinder
- Don't use a chisel that has been ground down so much that it has become too small to handle safely.

Cold chisels

Hammers

- Make sure the head is fitted correctly to the shaft
- Don't use a hammer with a defective shaft.

Lump hammer   Ball pein hammer   Cross pein hammer

Pipe grips and wrenches

- Keep the teeth free from jointing compounds. If they're clogged up, it could cause the tool to slip
- Once the teeth become worn, the tool should be replaced
- Check for wear on the ratchet mechanism when using pump pliers. These often slip when under pressure

Adjustable spanner

- Be careful when loosening a joint or pipe that is difficult to move. It might give suddenly, and you could damage your hands, or even pull a muscle.

Screwdrivers

- Keep flat-ended screwdrivers for slotted screws to a uniform thickness
- Never use a screwdriver with a defective handle.

Normal slotted and Philips screwdrivers

Stubby PZD/SDV, Philips and normal screwdrivers

Additional tools that commonly form a plumber's toolkit

- Basin key (spanner) – for tightening tap connector nuts in awkward locations.
- Blow torch assembly.
- Compact tube cutter – for cutting copper tube in tight locations.
- Files – smooth and coarse, including a rasp type for chamfering plastic pipework for push-fit connections.
- Flooring chisel – for lifting floorboards.
- Footprint type wrench – for applications such as pump valves.
- High-speed drill bit set – for drilling small holes in timber and metals.
- Hole saw set – for preparing holes in storage cisterns.
- Immersion heater spanner – for tightening immersion heaters to cylinder bosses.
- Insulated screwdrivers – for electrical connections.
- Junior hacksaw – for use on small copper pipe cutting.
- Knife – for trimming materials.
- Masonry drill set – for drilling holes in masonry wall surfaces.
- Pad saw – cutting small holes, for example through plasterboard.
- Pointing trowel – for making good to masonry wall surfaces.
- Radiator spanner – for making valves etc. into radiators.
- Spirit level (small and large) – for fixing to correct levels.
- Stillson wrench – a heavy duty tool used for tightening steel pipework etc.
- Tape measure – for measurement of materials.

- Water pump pliers – general purpose tool for tightening.
- Wood drill bit set – for the drilling of holes in timber to accommodate pipework and components.

Power tools

The main power tools used in plumbing are:

- power drills
- cordless powered screwdrivers
- combined cordless drills and screwdrivers
- power saws.

Power drills

There is a wide range of power drills on the market. Their power varies with the size of the motor, and this will have a bearing on what each drill can do. Some have a 'hammer action' which, when engaged, makes it easier to drill through masonry. Not only does the drill rotate, but it also moves fractionally backwards and forwards at high speed, giving the effect of hammering the drill into the material.

Typical power ratings range from 620 W to 1400 W. Most drills are variable speed and some have reversible action.

Standard rotary hammer drill 110V

Cordless drills and screwdrivers

The photograph shows a typical example from a very wide range. Most drills are combined so they can be used as a drill and screwdriver. They are powered by batteries and are usually supplied with two batteries and a charger, so that one is working and one is charging. These tools are popular because they are much safer than ones with cords, and you don't need to carry a transformer around. Also, if there is no electricity on site you don't have to revert to hand tools.

Cordless drill/screwdrivers are available in a range of voltages: for example 12 V, 14 V, 14.4 V and 24 V.

Battery-operated hammer drill

Power saws

Most powered circular saws run off 110 V. They are used by plumbers for taking up floorboards or sheets to install pipework under floors. Make sure that nails or screws are avoided when sawing. If possible, take up a board by hand to check what is beneath the area where you are going to use the saw. Always make the cutting depth the same depth as the floor thickness.

Battery-operated circular saws are also available, usually with an 18 V motor.

Another type of saw that plumbers often use for cutting wood is the jig-saw, as shown in the photograph. Similar care should be taken with this type of saw.

Powered circular saw

Jig-saw

Checklist of power-tool use

✓ You should have been properly instructed on the use of the power tool before using it on site

✓ Each power tool should be visually inspected for signs of damage prior to its use – if it is damaged it should not be used

✓ All electric tools should be double insulated or incorporate an earth cable

✓ Battery-powered tools are preferable to mains-operated as they are safer. If you use the mains or a temporary power supply, always use 110 V tools as opposed to 240 V

✓ If 240 V power tools are used then the supply to the tool must be protected by the use of a Residual Current Device (RCD)

✓ Check that electrical cables are not damaged or worn

✓ Check that plugs are not damaged

✓ equipment should be PAT tested in accordance with your employer's procedures. PAT tests are maintenance records of all portable electrical equipment to ensure it is in safe working order – the maximum interval between tests for equipment used for construction is usually 3 months

✓ Cartridge-operated tools are covered in Chapter 2 *Health and Safety*. You *must* receive proper instructions on their use before using them

Safety tip

Never use a circular saw without a guard

Safety tip

Remember to wear safety goggles when using drills and saws. These will protect your eyes from dust and any splinters of material that might fly off while working

Knowledge check 3

1   What is normally considered to be the safest type of power tool to be used on a construction site?

2   What is the main safety feature that must always be used on a portable circular saw?

3   What type of cutting device is normally used to cut copper tube in a confined location?

4   How are the tap connectors to a pedestal wash hand basin normally tightened?

5   What must screwdrivers for electrical use feature?

Fixing devices, pipe supports and brackets

In this section we will look at the various types of fixing devices used to secure pipework so that it looks neat and is kept in its proper position. Fixings should provide sufficient support to withstand possible accidental damage from people treading on pipework, children pulling at it and so on.

As a plumber you will be required to fix pipework, sanitary ware and appliances to various surfaces. You will also need to know how to re-fit boards and access traps in timber floors.

Fixing devices

These include:

- screws
  - brass wood screws
  - self-tapping screws
  - turn-threaded wood screws
  - steel countersunk screws
  - chipboard screws
  - mirror screws
- plastic plugs
- plasterboard fixings
- cavity fixings
- nails.

Find out

The range of sizes for fixing devices is vast, and too detailed to explain here. Have a look at a hardware catalogue so you can see the range for yourself

Screws

Screws are specified by their length in inches, and gauge. The usual lengths and gauges used in plumbing range from $\frac{5}{8} \times 8$ for fixing 15 mm saddle clips to $2\frac{1}{2} \times 12$ for fixing radiator brackets.

This is just a rule of thumb, and as a plumber you often have to make a decision about the length and gauge that is right for a particular situation. For example, if the wall you are fixing to is not in good condition, you might have to use a longer screw and thicker gauge. Trial and error and experience are factors here!

Brass or alloy screws are used internally where they could be affected by moisture; this would include sanitary appliances, for example. They are also used externally (mostly in alloy form due to costs) for soil and rainwater fixings and wall-plate elbows for hose union bib taps.

Self-tapping screws are used when fixing into metal sheet. This is not typical in domestic installations, but could apply if you needed to clip to a metal stud partition. This involves drilling a pilot hole in a smaller gauge than the screw, using a steel drill bit, and then screwing into it.

Steel countersunk screws are used for general purposes, such as fixing clips and radiator brackets.

Chipboard screws, as the name suggests are used for chipboard fastenings, for example if you had to fix an access trap in a chipboard floor following some installation work.

Mirror screws are used where appearance is important, and are often used for fixing timber or plastic bath panels.

Plastic plugs

Plugs come in a range of gauges that are appropriate to the gauge of the screw. They are colour coded for ease of selection.

Head   Slot
Shank
Gauge
Length

 Slotted
 Pozidrive

 Philips
 Countersunk

 Pan head
 Mirror screw head

 Roundhead
 Raised head

Figure 5.31 Some commonly used screws

The wall is drilled with the correct size masonry drill

The correct size plug is inserted into the hole

The right size screw is driven into the plug

Figure 5.32 Fixing operation using plug and screw

Screw size (gauge)	Drill size (mm)	Plug colour code
6–8	5	Yellow
8–10	6	Red
10–14	7–8	Brown
14–18	10	Blue

Figure 5.33 Size matches of screws, drill bit and plug

Plasterboard fixings

These are used when fixing to plasterboard stud partitions, where there is nothing solid to fix into at the back of the plasterboard. There are several types available.

Figure 5.34 Spring toggle, rubber-nut and cavity fixing

When using a spring toggle, drill a hole in the plasterboard big enough to take the toggle when folded. This is inserted through the hole, and when it is in the space behind the board, the toggle is pushed open by the spring.

Rubber-nut fixings work on the principle of drawing the nut mounted in the rubber towards the screw head. As it tightens, the rubber is squashed to form a flange at the back of the board.

Cavity fixings work on the same principle, but this time an aluminium body is 'squashed' to form the flange. When using this type of fixing, a small pilot hole is drilled into the plasterboard and the complete fixing is screwed into the board. The screw is then removed, leaving a fixing point similar to a plastic plug.

Other fixings

The fixings described so far should be adequate for day-to-day work. There will be occasions, however, when you need larger fixings, perhaps for a heavy appliance, or where the fixing surface is in poor condition.

Coach screws are normally supplied with their own purpose-made plastic plug, and are fixed like a normal plug and screw. They are tightened using adjustable grips or a ring spanner.

A wall bolt is an extremely strong masonry fixing. A hole is drilled in the masonry just large enough for the gauge of the bolt. The bolt is inserted and the nut tightened. This pulls the tapered end forward, expanding the segments against the masonry (similar to a cavity fixing). The nut is then removed, the item put in position, and the nut and washer reattached.

Wall bolt

Nails

Plumbers use a range of nails, particularly on maintenance, repair and refurbishment work. This includes procedures such as refixing floorboards (although you should use screws where access to pipework is required), pipe boxing, skirtings and boards for clipping pipe runs. If you are working for a small business, you may have to build your own support platforms for cold-water storage cisterns.

You will probably carry the following in your tool bag:

- panel pins
- masonry nails
- oval brad/lost head
- round head, plain and galvanised.

All are available in a range of lengths and diameters.

Clips and brackets

It is likely that the majority of your work will take place in domestic dwellings where the use of copper or plastic clips is adequate for supporting copper and plastic pipework. Once again, there is a range of sizes of clips available.

Selection of metal and plastic pipe fittings

Remember

You should always aim to fix pipework to meet these spacing requirements

However, there may be times when you will work on other buildings: schools, hospitals or small industrial units, for example, where the clips or brackets need to be strong and robust.

On large jobs, clips or brackets will be specified, but frequently the plumber has to decide what type of fixing to use.

Plumbers also use low carbon steel pipe (LCS) fixings, which are similar to the brass fixings shown for copper.

The table below lists the recommended spacings for use between internal pipework fixings:

Pipe size		Copper		LCS		Plastic pipe	
mm	in	horizontal (m)	vertical (m)	horizontal (m)	vertical (m)	horizontal (m)	vertical (m)
15	$\frac{1}{2}$	1.2	1.8	1.8	2.4	0.6	1.2
22	$\frac{3}{4}$	1.8	2.4	2.4	3.0	0.7	1.4
28	1	1.8	2.4	2.4	3.0	0.8	1.5
35	$1\frac{1}{4}$	2.4	3.0	2.7	3.0	0.8	1.7
42	$1\frac{1}{2}$	2.4	3.0	3.0	3.6	0.9	1.8
54	2	2.7	3.0	3.0	3.6	1.0	2.1

Figure 5.35 Recommended spacing between pipework fittings

Associated trade skills

At the end of this section you should be able to:

- describe the procedures for lifting floor surfaces
- state the requirements for cutting holes and notching timber joists
- describe the procedures for cutting holes through a range of building materials
- list and describe the tools used in associated trade work.

Crowbar

You will probably appreciate by now that a plumber's job is both interesting and varied. As a result, you will need to learn some additional skills to carry out work other than just plumbing. For example, you will have to run pipes under timber floors, which involves notching joists.

In existing properties, gaining access to work under the floor will mean lifting floor surfaces, so you will need to know how to do that. Often, both on new work and maintenance work, you will be required to cut or drill holes in brickwork, blockwork, concrete and timber, so that will also be covered.

Wood chisel

Pad-saw

Lifting floor surfaces

Lifting floorboards using hand tools

The hand tools shown are used for lifting floorboards.

Lifting a length of floorboard to run pipework through joists

This usually involves lifting a single length of floorboard. If it is a full length it is easier, because you won't have to cut across the board.

Bolster chisel

Using a hammer and sharp bolster, carefully cut the tongue-and-groove joint down either side of the floorboard. Alternatively, you could use a pad-saw.

Good practice requires punching down the nails to enable the board to be removed. Alternatively, a wrecking bar, or draw bar, is sometimes used to prize up the floorboard and nails. Once the board is partially lifted, if it is pushed down slightly the nail heads will be revealed, allowing their removal with a claw hammer.

If you need to lift only part of a board, you will have to make one, possibly two, cuts across a joist, so that when the board goes back it has a firm fixing point. If you cannot locate a joist, you will have to insert timber cleats. You can locate a joist by finding the floorboard nails.

Cross-cuts on a floorboard can be made using an extremely sharp wood chisel or a purpose-made floorboard saw.

Safety tip

Eye protection and dust mask may be required when lifting sections of flooring

Safety tip

Put the saw blade at a slight angle to help you to avoid the nails.

Pay particular attention to what is underneath the boards you are cutting (especially with a circular saw): you never know what is under them.

NEVER use a circular saw unless you have been trained to do so

Lifting floorboards using power tools

This is done using a circular saw.

Circular saw

Here are a few things to remember about using a circular saw:

- use 110 V supply only

- the depth of the cut must not exceed the depth of the floorboard

- if possible, remove a trial board using hand tools to check for electrical cables or hidden pipework

- a guard must always be fitted to the saw blade.

The saw can be used to cut down the full length of the tongue and groove on each side of the board. A cross-cut is made, again over a joist, making sure that the blade does not hit the nails.

Cutting traps in floorboards

Cutting a trap uses the same method as removing a single floorboard, using either hand or power tools. A trap requires you to cut more boards, but in shorter lengths.

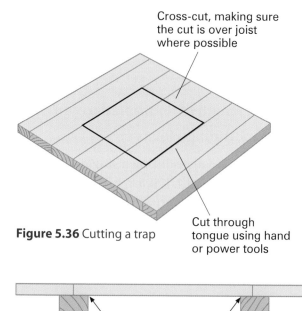

Cross-cut, making sure the cut is over joist where possible

Cut through tongue using hand or power tools

Figure 5.36 Cutting a trap

Replacing floorboards and traps

The floorboard length or trap should be screwed back into position to make future inspection or maintenance easier. Use countersunk wood screws. When refixed over pipework, the board surface should be marked accordingly, e.g. 'hot and cold pipework'.

If it is not possible to find a joist to refit the board or trap, cleats must be used to support the board end.

The illustration shows a trap or board replacement over joists and using cleats.

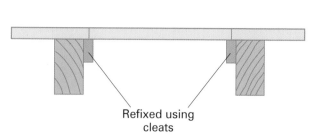

Cut over joist if possible

Refixed using cleats

Figure 5.37 Replacing floorboards

Removing chipboard

This is a more difficult job than floorboards, as chipboard is laid in wider sheets.

The best way to remove a chipboard section is by using a circular saw. If a power supply is not available, a section of board can be cut using a floorboard saw. The section of board to be removed should be marked out across the board, creating guidelines to follow for the cut. If a pad-saw is used to make the cut, it is helpful to drill holes in each corner of the area to be lifted in order to start the cutting process.

If this method is used, a new piece of chipboard will be needed to replace the removed section.

Holes drilled in each corner to start the cut with either pad-saw or jig-saw

Figure 5.38 Preparation for removal of chipboard

Replacing chipboard

Like floorboards, chipboard should be screwed back into position.

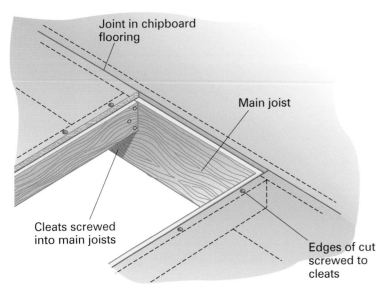

Joint in chipboard flooring

Main joist

Cleats screwed into main joists

Edges of cut screwed to cleats

Figure 5.39 Re-fixing chipboard

Pipe guards

We mentioned earlier that pipe runs under floors should be marked. A more effective way of protecting pipes that pass through joists is to use pipe guards. Figure 5.40 shows a typical example of their use.

Steel plate

Pipe runs

Figure 5.40 Pipe guards

Timber joists

It is inevitable that joists will have to be drilled or notched to permit pipe runs under timber floors. The preferred method would be to drill the joist in the centre of its depth, as this is where there is least stress. In practice, unfortunately, apart from when using plastic hot- and cold-water supply pipe, this tends to be impractical.

The main thing to remember, therefore, for either notches or holes is not to weaken the joists. This also applies to the distance from the wall where the joist is notched or drilled.

The Building Regulations set out requirements for notching or drilling joists, and these must be followed at all times.

A worked example for notching a joist

Remember

When drilling a joist, make sure that what you are doing will not weaken the joist

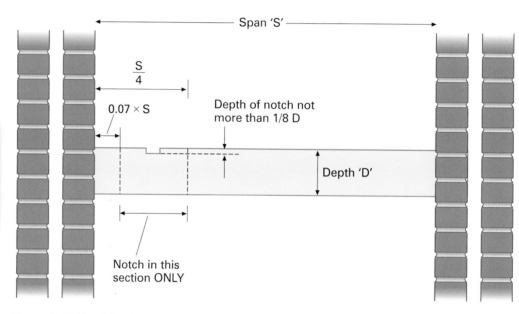

Figure 5.41 Notching joists

Consider a joist 200 mm deep (D) and 2.5 m long (the span, S).

Any notch must have a maximum depth of $\frac{1}{8}$ mm. Therefore, for our joist, the depth of the notch is $\frac{200}{8}$ = 25 mm.

The minimum distance from wall is $7 \times \frac{S}{100}$ mm. For our joist, this is $\frac{7 \times 2500}{100}$ = 175 mm.

The maximum distance from wall is $\frac{S}{4}$ mm from its bearing, giving a maximum span for our joist of $\frac{2500}{4}$ = 625 mm.

Joists are normally cut by hand, using a hand or floorboard saw. It is cut to the required depth and width, and the timber notch removed using a hammer and sharp wood chisel.

The width should be enough to give freedom of movement, in order to allow for expansion and contraction, particularly in the hot-water supply.

Drilling a hole in a timber floor joist

There are similar requirements laid down for drilling holes in joists, although the calculation figures are slightly different:

- The maximum hole diameter is one quarter of the depth of the joist.

- Holes should be positioned no closer together than 3 times their diameter.

- The minimum distance from the wall is $0.25 \times$ the span.

- The maximum distance from the wall is $0.4 \times$ the span.

Figure 5.42 Acceptable positions for drilling holes in timber floor joists

Cutting holes in the building fabric

By building fabric we mean:

- brickwork
- concrete
- blockwork
- timber.

Hand or power tools can be used to cut holes.

Power tools

The selection of drill will depend on the job in hand.

Rotary hammer drill	500 watt	850 watt	1400 watt Diamond core drill
Brick/block	30 mm	42 mm	152 mm
Concrete	24 mm	30 mm	120 mm

Figure 5.43 Choosing the appropriate drill bit

Drill bits are designed for specific tasks and are purpose-made for brick, block, concrete, steel or wood. A diamond-core drill uses either diamond- or tungsten-tipped bits.

Core drills are excellent for drilling through brickwork, blockwork or concrete where you need to pass large-diameter waste or soil pipes, or flue pipework. They provide a much neater finish to the job, and leave less **making good** than hand tools.

Selection of wood bits

Hole saw set

Wood bits or wood-boring bits

These are useful for drilling holes in joists or other timber constructions for the passage of pipes.

Hole saws

These are handy when drilling through kitchen units in order to pass waste pipes through the side or back of cabinets. They are also used to drill holes in plastic cold-water storage cisterns.

Hand tools

Tools for brick, block and concrete include cold chisels, brick bolsters (preferably with guard), plugging chisels and club hammers. Tools for timber include claw hammers and wood chisels, and brace and bits for cutting or boring holes in timber – although power tools have generally replaced the brace and bit.

Hand tools, such as club hammers, cold chisels and bolsters are used to cut holes and chases in brickwork, blockwork or concrete. Care must be taken to cut out only the minimum of material required for the passage of the pipe, in order to keep making good to a minimum. Plugging chisels are used to chase out mortar between brickwork, and the plumber does this to let in sheet lead flashings, or for timber-fixing plugs.

A brace and bit, or cordless drill, is used to bore holes through timber in the same way as power tools are used. However, the power drills can use much larger-diameter bits, and do not require as much hard work.

Making good

> **Definition**
>
> **Making good**
> – repairing and finishing off the brickwork, blockwork and concrete

For most jobs, a mortar mixture of 4 parts sand to 1 part cement will be adequate for pointing any brickwork joints disturbed while doing the job, and for **making good** the gap around the pipe penetration. Mastic sealant, either clear or close to the colour of the pipework, can be used as an alternative.

Generic systems knowledge

Identification of pipework and use of plumbing symbols

Identification of pipework

British Standards are produced to standardise the colour coding of pipework. This is particularly relevant in larger installations where a number of services are used. In domestic installations, pipework colour coding is not essential in most situations, but it is used on service pipes below ground, as follows:

- cold-water service pipes = blue

- domestic natural gas = yellow

- central heating pipework for use in solid floors is covered in white plastic.

Plumbing symbols

In order to provide some standardisation of symbols used on drawings, the BSI has produced a specification for pipework symbols; the main symbols are shown in Figure 5.44 below.

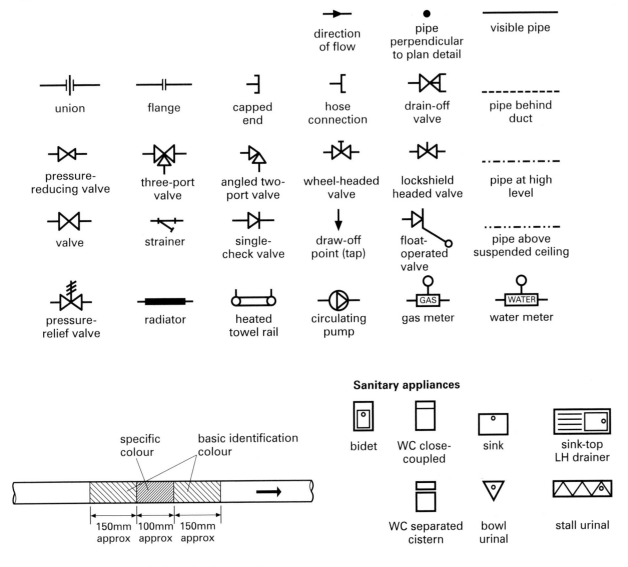

Figure 5.44 Pipework symbols and colour-coding

Measuring jobs and estimating materials

This topic includes:

- estimating from a drawing
- estimating from site.

Estimating from a drawing

It is unlikely that you will be required to carry out an estimate from a complex services drawing, but you may be required to estimate from a detail such as the one shown here:

Figure 5.45 Example of line drawing, not to scale

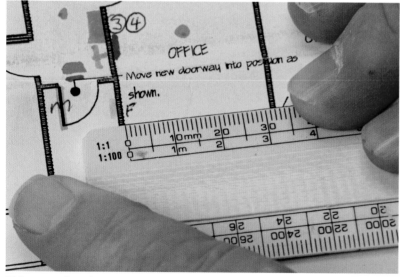

Using a scale rule

On this drawing the dimensions are given. The detail could have been given drawn to scale, in which case you would need to use a scale rule to measure the dimensions.

The photograph shows an example of a scale rule, used to measure dimensions on a drawing.

Estimating from site

Some of your estimating will take place on-site, without drawings or specifications. On one-off new installations, it will be carried out in two stages:

1   first fix

2   second fix.

The **first fix** involves installing all the pipework prior to floor covering going down and plasterboard finishing (sometimes referred to as carcassing). The **second fix** is the completion of the installation – fitting the bathroom suite, other sanitary appliances or radiators.

You will need to have a thorough understanding of the system requirements and where the intended pipe runs are. The next stage is to take measurements, using as a minimum a 3 m, and preferably a 5 m tape, and then recording your measurements in a notebook. An estimate of fittings can also be done at the same time.

Once completed, the dimensions can be totalled to give an estimate of the pipe requirements.

On larger contracts, a large housing project for example, the dimensions, pipe and fitting requirements will probably already have been worked out. These are often provided to the plumber in the form of a specific materials pack, booked out from the stores on a house by house basis using a requisition order, so that the stock can be monitored.

Find out

Estimate the material requirement for a typical installation – maybe the installation in your own home

Documentation

You will come across the following general documentation:

*   Water Regulations

*   requirements of British Standard specifications and in particular BS 6700

*   Building Regulations, which we mentioned earlier with regard to drilling and notching joists

*   Health and Safety regulations.

As well as these documents, which are the 'yardstick' of everything you do, there is other documentation that you need to be aware of. The other documentation that is used in support of plumbing work will vary from job to job, depending on what is being done.

Job specifications

These usually accompany site and services drawings and details, and are mainly used on larger contracts. Job specifications provide details such as:

*   type and quality of components, materials and fittings to be installed

*   type of clips or brackets to be used

*   system test specifications

- any specific installation requirements, e.g. who is to carry out any associated building work, such as cutting holes for pipework, drilling or notching joists and making good.

Remember: specifications form part of the contract documentation; any alteration required to the specification should not be done by you. If, for any reason, a part of the job cannot be done to the specification, you should inform your supervisor or employer immediately.

Quotations

On a smaller job, for example renewing a customer's bathroom suite, it is unlikely that a detailed specification will be produced. Your employer, however, will have issued a detailed quotation that the customer will have agreed. This in itself is a form of contract. Again, any revisions to the original quote must be passed on to your employer, so that they can be agreed with the customer.

Quotations are also used by plumbers' merchants to supply a price for the materials required to carry out the job. This would be followed by an invoice, which is a statement of the materials supplied, requiring payment by the plumber.

An example of the paperwork system

Some of the other important documentation can be illustrated by looking at how the paperwork system works for a bathroom installation.

1 Customer asks plumber for a quote.

2 Plumber visits customer, measures up, estimates materials.

3 Plumber gets a quotation for the materials from the merchant.

4 This enables plumber to finalise quote to customer.

5 In this case, the customer accepts and the quote is given to customer in writing.

6 Plumber orders materials from merchant after the materials quotation is confirmed in writing. This is an order (it may not be confirmed in writing if the plumber has an account with the merchant). Plumber receives delivery of materials, confirmed by a delivery note.

7 Job is carried out. Plumber receives invoices for payment of materials and also invoices the customer for the work carried out. A remittance advice (record of payment) is sometimes issued. A receipt may be given to the customer.

On any job, large or small, if you are advised that the delivery of materials will be delayed, or a particular item is not the one ordered, notify the site manager or your immediate supervisor/employer (so that he or she is aware of potential delays).

Work programmes

Work programmes will be covered in greater depth at Level 3. On a small job, you will have the work programme inside your head, based on agreed start and finish dates. On larger contracts, the approach is more scientific, and a contract programme will have been provided. This could consist of an overall programme for all site trades, as well as a separate programme for each trade. There could be a number of variations on this theme, but the principle of programming is basically 'activity against time'.

Preparing for the installation

Depending on the size of the job, the following checklist will cover most of the preparation items:

- Have you confirmed with the client the start time and date?
- Are all the tools, materials and equipment required to do the job on site?
- Are electrical tools safe to use?
- Is the work area clean and safe to work in?
- If using access equipment, ladders etc., are they safe to use?
- If you are working in a loft, do you have adequate lighting to see what you are doing (torch, inspection lamp etc.)?
- If working in an occupied dwelling, have you checked the work area for any pre-work damage? If you have noticed any damage, have you notified the customer prior to starting the work?
- Again, in an occupied dwelling, have you made sure you have protected the customer's property before you start?
    - Use dust sheets (but not where they could be dangerous, e.g. on stairs).
    - Request that vulnerable furniture is moved.
    - Request that carpets are taken up if necessary.
    - Cover up sanitary appliances if working in the bathroom area.

Safety tip

Never leave tools such as hammers or spanners where they could be accidentally knocked – for example, into sanitary appliances

FAQ

How important is access to pipework and fittings installed in the structure of a building?

It is essential to have reasonable access to plumbing services, otherwise expensive, time-consuming work is needed, much to the annoyance of the customer. A lack of access may be in breach of Regulations.

Knowledge check 4

1   What type of screws are normally used to make fixings into sheet metal?

2   What is the maximum clip spacing distance for a 28 mm horizontal copper pipe?

3   What is the maximum hole diameter that can be drilled in a 250 mm deep timber joist?

4   Where is a fixed price for a job normally detailed?

5   What is the term that describes the installation of pipework in a new build property?

Additional information sources

Sources of additional information on the topics in this chapter include:

Copper Development Association – www.cda.org.uk
Yorkshire Copper Tube, tube manufacturer – www.yorkshirecopper.com
Yorkshire Fittings, copper fittings manufacturer – www.yorkshirefittings.co.uk
John Guest, plastic pipe and fitting manufacturer – www.johnguest.co.uk
IBP Conex, compression fitting – www.ibpconex.co.uk
George Fischer, LCS tube and fittings – www.georgfischer.co.uk
Hepworth Building Products, plastic pipe manufacturer (sanitation, hot- and cold-water systems, central heating) – www.hepworth.wavin.com
Geberit, plastic pipe manufacturer (sanitation, hot- and cold-water systems) – www.geberit.co.uk
Hunter Plastics, plastic pipe (sanitation) – www.hunterplastics.co.uk
Wavin Plastics, plastic pipe (sanitation, hot- and cold-water systems) – www.wavin.co.uk
Ridgid Tools, hand and specialist tool manufacturer – www.ridgid.com
Rothenberger Tools, hand and specialist tool manufacturer – www.rothenberger-tools.com
Dewalt, manufacturer of power tools – www.dewalt.co.uk

chapter 6

Cold-water supply

OVERVIEW

The supply of fresh wholesome cold water to people's homes is a basic human need. As a plumber, it is your job to get the water to the taps so that it is clean and fit for human consumption, otherwise severe illness can occur. Most people take for granted the supply of cold fresh water to their homes. Few would probably appreciate what goes into providing this service. Although water supply legislation is fully covered at Level 3, it is important that you gain a general understanding relating to all areas of your work at Level 2. In this chapter you will cover:

- **Cold-water supply and treatment**
  - Water Regulations
  - the origin, collection and storage of water
  - water treatment
  - water supply and distribution
- **Cold-water systems – inside the building**
  - cold-water storage cisterns
  - frost protection
- **The basics of backflow prevention**

Cold-water supply and treatment

Water Regulations

The purpose of the Water Supply (Water Fittings) Regulations 1999 is to prevent contamination of a water supply, prevent the waste of water, prevent misuse of a water supply, prevent undue consumption of water and prevent erroneous measurement (fiddling the meter).

The Regulations are also designed to permit the introduction of new products and ideas, as well as supporting environmental awareness.

Enforcing the Regulations

The water supplier is responsible for enforcing the Regulations. Under the Water Regulations, water companies are encouraged to set up approved contractors schemes. These require that approved contractors certify to the water company that water fittings installed are in compliance with the Regulations.

Although there is no legal requirement for a person working on water services to be qualified, anyone who carries out such work can be prosecuted for an offence against the Water Regulations and, if convicted of an offence, can be liable to a fine.

The origin, collection and storage of water

Water is clear, tasteless and colourless. It is defined in chemical terms as a compound of the two gases, hydrogen and oxygen, in the proportion two parts hydrogen and one part oxygen (H_2O).

The classifications of water

Probably the most important property of water is its ability to dissolve gases and solids to form solutions; this is referred to as its **solvent power** and has a bearing on how soft or hard the water eventually becomes. Water is classified according to its hardness. Hard water can result in limescale build-up, as seen in the photograph. This can be tackled by using water treatment methods, which we'll look at later.

Limescale build-up in pipework

This diagram explains the classification of water.

Rainfall

CO_2 CO_2 O_2

dust and
gases added
to rainfall

CO_2 SO_2

Rainwater falls to the ground as a dirty and acidic solution
Below: typical ground conditions

| moorland
CO_2 added | sandstone
CO_2 retained | salt
(calcium sulphate) | limestone and chalk
(calcium carbonate)
CO_2 dissolving
limestone |

| very soft
water
(acidic) | soft
water
(acidic) | permanent
hard water
(alkaline) | temporary
hard water
(alkaline) |

Figure 6.1 Classification of water type

Find out

There's more on water
hardness in Chapter 4

Did you know?

Solvent power can be
a negative effect for the
plumber to deal with as
it causes corrosion and
'blocking up' of systems
and components

The measurement of the hardness of water is shown as parts per million or alternatively milligrams per litre. This is how hardness is classified:

Water type	Dissolved gases
Soft water	0 to 50 mg/litre
Moderately soft	50 to 100 mg/litre
Slightly hard	100 to 150 mg/litre
Moderately hard	150 to 200 mg/litre
Hard	200 to 300 mg/litre
Very hard	over 300 mg/litre

Figure 6.2 Water hardness

Where does water come from?

Water evaporates from the sea, rivers, lakes and the soil. It forms clouds containing water vapour, which eventually condenses and falls as rain. When it hits the ground, some of the rainwater runs into streams, rivers and lakes, some soaks into the ground, where it will collect temporarily and evaporate, or it will soak away and form natural springs or pockets of water to be accessed by wells (see figure 6.3).

Water companies, for example Severn Trent, Thames Water, etc., obtain their water for public consumption from two main sources:

* surface sources, such as:
  * upland surface water
  * rivers and streams
* underground sources, such as:
  * wells
  * artesian wells
  * springs.

Surface sources

Upland surface water

This category covers **impounding reservoirs**, lakes and natural reservoirs. This source of water is mostly found in the northern part of the UK, as the landscape there is hilly or mountainous, allowing lakes to form naturally, or the damming of streams to form impounding reservoirs. The water quality from this source is good because it is generally free from human or animal contamination. It is usually classified as soft as it runs directly off the ground surface and into the water source, so it is not affected by passing through a particular soil type. Where water comes into contact with peat, it can become acidic.

Remember

All surface and underground sources are dependent upon rainfall

Definition

Impounding reservoir – a man-made reservoir

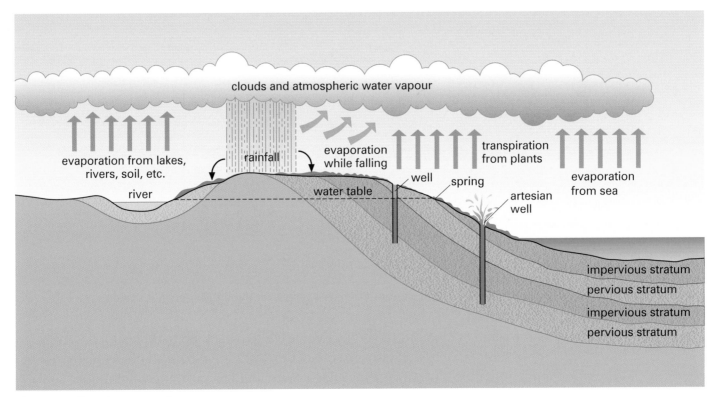

Figure 6.3 The rainwater cycle

Artificial reservoirs

Artificial reservoirs are constructed to meet an ever-increasing demand for water in both domestic and industrial sectors. They are also used in areas where insufficient natural resources exist. Reservoirs are created by flooding low-lying areas of land, normally by damming a water course. The water is classified as soft.

Rivers and streams

The quality of water from rivers depends on the location. Water from moorland rivers and streams tends to be relatively wholesome compared to further downstream, where it could become polluted by natural drainage from farmyards, road surfaces and industrial waste. The quality of river water varies in hardness depending on the nature of the ground where it originated. Water from upland river sources is generally soft compared with that of the lower reaches, which is usually hard.

Underground sources

Wells

In the past, before the existence of Water Authorities (water companies), wells were used to provide the water supply to dwellings or small communities. There are two types of wells, shallow and deep. Surprisingly, this classification does not necessarily refer to the actual depth of the well, but whether it penetrates the first impervious stratum of the earth.

Definition

Wholesome water – good quality water that is required to be provided for human consumption

Shallow well

Deep well

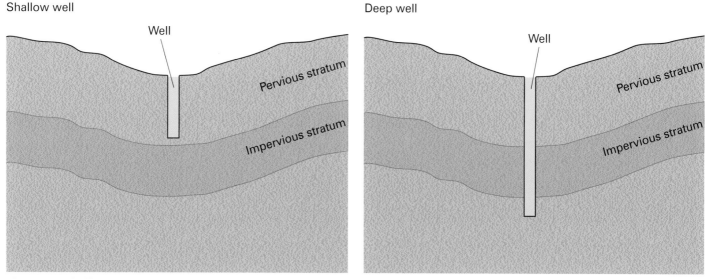

Figure 6.4 Shallow well does not penetrate impervious stratum. Deep well penetrates through impervious stratum.

Some water companies still retain deep wells as a back-up to their supplies from other sources, or as a standby in case of drought. A shallow well has a greater risk of contamination, but water from a deep well should be pure and wholesome.

Artesian wells

An artesian well penetrates the impervious stratum and enters a lower porous zone containing water. The outlet of the well is situated below the **water table**, so the water is forced out by gravity through the mouth of the well.

> **Definition**
>
> **Water table** – the natural level of water under the earth

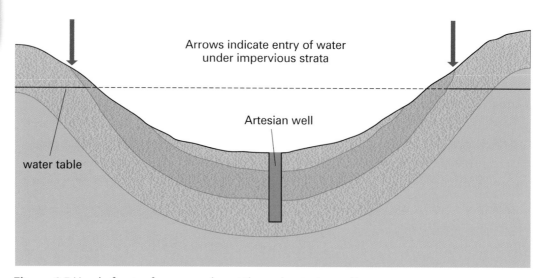

Figure 6.5 Head of water forces supply out through artesian well

Springs

The quality of spring water varies depending upon the route that the water has taken from underground to the surface. If it has travelled for a long distance through rock formations, it will probably be free from contamination, but it is likely to be hard.

The following table summarises the various water sources and their level of contamination before treatment.

Classification	Source	Contamination level
Wholesome	1 Spring water 2 Deep well water 3 Upland surface water	Very palatable
Suspicious	4 Stored rainwater 5 Surface water from cultivated land	Moderately palatable
Dangerous	6 River water 7 Shallow well	Non palatable

Figure 6.6 Summary of water source classification

Water storage

Water is stored by water companies either in its untreated state in impounding reservoirs or lakes, or as wholesome water in service reservoirs. It is worth noting that water companies usually aim to store enough drinking water in their service reservoirs, for emergencies, to maintain supply for about 24 hours. This safeguards against failure of pumps or mains and allows time to repair any faults before supplies run out.

Water treatment

It is the responsibility of the water companies to ensure water is fit to drink. All water must be treated before it is put into the supply system. How it is treated will depend on its source and what impurities it contains. Some impurities are actually essential to our health and will be retained; others are harmful and must be removed during the treatment process.

Deep wells and bore holes

The quality of water from this source is already quite good, due to the natural filtering process as the water passes through the rock strata. In this case, the only treatment needed is sterilisation. Sterilisation serves to keep both the water and supply pipework free from bacteria as the water is piped to our homes.

Rivers, lakes and impounding reservoirs

These are our main sources of supply, but generally speaking this water is dirty and polluted, so treatment needs to be extremely thorough. The treatment process usually involves the water being strained, filtered, chemically treated and sterilised.

Definition

Sterilisation – purification by boiling, or, within the industry, dosing the supply with chlorine or chlorine ammonia mix

Water supply and distribution

Definition

Water mains – the network of pipes that supply wholesome water to domestic and commercial properties

Water is supplied to our homes via a network of pipes known as **mains**. These will vary in diameter depending on the purpose of the main and the likely demand of the supply. The diagram shows an example of a typical water-supply system layout.

Figure 6.7 A typical mains water distribution system

Figure 6.8 Section through a mains tapping machine

The local mains shown will provide the 'final leg' of the supply of water to your home.

Mains connection

Water mains are constructed of asbestos cement, steel, PVC or cast iron. PVC is now used extensively on new installations and mains replacements. The connection to the mains is the responsibility of a water company, and new connections to existing mains, operating under live pressure, require the use of specialist equipment. The illustration shows an example of a connection to a cast iron main using a mains tapping machine where the new connection can be made without the need to isolate the live main.

The connection to the main is made using a **ferrule**. The ferrule is in fact an isolating valve, so once the threaded section has been inserted into the mains, the valve can be isolated, permitting the mains tapping machine to be removed and connection of the service pipework to the ferrule outlet.

Figure 6.9 Section through a brass ferrule

Definition

Ferrule – a metal fitment used to connect to a main allowing isolation

The supply from the mains to the building

The next diagram shows a full installation from the main to the stop and drain valve wherever it enters the building. A plumber's work usually starts from the external stop valve.

Figure 6.10 Supply pipework to the building

There are some key points to remember about service pipes:

- In order to protect against frost damage, mains and service pipes should be at least 750 mm beneath the surface of the ground.

- The maximum depth of cover should not exceed 1.35 m, as this would prevent ease of access.

- Blue-coloured medium-density polyethylene pipe (MDPE) is normally used for the underground service pipework into the property. Plastics are now widely used to make service pipe connections as they tend to be jointed easily and are not subject to corrosion. Care needs to be taken, however, when installing standard plastic pipes in ground containing contaminates, such as petrol or oil, as they are susceptible to permeation, causing the water to be contaminated. When installing service pipes across contaminated land, a special barrier plastic pipe, or metallic pipe such as copper, should be used.

- Metal pipes should be protected against possible corrosion from the soil, particularly acidic soils. This can be done by:
  - use of plastic-sheathed pipe
  - wrapping the pipe with an anti-corrosive tape
  - installing the pipe inside a duct.

- The minimum size allowed for a cold-water service pipe to a dwelling is 15 mm. Service pipes to new dwellings are now usually provided with a 25 mm internal diameter in order to meet the flow rate demands required by higher pressure mains plumbing systems.

Installation of external stop valves

As you can see from figure 6.10, the service pipe to the building can be isolated from the main by using the external stop valve. Ease of access to the stop valve is very important, so it should be located in a stop valve chamber, constructed from 150 mm PVC, or earthenware pipe, sited on a firm base, and finished off at the top with a stop valve cover. The stop valve cover is usually made from steel plate, or a combination of steel plate and plastic.

Figure 6.11 shows a PVC stop valve chamber; these are now widely used on new installations. The guard pipe is cut to the desired length to suit the installation as part of the fitting process.

Installation of water meters

The installation of water meters for domestic premises is becoming more widespread, particularly on new housing developments. They measure the amount of water used by the householder, who then only pays for the water used. Water meters can be installed either internally or externally.

External water meters (an example is shown in Figure 6.12) usually incorporate a service stop valve and are found on new domestic installations. The parts come already assembled, with the guard pipe being cut to length as part of the fitting process.

The next two diagrams show the installation of internal water meters. They are normally seen in commercial or larger domestic properties.

Figure 6.11 PVC stop valve chamber without meter

Figure 6.12 External water meter installation

Figure 6.13 Exposed internal water meter

Figure 6.14 Concealed internal water meter

Knowledge check 1

1   What is the minimum depth of cover over an external cold-water service pipe?

2   What is untreated water drawn from a shallow well classified as?

3   Which organisations are responsible for enforcing the requirements of the Water Regulations?

4   What type of component is normally sited inside an external stop valve box together with the stop valve?

5   What must a service pipe sited in contaminated land be manufactured from?

Cold-water systems – inside the building

Stop and drain valve

This section deals with the supply from the stop and drain valve inside the building.

Remember

The supply pipe must contain an isolation valve as it enters the property, together with a drain-off facility at the lowest point. These arrangements can be made up using separate fittings as an alternative to the combined valve shown

Figure 6.15 Combined stop and drain valve

The purpose of the stop and drain valve is to:

- turn on/off the water supply to pipework fittings and components to enable system maintenance

- drain down all the system pipework fittings and components to enable system pipework repair/replacement.

System pressure and flow rate

The incoming water pressure and flow rate supplied via the main is vital as it has a key bearing on the size of pipework and fittings used in the system. We will look at pipe sizes in more detail at Level 3, however you may be required to assist in taking readings of incoming water pressure and flow rate as part of the installation process.

Water pressure measurement

This is carried out using a pressure gauge as shown below. The gauge usually reads in bar pressure.

Water pressure reading being taken

Water flow rate reading being taken with a weir cup

Remember

Keep the weir cup *level* when taking a water flow rate reading

Water flow rate measurement

This is carried out using a water flow measuring device as shown above. The reading is usually quoted in litres per second.

A common fault with domestic systems installation is that water pressure readings are not taken at the right stage in the job – that is at the design or pricing stage. Insufficient supply pressure and flow rate can result in big problems: for example, if the wrong system components, such as combination boilers, are installed, they will not work correctly. If the pressure to an existing dwelling is poor, check first that there is no burst on the service from the external stop tap. You can usually tell by putting your ear against the stop tap key while it is on the stop tap and listening for a hissing noise.

Use of water conditioners

One way to prevent damage to water systems by hard water is to install a water conditioner in the cold water supply. This has the effect of softening the water prior to it being heated, and so reduces scale formation.

There are three main ways to treat water hardness:

- base exchange softeners
- scale reducers
- magnetic water conditioners.

Base exchange softeners

These work by passing the hard water through a tank containing resin particles. The resin attracts and absorbs the hardness salts – mainly calcium and magnesium – from the water. At the same time it replaces them with sodium from the resins. After a while, the resin becomes saturated with hardness salts and needs to be regenerated, using salt solution to put sodium back into the resin. The hardness salts are released from the resin and washed down the drain.

The unit requires regular maintenance and checking, an annual service and a check on water hardness needs to be carried out by a service engineer. The salt, however, can usually be topped up by the end user. BS 6700 makes recommendations for the installation of base exchange water softeners. Its main consideration is the prevention of backflow and contamination of the water supply. As there is a backflow risk associated with the use of a base exchange water softener, where salt additives are put into the cold-water system, the installation must be protected by the use of a single check valve. For further details on the basics of backflow prevention see pages 213–216. BS 6700 is the key reference source, together with the manufacturer's instructions, when installing the water softener.

Manual override

Timing device

Salt compartment lid

Drain

Salt compartment

Overflow

Maximum salt level

Resin

Salt storage compartment

Resin pressure vessel

Brine well

Brine filter

Figure 6.16 Section through a typical base exchange water softener

Scale reducers

Scale reducers are also examples of chemical conditioners designed to reduce limescale in water; they dispense chemical additives into the system to treat the water. The chemicals alter the molecular structure of the calcium salts in the water, preventing them from bonding together. The scale reducer will need periodic replenishing with chemical additives. It is commonly used to provide protection to individual system components such as multipoint water heaters or combination boilers.

Figure 6.17 Scale reducer

Magnetic conditioners

The molecular structure of the salts in the water is changed by a small electric field set up by the magnetic device that is clamped to the pipework. There are many variations on magnetic conditioners, some simply comprising of a strong magnet, others connected to the mains via a transformer to generate the electric current.

Figure 6.18 Magnetic conditioner

Types of cold-water system

There are two types of cold-water system: direct and indirect. In a direct system, all the pipes to the draw-off points (sink, bath, hand basin, WC, etc.) are taken directly from the rising main or service pipe, and operate under mains pressure. In an indirect system, one point – usually the kitchen sink – is fed directly from the rising main, which then supplies the cold-water storage cistern. The remaining draw-off points are fed from the cold-water storage cistern – hence the term indirect.

The direct cold-water system

The installation of direct systems is permitted by water companies in domestic properties in medium to high-pressure areas where the supply can provide adequate quantities of water at sufficient pressure to meet the building's needs.

The advantages of a direct system are that it is:

- cheaper to install because
  - less pipework is required
  - the storage cistern is smaller (100 litre minimum); BS 6700 recommends that the storage cistern to a direct system should be at least the same size as the hot-water storage cylinder.

- drinking water is available from all draw-off points

- less risk of frost damage due to a smaller amount of pipework

- less structural support required for smaller cistern.

However, there are disadvantages:

- higher pressure may make the system noisy

- there is no reserve of cold water if the mains or service supply is shut off

- more wear and tear on taps and valves due to high pressure

- higher demand on the main at peak periods.

The direct system is the most commonly installed type of cold water system in domestic properties because its installation is cost effective and there is usually relatively high pressure of supply available. It is a form of direct system that normally provides the cold-water supply to an instantaneous water heater or a combination boiler.

Figure 6.19 Pipework layout for a direct cold-water system

The indirect cold-water system

The draw-off points in an indirect system are fed indirectly from the cold-water storage cistern, one outlet being fed directly from the supply pipe. The system is designed to be used in low-pressure water areas where the mains supply pipework is not capable of supplying the full requirement of the system. This type of system also has a reserve of stored water in the event of mains failure.

The advantages of indirect systems are that there is:

- a reserve of water should the mains supply be turned off
- reduced risk of system noise due to lower pressures
- reduced risk of wear and tear on taps and valves, again due to lower pressure
- lower demand on the main at peak periods.

Figure 6.20 Pipework layout for an indirect cold-water system

The disadvantages are:

- an increased risk of frost damage
- the space occupied by the larger storage cistern (230 litre minimum – the minimum size for supplying both hot and cold outlets in dwellings as laid down by BS 6700)
- the additional cost of the storage cistern and pipework
- before the 1999 Water Regulations water might not have been potable.

Knowledge check 2

1   What is the normal size of cold-water pipework inside a small property supplying cold water via a direct system?

2   What is the minimum size of a cold-water storage cistern in a direct system (feeding hot water only)?

3   A water conditioning device that uses an electric current is described as what type of device?

4   What type of device is used to take water flow rate readings in a system?

5   What is an easy method to identify whether a cold-water system is direct or indirect?

Cold-water storage cisterns

Under the Water Regulations, a storage cistern supplying cold-water outlets, or feeding a hot-water storage system, should be capable of supplying wholesome water, and therefore various protection measures are included in the design of the cistern to ensure the water supply does not become contaminated.

Cisterns must therefore be:

- fitted with an effective inlet control device to maintain the correct water level
- fitted with service valves on inlet and outlet pipes
- fitted with screened warning/overflow pipes to warn of overflow
- covered to exclude light or insects
- insulated to prevent heat loss and undue warming
- installed so that the risk of contamination is minimised
- arranged so that water can circulate preventing stagnation
- supported to avoid distortion or damage leading to leaks
- readily accessible for maintenance and cleaning.

Figure 6.21 Protected cold-water storage cistern

Materials for cisterns

In the past, galvanised low carbon steel was the main material used to make cold-water storage cisterns. You might still come across these on maintenance jobs, but most new installations use cisterns made from plastic such as polyethylene, polypropylene and polyvinyl chloride.

The majority of cisterns are polypropylene, because this material allows them to be:

- light
- strong
- hygienic
- resistant to corrosion
- flexible enough to be manoeuvred through small openings.

Cisterns are available in square, rectangular or circular shapes, and are produced in black to prevent the growth of algae. However, because they are flexible, the base of the cistern *must* be fully supported throughout its entire length and width. Holes for pipe connections should be cut out using a hole saw. The joint between the cistern wall and fitting should be made using plastic or rubber washers.

On no account must the hole in the cistern be made by heating a section of pipe and using it to make a hole in the cistern – *this degrades the plastic and will result in cistern failure.*

Safety tip

Never use oil based pastes on a plastic cistern – they cause the plastic material to degrade

Connections to cisterns and control valves

Inlet controls

Water Regulations require that a pipe supplying water to a storage cistern be fitted with an effective adjustable shut-off device, which will close when the water reaches its required level. For most domestic applications, a float-operated valve is used. These must comply with BS 1212 (parts 1–4), and the following types are available:

- Portsmouth type
- diaphragm valve made of brass
- diaphragm valve made of plastic.

Piston

Figure 6.22 Portsmouth float valve to BS 1212 Part 1

Seating

Washer

Split pin

Figure 6.23 Diaphragm float valve to BS 1212 Part 2 or 3

The Portsmouth valve is not widely used on new installations as it does not provide an effective air gap between the water level and the point at which the valve discharges, but you may see this valve on existing installations. Brass or plastic diaphragm float valves to BS 1212 Part 2 or 3 are commonly used in new storage cisterns, as an effective air gap is provided between the water outlet from the valve and the water level in the storage cistern. BS 1212 (Part 4) refers specifically to diaphragm equilibrium float valves (not illustrated) designed primarily for use in WC flushing cisterns; the Portsmouth and diaphragm valves work by the principle of leverage; the equilibrium valve uses the water pressure. However, equilibrium valves are available with a similar external appearance to the Portsmouth and diaphragm valves. Inlet pipes to cisterns must have a servicing valve fitted immediately before connection to the cistern, in order to enable any maintenance to be carried out on the cistern without turning off all the water supply. This also applies to WC cisterns. The valve usually installed is a spherical type plug valve, suitable for use with higher pressure cold-water supplies.

Outlet pipes

The outlet pipes should be connected as low in the cistern as possible – there is a move in the new Water Regulations to preferred connections being in the bottom of the cistern rather than the side. This prevents the build up of sludge in the bottom of the cistern. Outlet pipes, such as cold feed and distribution pipes, should be fitted with servicing valves, and these should be located as near to the point of connection to the system as possible. The valve type used in this position is usually a wheelhead gate valve, suitable for isolating lower pressure cold-water supplies.

Overflow and warning pipes

When water in a cistern rises above a pre-set level, usually due to a faulty float-operated valve, the water is allowed to flow through a pipe away from the cistern. An **overflow pipe** is used to discharge water where it will not cause damage to the building. A **warning pipe** is a pipe used to give warning to the occupiers of a building that a cistern is overflowing and needs attention.

Definition

Overflow pipe – discharges excess water safely

Warning pipe – alerts occupants to a problem with float valves

Small cisterns of up to 1,000 litres (that is most domestic cisterns), must be fitted with a warning pipe and no other overflow pipe.

Larger cisterns (between 1,000 and 5,000 litres), are fitted with both.

Location of a warning pipe in a small cistern

The warning pipe must be located in the cistern so that a minimum air gap is maintained between its point of discharge and the normal water level in the cistern. The position of the float-operated valve is also crucial, to ensure that a minimum air gap is maintained between its outlet and the spill-over level of the cistern at the warning pipe. Note also:

- If the float-operated valve becomes defective, the warning pipe should be capable of removing excess water without becoming submerged.

- The minimum size pipe is 19 mm internal diameter, although the size may need to be larger than this with higher inlet flow rates to the cistern.

- The warning pipe should fall continuously from the cistern to the point of discharge.

Figure 6.24 Warning pipe positioning

- Warning pipes should discharge where the water will be noticed, usually outside the building.

- Warning pipes should be fitted with a screen or filter to exclude insects.

Connecting two or more cisterns together

We mentioned earlier the need to avoid contamination and stagnation in cisterns. This can occur when water cannot fully circulate throughout the cistern and is permitted to stagnate. This problem is more likely to occur when two or more cisterns are joined together – perhaps because the access hatch was not big enough. When jointing cisterns we must ensure proper water movement throughout the cisterns to avoid stagnation, as illustrated in figure 6.25.

Filtered overflow
(must be in same
cistern as ball valve)

28 mm connecting pipe

Supply pipe

Cold feed/distribution pipe

Figure 6.25 An example of linking two or more storage cisterns together

When linking larger cisterns in domestic properties attention must be given to:

- the provision of inlets to both cisterns fed via float-operated valves
- the method of connecting the distribution pipes via a manifold arrangement taking equal draw-off from both cisterns simultaneously
- siting outlet connections at the opposite end to inlet connections to allow effective water distribution across the cistern
- the method of linking the cisterns using lateral connections between the cisterns.

Frost protection

The major problem associated with freezing temperatures is the dramatic effect it can have on plumbing systems. Be aware that when water freezes it expands by about 10 per cent, causing damage to unprotected pipes and fittings.

When water freezes within the confines of, for example, a copper pipe, it will expand. The water either side of the 'ice plug' cannot be compressed, so the expanding ice will cause the wall of the pipe to split. Once the ice thaws, you have a burst pipe. Water damage in domestic properties can prove to be very expensive. It also causes an undue waste of water, which goes against the requirements of the Water Regulations.

The Water Regulations and BS 6700 both deal with frost protection.

Solid and timber floor installations

Where a supply pipe enters the building through a solid floor, it should be ducted, and the ends sealed to prevent moisture entering the pipe. Depending how close the pipe is to the outside wall as it rises to the property, it may need to be insulated as well. If it enters the building through a suspended timber floor, it should be ducted, and the ends sealed, and insulated as well. This is because the space between the suspended timber floor and the oversite concrete is exposed.

FAQ

When does a pipe burst due to freezing? Is it when it thaws out?

No, the pipe actually bursts (splits open) when it freezes due to the expansion of water changing to ice, it is only when it thaws out that we realise it has burst by seeing the water leak out.

Figure 6.26 Suspended timber supply pipe entry

Figure 6.27 Solid floor supply pipe entry

Other outside applications

Outside taps or standpipes for garden hoses are common. If the tap has a dedicated underground supply, it requires protection.

The diagram shows a typical garden tap installation. Note that a drain valve (see Figure 6.28) is located below the pipework. This enables the pipework to be drained once the stop tap is turned off. A double check valve is also fitted to protect against back flow.

Location of pipework

When planning pipe runs, you should try to avoid areas that are hard to keep warm, such as:

- any outside locations or outside walls
- roof spaces, cellars and under-floor spaces
- garages and other outbuildings
- near to windows, air bricks or ventilators.

Obviously it is not always possible to avoid these locations, so on these occasions you will need to provide some protection to the pipework.

Figure 6.28 Installation requirement of externally sited and fed hose union tap

Protection of pipes and fittings

Where pipework is at risk of freezing, it should be protected using pipe insulation material. The thickness of the insulation used will depend on the type of insulation used and its insulation properties.

The table gives the minimum recommended thickness of insulation in millimetres for 15 mm, 22 mm and 28 mm water pipes, to comply with BS 5422 and Water Regulations schedule 2, paragraph 4.

External diameter of pipe (mm)	Thermal conductivity of insulation materials (W/mK)				
	0.02	0.025	0.03	0.35	0.04
15	20	30	25	25	32
22	15	15	19	19	25
28	15	15	13	19	22

Figure 6.29 Minimum recommended insulation thickness (mm)

The table shows how thick the insulating material must be for products with varying insulation properties. Note that all of the entries in the table require an insulating material that is thicker than the typical 10 mm wall thickness you buy off the shelf at the merchant – so be careful.

Generally, the insulating material must be resistant to, or protected from:

- mechanical damage
- rain
- moisture
- subsoil water
- vermin.

Trace heating

Trace heating involves attaching a low-temperature heating element to the outside of the pipe, controlled by a thermostat which activates the heater when the temperature is low, thereby preventing the pipe from freezing. It is more common on industrial installations, but there are domestic products available.

Figure 6.30 Trace heating element protecting pipework against frost

Pipes and cisterns in roof spaces

The illustration shows the correct installation of a cold-water storage cistern and associated pipework in a roof space. Note:

- the provision of a 350 mm gap to the roof surface to allow access for maintenance
- both pipework and cistern are fully insulated
- the space under the cistern is left uninsulated to allow heat from the property to warm the cistern.

Figure 6.31
Cold-water storage cistern installation in roof space

Water Regulations also require that cold-water systems should not exceed 25°C. The use of insulation on pipework and components acts in hot weather to prevent stored water from becoming over-heated.

FAQs

Why does the water come out of some cold water taps with so much force?

This is due to high pressure in the cold-water supply to the taps which have been fed directly from the main.

The cold-water system in my house is quite noisy, when someone flushes the loo the water screeches through the pipework until the loo cistern is full again.

High water pressure is undoubtedly the cause of the noise; appliances fed from a storage cistern are much quieter in operation although it is possible to fit a pressure-reducing valve to the incoming supply.

Does water in storage cisterns and pipework go stale?

Yes it does. When you return from holidays, or if the property has been empty then it is wise to run all taps for a while to replenish any stale water with fresh.

Knowledge check 3

1   Which type of float-operated valve should not normally be used on a new cold-water storage cistern?

2   What component must be fitted to a warning pipe to a protected cistern to guard against the ingress of contaminants such as insects?

3   The outlets from cold-water storage cisterns are normally fitted with which type of servicing valve?

4   What is the minimum thickness of pipework insulation to a 15 mm cold-water pipe if the insulating material has a thermal conductivity of 0.03 W/mK?

5   What is the minimum recommended distance for servicing purposes between the nearest point of a cold-water storage cistern and the roof timbers?

The basics of backflow prevention

When installing a hot and cold plumbing system we must ensure that the system remains hygienic at all times and is not subjected to contamination – do not forget that we are going to drink from these systems, so getting contaminants inside the system could be dangerous to health.

There are two major ways in which water in a system may become contaminated:

1 The use of a non-approved material such as lead in the supply system which can lead to it becoming contaminated.

2 Backflow in the system can take place causing contaminated water to be drawn off inside the system.

The first point is dealt with by always installing approved pipework, components and fittings in hot- and cold-water systems.

The second point is dealt with by installing what are called **point-of-use backflow prevention devices** at all water outlets and fitting connections.

The subject is quite complicated and will be covered in detail at Level 3. Here the basics and how they relate to typical domestic plumbing systems are covered.

There are two ways in which backflow can occur in a plumbing system:

- Back-pressure – reversal of water flow in a pipe caused by a pressure increase in the system.

- Back-siphonage – siphonage from a cistern, appliance or water storage vessel back into the pipe feeding it.

There is an extensive range of backflow prevention devices on the market which provide backflow protection at the point of use, for example a tap outlet or a plumbing component. The selection of the correct type of device is based on an assessment of the risk presented by the fluid that the plumbing fitting may come into contact with.

Did you know?

Backflow is water flowing in a reverse direction from its intended direction of flow

Fluid risk categories

Before selecting a suitable backflow prevention device, an assessment of the fluid risk category that the water fitting may come into contact with needs to be made. A backflow prevention device can then be selected based on that assessment of risk. In the UK a five risk level system is used to undertake the assessment. This is summarised in the table on the next page.

Fluid category	Description	Examples in domestic properties
1	Wholesome water provided by the water undertaker	
2	Water that has had its aesthetic quality changed, e.g. by a change in temperature, taste or odour	• Mixed hot- and cold-water discharged from mixer taps • Water discharged from a base exchange water softener
3	Fluid that represents a slight health hazard	• Water in primary central heating circuits • The water in the bowl of a bath, basin or shower tray • Hand-held garden hose used with a hose union tap • The contents of a washing or dishwashing machine
4	Fluid that represents a significant health hazard	• Water used in a mini-irrigation system in a house, such as a porous hose or a sprinkler system with sprinkler heads less than 150 mm above ground level
5	Fluid that represents a significant health hazard	• The contents of a sink bowl, WC pan or bidet

Figure 6.32 Fluid risk categories

Examples of typical backflow prevention devices used in domestic properties

Taps and shower roses to baths, basins and shower trays

The contents of the bowl, a bath basin or shower tray are assessed as a fluid category 3 risk. To protect against the contents of the bowl, bath basin, etc. being drawn back into the hot- or cold-water system a Type AUK2 backflow prevention device using a simple air gap between the outlet of the water fitting and the spillover level of the bowl of the appliance is normally installed. The fixed air gap prevents the contaminated water from being drawn back into the system.

The air gap dimension must be a minimum of:

* 15 mm or less diameter outlet – 20 mm

* Between 15 mm and 25 mm outlet – 25 mm

* Over 25 mm outlet – 70 mm

Figure 6.33 Air gap above water fitting supplying domestic sink

Figure 6.34 Air gap above water fitting supplying domestic bath, basin or shower tray

In the event that the air gap cannot be provided to the dimensions highlighted above, then fluid category 3 protection in the form of a double check valve will need to be installed to both the hot and cold pipes feeding the individual appliance.

Taps and shower roses to domestic sinks

Here the contents of the sink bowl are regarded as a fluid category 5 risk – the most serious, owing to the bowl's potential contents, for example the washing of animal blood products, washing of vegetables, etc. Protection against the fluid category 5 risk is provided by means of a Type AUK3 air gap. The minimum air gap dimensions are 20 mm or twice the diameter of the inlet pipe supplying water to the fitting. So if it's a 15 mm pipe then the air gap should be 30 mm.

Here we must ensure that the air gap is achieved as we cannot fit another type of device if the gap is not sufficient, for example a double check valve connected to the inlet pipes only provides up to fluid category 3 protection.

Other examples of simple backflow protection in domestic properties

- A close coupled WC suite includes a Type AUK1 air gap built into the design of the appliance to ensure that the contents of the WC pan cannot be drawn back in the pipe supplying cold water to it. This air gap is built into the design of the WC suite so there's nothing more for us to do with this one!

- A new outside hose union bib tap requires fluid category 3 backflow protection to protect against the end of the hose being left in potentially contaminated water. Protection is provided here in the form of a double check valve fitted to the tap inlet (sited inside the building preferably) providing fluid category 3 protection. This is shown earlier in this chapter on page 209.

Remember

A double check valve only provides protection up to fluid category 3

Figure 6.35 Backflow prevention applied to mixer taps with mixing of water in the valve body

Hot-water supply

Cold-water supply

Single check valves

- The water supplied to a pressurised (sealed) central heating system, for example a combination boiler system, is fed through a double check valve built into the filling loop. This provides fluid category 3 protection against contamination from the contents of the central heating system.

- Mixing of hot and cold water in a mixer tap. Here the mixed water in the tap is identified as a fluid category 2 risk. The hot and cold supplies to the mixer tap must therefore be fitted with single check valves (protecting against a fluid category 2 risk) – unless the supplies to the appliance are from dedicated distributing pipes from a storage cistern.

- The cold-water supply to a water conditioning device such as a base exchange water softener is protected by installing a single check valve to provide fluid category 2 protection against mild contamination of the water supply through backflow from the salt additives in the device.

Knowledge check 4

1   What type of air gap is applied to a mixer tap to a kitchen sink?

2   What is the minimum air gap distance for a 25 mm cold-water tap to a bath?

3   What type of backflow prevention device is normally used with a hose union bib tap in a domestic property?

4   What type of backflow prevention device must be fitted to the inlet pipework to a twinflow mixer tap (mixing of hot and cold water cannot take place in the valve body)?

5   The contents of a bidet bowl are classified as what type of risk?

Additional information sources

Sources of additional information on the topics from this chapter include:

Water Regulations Advisory Scheme (WRAS) – www.wras.co.uk
Pegler Ltd. – www.pegler.co.uk
Department of the Environment, Food and Rural Affairs (DEFRA) – www.defra.gov.uk (includes a guide to the Water Regulations)
Scalemaster (water conditioning devices) – www.scalemaster.co.uk
Arrow valves (backflow prevention device) – www.arrowvalves.co.uk
Bathroom Manufacturers Association (member organisations have learning materials on plumbing systems) – www.bathroom-association.org

chapter 7

Hot-water supply

OVERVIEW

A supply of hot water is an essential requirement for anyone. We use it for personal washing and general cleaning purposes. There is a variety of hot-water systems available. In this chapter you will learn about:

- **Types of hot-water system**

- **Vented hot-water storage systems**
  - direct vented systems
  - indirect vented systems
  - vented cylinder details
  - combination storage systems
  - unvented storage systems
  - water jacketed heater

- **Storage heaters**
  - outlet controlled
  - inlet controlled

- **Instantaneous hot-water systems**
  - multipoint
  - single-point

- **Connection requirements of showers**

- **Dead legs and secondary circulation**

- **Connection requirements of bidets**

- **Pipework insulation requirements**

It would be hard to imagine life without hot water: no hot showers, baths or water for cleaning. It is the plumber's job to ensure that hot water is available as required.

A supply of hot water to domestic dwellings only became commonplace after the Second World War, and systems design has been improving since then. Hot water system design is also covered by the Water Regulations and BS 6700. The Building Regulations under part L1 lay down legal requirements for the energy efficient installation of hot-water systems. These details are covered at some length in the document *Domestic Heating Compliance Guide* provided by the government department Communities and Local Government. The document is freely available at www.planningportal.gov.uk

The design of a hot-water system will depend on the type of building. For example, when installing a toilet in a factory canteen, an instantaneous water heater is a more economical option than a storage system. In a domestic situation, there are instantaneous methods of supplying hot water as well as systems fed from storage.

Types of hot-water system

Connection

The connection between main and service pipework will depend on the type of hot water system that is to be installed, but generally will be categorised as direct or indirect.

Direct

Appliances such as instantaneous water heaters, when supplied directly should be fitted with a servicing valve (ball type) as close to the appliance as possible. Direct hot-water storage systems are fed via the cold-water storage cistern (CWSC).

Indirect

Instantaneous water heaters can also be supplied indirectly, and again should be fitted with a servicing valve. Again, on indirect hot-water storage systems the cold-water supply is via the CWSC and the installation details were covered in Chapter 6, *Cold-water supply*.

There is a wide range of hot water systems available for domestic properties. The following factors should be considered in their selection and design:

- the quantity of hot water required
- the temperature during storage and at outlets
- the cost of installation and maintenance
- fuel energy requirements and running costs
- any wastage of water and energy
- safety for the user.

Did you know?

There are many types of hot-water system but, generally speaking, they are either instantaneous systems (supplying hot water instantly) or storage systems (where water is heated and stored in a vessel until needed)

Safety tip

Catering for the expansion that occurs during the heating process is vital to ensure that the system works correctly and safely

Methods of heating hot-water systems

The following energy sources or fuels are used for heating hot water.

- Electricity
  - immersion heater
  - instantaneous heater
  - storage heater

- Gas
  - boiler
  - water circulator
  - instantaneous heater
  - storage heater

- Solid fuel
  - boiler
  - combined cooker and boiler

- Oil
  - boiler
  - combined cooker and boiler

- Solar (usually used to supplement other systems)
  - solar panel or collector.

Choosing a system

Systems can range from a simple, single-point arrangement supplying one outlet to a more complex centralised boiler system supplying hot water to a number of outlets. BS 6700 sets out a number of ways of supplying hot water, as detailed in Figure 7.1 opposite.

The chart is divided into **centralised** and **localised** systems. A centralised system is one where water is heated and can be stored centrally within a building. The heating of the water can be controlled by a thermostat. A system of pipework supplies the heated water to various draw-off points. A localised system is one where water is heated where it is needed – for example, a single-point water heater sited over a sink. Localised systems are often used in situations where long distribution pipe runs would involve a waste of water and energy.

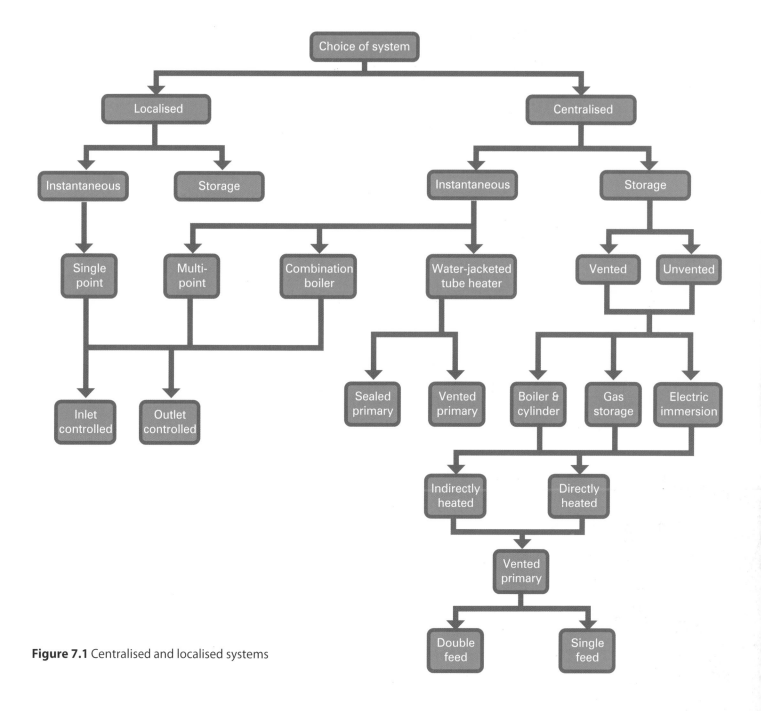

Figure 7.1 Centralised and localised systems

Vented hot-water storage systems

Definition

Gravity circulation – cold water is heavier than hot water and gravity therefore exerts a stronger pull on it, drawing it down and allowing the hot water to rise through the system

Immersion heater – an electric element fitted inside the hot-water storage vessel. It can be controlled by a switch and thermostat

Direct hot-water system (vented)

You are only likely to work on the system shown in the illustration when carrying out maintenance or repair, as it is no longer widely used. This direct system works by water being heated in a boiler (by gas, oil or solid fuel) and then rising due to the principle of convection – otherwise known as **gravity circulation**. It rises through the primary flow pipe and into the hot water storage vessel, heating the contents of the vessel directly. The hot water from the boiler is replaced by the cooler and heavier water moving in the primary return from the lower area of the storage vessel. The system does not always have to be heated by a boiler via primary flow and return pipes. It can be directly heated by means of an **immersion heater**; gas circulators are also used on direct systems, connected directly to the storage vessel.

There are a number of points to remember about direct systems using gravity circulation.

- There are minimum pipe sizes for primary circuits to hot storage vessels:
  - 22 mm for short pipe runs
  - 28 mm for longer pipe runs (or from continuous burning appliances).
- Vent pipes should not be less than 22 mm in diameter.
- All pipes to be laid to falls to prevent air locks and help system drain down.
- The vent route from the boiler, primary flow and open vent should not be valved.

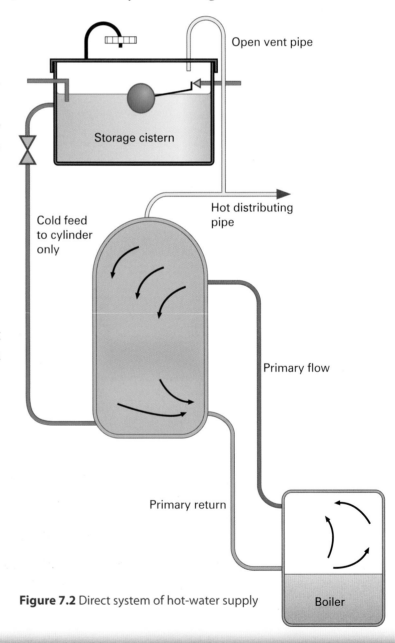

Figure 7.2 Direct system of hot-water supply

- The cold-feed pipe (normally sized at 22 mm in a small domestic property) should be sized in accordance with BS 6700 – the cold feed is the key route in which expansion water is taken up from the cylinder when it is heated, i.e. the heated water from the cylinder moves through the cold-feed pipe and the water level rises in the storage cistern.

- The open-vent pipe cannot be taken directly from the top of the hot-water storage vessel. The hot draw-off pipe should incorporate a 450 mm offset between the storage vessel and its point of connection to the open-vent pipe to prevent one-pipe circulation.

- Corrosion inhibiters should not be used, as the water in the boiler is fed directly to the appliances.

- No other supplies or draw-offs should be connected to the cold feed.

The next illustration shows a direct system heated by an immersion heater. The immersion heater should be controlled by a thermostat.

Did you know?

The vent pipe is where the term vented system comes from; the pipe is a key safety component in the system

Figure 7.3 Direct hot-water system heated by an immersion heater

A new immersion heater will now be supplied with a secondary thermostat or overheat thermostat to guard against overheating in the event that the normal control thermostat fails. Following a number of recent high-profile scalding incidents with immersion heaters, it is now advisable to replace any immersion heaters with a single thermostat with the safer dual thermostat variety.

Hot-water cylinders solely heated by electricity are often designed to take advantage of using low tariff electricity supplied in the evenings. This means a much larger cylinder than normal is installed together with two immersion heaters:

- The bottom heater works on the low electricity tariff and is used to heat the entire contents of the cylinder during the evening.
- The top heater uses normal tariff electricity and provides daytime top-up in the event that there is a shortage of hot water.

The purpose of the vent pipe

The purpose of the vent pipe is to maintain atmospheric conditions in the pipework. It permits any air entering the system to escape and, should the water in the system become over heated, it allows it to expand up the vent pipe and discharge into the cistern. The use of better system controls has reduced the risk of overheating, but it can occur on direct systems, particularly ones using solid-fuel back boilers.

Indirect hot-water system (vented)

This is the most common form of vented domestic hot-water system. It allows the boiler to be used for the central heating circuit as the system permits the use of a variety of different metals because the primary circuit is totally separate from the secondary circuit. The system is called indirect because the water contained in the

Figure 7.4 Double-feed indirect system of hot-water supply

> **Remember**
>
> The thermostat controlling the water temperature in any type of hot-water storage cylinder should normally be adjusted to a maximum of 60-65°C

storage vessel is heated indirectly through a heat exchanger. The diagram shows a vented double-feed indirect system.

Key points of the double-feed type of cylinder:

- The open-vent and cold-feed pipes may be connected to the primary flow pipes as shown, or fed separately into the boiler.

- Where the vent pipe is not connected to the highest point in a primary circuit, an air release valve should be fitted.

- A separate feed and expansion cistern needs to be provided to feed the primary circuit – this ensures that where a double-feed cylinder is used, the primary water is kept totally separate from the secondary hot water.

Indirect single-feed system

This system uses a self-venting cylinder and does not require a separate feed and expansion cistern. The water in the primary and secondary circuits is separated by means of an air bubble. The installation of the cylinder needs to be carefully made in accordance with the manufacturer's requirements, in order that the air bubble is

Remember

Hot water storage vessels must be insulated. New cylinders are now pre-lagged in the manufacturing process, using expanded foam, but lagging jackets can also be bought

Figure 7.5 Single-feed vented hot-water storage system

not dislodged in the cylinder permitting the two waters to mix. It is for this reason that this type of cylinder does not tend to be widely used in modern vented types of hot-water system.

Key points of the single-feed type of cylinder:

- Water enters the primary circuit via a number of holes at the top of the vertical pipe immediately under the upper dome.

- The system is self-venting through the air vent pipe while the primary circuit is filling.

- Once the primary circuit is filled, the filling of the secondary supply (the one which feeds the appliances) continues.

- When the secondary supply is full, two air seals are formed and a permanent seal is maintained.

- Once the water is heated, expansion of the water in the primary circuit is taken up by forcing the air from the upper dome.

Stratification

In a hot-water storage vessel, layers of water form from the top of the vessel where it is hottest to the base of the vessel where it is coolest. **Stratification** must take place if the vessel is to function to its maximum efficiency.

Manufacturers of storage vessels build stratification into the design. The following design rules enable stratification to take place:

- the vessel should be cylindrical in shape

- the cylinder vessel should be installed in a vertical position rather than horizontal

- the cold-feed connection should be in a horizontal position.

Figure 7.6 Stratification in a hot-water cylinder

Vented cylinder specification

Vented hot-water storage cylinders are available 'off the shelf' in a number of sizes to suit a particular circumstance, such as airing cupboard dimensions:

- standard widths vary from 300 mm to 600 mm

- standard heights vary from 675 mm to 1,800 mm.

The most common size for a standard one bathroom property tends to be the 450 mm × 900 mm, storing approximately 120 litres of water. Some manufacturers also produce 'one-off specials' made to the required dimensions for the job.

Cylinders are available in three grades. The grade of cylinder is based on the overall head of water generated by the cold-water storage cistern on the base of the hot-water storage cylinder:

- Grade 3 maximum head – 10 m

- Grade 2 maximum head – 15 m

- Grade 1 maximum head – 25 m.

Grade 3 cylinders are used in the majority of domestic properties.

Any new, or replacement, vented hot-water storage cylinders must comply with the requirements of the *Domestic Heating Compliance Guide*. The main points are:

- A direct, indirect single or double feed cylinder must be manufactured to the standards laid down in BS 1566.

- The cylinder must be labelled as Building Regulations Part L compliant with the following information:
  - type of vessel
  - nominal capacity in litres
  - standing heat loss in kWh/day
  - heat exchanger performance in kW
  - detail showing compliance with BS 1566

In relation to energy efficiency BS 1566 lays down minimum requirements for the insulation applied by manufacturers to vented cylinders, as well the surface area of the heat exchanger (coil) in an indirect cylinder. The greater the coil surface area the more energy efficient the cylinder tends to be.

The *Domestic Heating Compliance Guide* requires that all the pipework connecting to a hot-water cylinder should be insulated:

- to a point not less than 1 m from the cylinder, or

- to the point at which the pipework becomes concealed.

The storage capacity of a hot-water cylinder is based on an assessment of the consumption of hot-water in the property. The recovery rate (the time needed for the cylinder to be re-heated) is also a factor that needs to be considered. BS 6700 outlines the methods of calculating the hot-water cylinder capacity. Commentary in BS 6700 identifies that the minimum storage capacity in a small property should be no less than 100 litres.

Remember

The grade of cylinder refers to the thickness of copper used in the manufacture of the cylinder, i.e. the thicker the copper the greater the water head

Remember

The *Domestic Heating Compliance Guide* laid down by government identifies the minimum energy efficiency requirements of hot-water systems

Did you know?

Some manufacturers produce 'superduty' type cylinders which have a higher performance rating than BS 1566 and so are more energy efficient

Combination storage systems (vented)

Under the requirements of the *Domestic Heating Compliance Guide* combination type hot-water cylinders should be manufactured to BS 3198. This lays down the quality standards (including energy efficiency requirements) for this type of cylinder. The main feature of a combination type hot-water storage cylinder is that it contains a cold-water storage cistern integral to the unit. The base of the cold-water storage cistern must therefore be positioned at a higher level than the level of the highest water outlet (tap) in the property. This height (head) must also ensure that sufficient water flow rate is available at the outlets. These systems do not tend to be that widely used in newer installations as the relatively low head of water generated tends not to provide the required water flow rate at the outlets, unless the system is boosted with a water pump.

The advantages of a combination storage cylinder are:

- low installation costs
- they can be useful in flats, provided minimum water flow rates are achieved.

Disadvantages include:

- cannot normally be used with gravity fed showers (unless the system is pumped)
- tends to generate lower pressure and hence flow rate at outlets
- the cold-water storage space is often limited.

Figure 7.7 Combination storage cylinder

Unvented hot-water storage systems

This is a system type that is primarily covered at Level 3. Below is an appreciation of the system.

An unvented hot-water storage system, as the name suggests, does not include an open vent pipe nor a cold-water storage cistern. Essentially it is a type of hot-water storage system that operates under the influence of mains/boosted cold-water pressure, therefore providing higher water flow rates at taps (outlets).

As the hot-water storage cylinder operates as a closed vessel (at high internal pressure) the Building Regulations require:

- The cylinder to be fitted with a range of safety (temperature) controls to ensure that the stored hot-water temperature never exceeds 100°C.

- The discharge pipes from the cylinder to be located in a safe place where they are not going to scald anyone.

- The cylinder to be installed/maintained only by operatives who have passed an unvented systems competency test.

- The cylinder installation to be notified to the local authority before installation or to be self-certified by a Competent Persons Scheme member company.

An unvented hot-water storage cylinder must be manufactured to the requirements laid down in BS EN 12897. This standard also details the insulation and heat exchanger performance requirements for the cylinder that are necessary to comply with Part L1 of the Building Regulations.

Under the requirements of the Water Regulations, the cylinder installation must include a range of functional controls preventing:

- over-pressurisation of the installation and therefore wastage of water

- contamination of the cold-water supply.

Figure 7.8 shows the typical layout of an unvented hot-water storage cylinder.

The following table outlines the purpose of the temperature related control devices:

Did you know?

Companies who are in membership of an approved Competent Persons Scheme can self-certify certain types of work providing exemption from notification under the Building Regulations

Safety item 1	Control thermostat	Controls the water temperature in the cylinder to between 60 and 65°C
Safety item 2	High-limit thermostat (energy cut out devices)	A non self-resetting device that isolates the heat source at a temperature of around 80–85°C
Safety item 3	Temperature-relief valve (component 6 in Figure 7.9)	Discharges water from the cylinder at a temperature of 90–95°C (water is dumped from the system and replaced by cooler water to prevent boiling)

Figure 7.8 Temperature control safety devices

Figure 7.9 Typical indirect unvented hot-water storage cylinder

Figure 7.10 outlines the purpose of the functional control devices fitted to the cylinder.

Line strainer (Component 1)	Prevents grit and debris entering the system from the water supply (causing the controls to malfunction)
Pressure-reducing valve (on older systems may be pressure-limiting valve (Component 2)	Gives a fixed maximum water pressure
Single check valve (Component 3)	Stops stored hot water entering the cold water supply pipe (a contamination risk)
Expansion vessel or cylinder air gap (Component 4)	Takes up the increased volume of water in the system from the heating process
Expansion valve (Component 5)	Operates if the pressure in the system rises above design limits of the expansion device (i.e. the cylinder air gap or expansion vessel fail)
Isolating (stop) valve (not shown in Figure 7.8)	Isolates the water supply from the system (for maintenance)

Figure 7.10 Functional control devices fitted to unvented hot-water storage cylinder

Water-jacketed heater

This is also known as a thermal storage system.

Figure 7.11 Water-jacketed tube heater

When the hot tap is turned on, cold water – from either the mains or storage cistern – passes through a heat exchanger, which is situated in a heat store of primary hot water. The size of this heat store will be calculated based on the volume and rate of flow that can be delivered without an unacceptable drop in temperature. The primary water-flow from the boiler is programmed by the cylinder thermostat. Hot water is pumped to the radiator heating-circuit and is returned to the heat store; the cooler water from the heat store is then returned to the boiler where it is reheated. This is similar to an indirect domestic hot-water system, only in reverse.

FAQ

Why is it that a hot-water cylinder is always full of water even when the hot tap runs dry?

A hot-water cylinder has the cold-water inlet at the bottom and the hot outlet at the top therefore it always remains full even if the cold supply is disrupted and hot taps are opened until they run dry. Furthermore it is safe for immersion heaters and boilers to remain working because the cylinder is always full.

Knowledge check 1

1   What thermostatic controls are contained in a new immersion heater?

2   What is the minimum size of a secondary open vent pipe to a hot-water system containing an indirect hot-water storage cylinder?

3   What types of heat exchanger may be found in a double feed indirect hot-water storage cylinder?

4   What is the main disadvantage of installing a combination type hot-water storage cylinder?

5   Which device controls the water pressure in a hot-water system containing an unvented cylinder?

Storage heaters

Outlet controlled

Figure 7.12 Direct-fired gas storage hot-water heater

These are more common in large domestic or small commercial/industrial buildings.

Figure 7.12 shows a typical storage heater. In this case, it is heated by gas, but electrical storage heaters are also available. A gas storage heater is basically a self-contained boiler and storage system. This system also includes an open flue, which must be terminated externally. Often referred to as pressure-controlled water heaters, they are usually designed to be fed by a cistern (indirect) or mains (direct) supply, in which case it would be included in an unvented system. This type of storage heater is classified as outlet controlled, as the supply is controlled at the appliance outlet, e.g. hot tap. It will also serve multiple outlets.

Inlet controlled

As Figure 7.13 shows, these are generally seen as single-point heaters, fitted either above the appliance with a swivel outlet spout, or under the appliance.

The heater is fed from the supply pipe, which has an inlet control. It is important that the outlet, or any connections made to it, must not be obstructed as the open outlet allows for expansion of the water on heating. If an under-sink model is used, then a special tap will be required to allow venting of the water heater. They can be heated by either gas or electricity.

Electric or gas single point heater

Inlet control valve

Alternative heater position (special tap required)

Basin

Supply pipe

Figure 7.13 Small, inlet controlled single-point storage heater

Instantaneous hot-water systems

Instantaneous systems work by passing cold water from the service pipe through a heat source, which heats the water by the time it comes out at the application end. The heat source can be either gas or electric.

The speed that the water can be heated is limited, so the flow rate of the water needs to be controlled so it can be heated properly. Because of the reduced flow rate, it is not possible to supply a large number of outlet points all at once, so they would not be installed in situations where there is high demand. For example, you might find a multipoint in a small property, or a single point in an office kitchen area or WC.

Types of instantaneous heater

Multipoint – gas fired

As you can see from Figure 7.14, this consists of a gas burner sited beneath the heat exchanger.

When the hot tap is opened, it allows water to pass through the heater. This causes the gas valve to open as a result of the drop in pressure in the differential valve. This drop in pressure is caused by water passing through the **venturi**, which creates a negative pressure as it sucks the water from the valve. The diaphragm is connected to a push rod and, as it lifts, it opens the gas line. The gas is then ignited by the pilot light. When the hot-water tap is turned off, the pressure in the differential valve is equalised, the diaphragm closes and the gas supply is turned off.

Figure 7.14 Typical gas fired multi-point water heater

Single-point instantaneous water heater

This type of heater uses electricity or gas. Sited directly above the appliance, it is usually inlet controlled, with the hot water delivered via a swivel spout. The electric multipoint is a small tank of water with an electric heating element inside. Because of the low volume of water, it quickly heats up as it is drawn through the heater. The temperature at the outlet will be related to the water flow rate and the kW rating of the heater.

You will find single-point heaters used in situations where a small number of hot-water draw-offs are fed by individual heaters in a non-domestic type building, and where the use of a centralised hot water system would be uneconomical. You might find them in the WC of small cafés, etc.

Instantaneous electric shower

The electric heater shown is designed for mains connection, although some can be fed indirectly. The electrical rating can be in excess of 10 kW, so it is important that the supply is adequate, and wired directly from the mains distribution unit (MDU). For a rating of 9.6 kW, the circuit protection device requirements would be 45 amp and the cable 10 mm². The shower should also be isolated with a switch; this should be located within easy access outside the shower room. Any pressure variations in the cold water supply to the shower will be handled by the flow governor. Most electrical instantaneous showers are fitted with a flexible hose outlet – they will require a **check valve** to be fitted to protect against possible contamination of the supply, as required by the Water Regulations. This will usually be provided by the manufacturer.

Single-point vented electric water heater

Instantaneous electric shower

Knowledge check 2

1 In which type of property are you more likely to come across a direct fired gas hot-water storage heater?

2 Is a single-point water heater that has a swivel spout classified as an inlet or outlet controlled heater?

3 A water heater that provides heated water on demand with limited storage is described as what type of heater?

4 What is an instantaneous water heater serving more than one hot-water outlet described as?

5 The electricity circuit to an electric instantaneous shower should be fed from which electrical component?

Definition

Check valve – a device used to protect the water supply from any contaminants that may get into it

Connection requirements of showers

Here we will consider the layout and connection requirements of shower mixing valves. A shower can be fed by mains pressure, or it can be storage-fed by gravity or storage-fed using booster pumps.

Mains-fed

Water Regulations require that provision be made to ensure that backflow cannot occur. This is done either by installing a double check valve, or by using a rigid connection to the showerhead. Mains connection can only be used on thermostatic mixing valves, using the manufacturer's valves designed specifically for this purpose.

Shower mixer

Storage-fed

The hot and cold supplies need to be of equal pressure, and the showerhead a minimum of 1 m below the bottom of the cold-water storage cistern to ensure that adequate pressure from the showerhead is achieved.

On the job: Dilemma for Craig

During alterations to a plumbing system on a house extension, Craig, a final-year apprentice, has been given an ordinary shower mixer by the customer for fitting, although the cold water is mains fed. Craig is aware from his studies that the shower should not be fitted directly to the mains, and he explains this to the customer. The customer accepts the explanation that it is not good practice and against the Water Regulations to fit it, but insists that this is his house and that it is acceptable to him.

'Anyway,' he says, 'thermostatic showers are three times the price – no chance, just fit it'.

1   What should Craig do now?

2   Who do you go to for expert advice?

FAQs

Why is the water flow from my shower poorer than it used to be?

Often the cause is just a build-up of debris in the holes of the shower rose/spray head. Removing it and cleaning it out is sometimes all that is needed to rectify the problem.

My friend complains that the water temperature in their inexpensive shower is very difficult to control and to set at a comfortable level. Why might this be?

The symptom is typical of a manual shower that has been installed with unequal pressures – that is, the cold from high mains pressure and the hot from low-pressure storage. Only special thermostatic mixer-type showers can cope with different pressures, and even then there are regulations to comply with to prevent backflow risks.

Figure 7.15 Pipework feeding a gravity shower

Boosted supply

In the system illustrated below, water pressure is increased using a booster pump. The diagram shows a single-impeller booster pump and a double-impeller booster pump. The pump increases the pressure, which means that the minimum static head of 1 m is no longer needed. However, a minimum of 150 mm head is necessary to allow the flow switch to open when the supply is turned on.

Figure 7.16 Typical shower pump installations

Dead-legs and the use of a secondary circulation system

Dead legs are long lengths of pipe from the hot-water storage vessel to the appliance (see Figure 7.17).

Generally speaking, the length of pipe measured from the hot-water storage vessel or heater to the tap should be as short as possible.

Dead legs should be avoided for two reasons:

1 they waste water by having to run off cold water before it turns hot

2 they waste energy by heating up the volume of water contained in the dead leg, which then cools.

One method of overcoming the problem of dead legs is **secondary circulation**. This can be demonstrated with the typical installation layout using a pumped secondary circulation system, shown in Figure 7.18.

A flow-and-return loop is installed, which feeds all the appliances. The water is kept circulating, either by gravity or by a non-corrosive circulating pump (bronze manufacture).

Figure 7.17 Dead-leg runs to individual plumbing appliances

Figure 7.18 Secondary circulation system designed to overcome dead-legs in systems

The return pipe is connected in the top third of the cylinder. This prevents cooler water lower down the cylinder from mixing with the hot water. The return leg of the cylinder normally includes a single check valve to prevent reversal of the water flow in the return pipe when a tap is opened. The operation of the pump is normally controlled by a simple time clock which is set to turn off the pump and prevent wasteful circulation of heated water during periods when the building is not in use.

Bidet connection requirements

Definition

Ascending spray that the water discharges below the top of the appliance, i.e. directly into its bowl

Bidets can be divided into two groupings:

- **ascending spray**, or over the rim taps, connected to a flexible hose
- over the rim with discharge via tap outlets.

Figure 7.19 Type AUK2 air gap to over the rim bidet with no shower attachment fitted

Did you know?

Dedicated distributing pipework refers to a hot/cold pipework system that supplies water solely to an individual appliance

Ascending spray bidets, or over the rim taps, with shower attachments should only be fed from a storage system which normally includes dedicated hot/cold distributing pipework connecting to it (see Figure 6.19 for the pipework layout). Remember these types of bidet cannot be directly fed with water from a combination boiler or an unvented hot-water system.

Over the rim types with standard tap outlets can be supplied with cold-water direct from the mains supply pipework or from a storage cistern without the need for dedicated distributing pipework.

Figure 7.20 Pipework to ascending spray or over the bidet with shower attachment

Pipework insulation requirements of hot-water systems

The following outlines the key requirements laid down in the *Domestic Heating Compliance Guide* relating to the insulation of hot-water system pipework in order to conserve energy.

- Primary circulation pipework should be insulated wherever it passes outside the heated living space or through voids in a building which are ventilated from unheated spaces.

- Primary circulation pipes for hot-water services should be insulated throughout their length with the exception of where they need to penetrate joists or building structural elements.

- The pipes connected to a hot-water storage cylinder should be insulated up to 1 m from the cylinder or to the point where they become concealed.

- If a secondary circulation system is used the pipework included in that circuit should be fully insulated.

- Pipework in insulated areas should comply with the requirements of BS 5422.

- Insulation that complies with the *Domestic Heating Compliance Guide* must not have a greater heat loss than the figures quoted in the table below:

Pipe diameter	Maximum permissible heat loss (W/m)
15 mm	7.89
22 mm	9.12
28 mm	10.07

Figure 7.21 Maximum permitted heat loss for insulation

Knowledge check

1   A gravity-fed shower will normally require what minimum head between the base of the storage cistern and the shower rose position?

2   A shower pump that boosts the mixed hot- and cold-water supply is said to be what type of pump?

3   What is a section of hot-water pipework in which the hot water is not continually circulated described as?

4   Which type of bidet should not be fitted to a mains water supply pipe?

5   Which British Standard details the requirements for pipework insulation in locations subject to freezing?

Additional information sources

Sources of additional information on this chapter include:

Water Regulations Advisory Scheme (WRAS) – www.wras.co.uk
Pegler Ltd – www.pegler.co.uk
Department of the Environment, Food and Rural Affairs (DEFRA) (includes a guide to the Water Regulations) – www.defra.gov.uk
Communities and Local Government (includes a link to the *Domestic Heating Compliance Guide* and Building Regulations Approved Documents) – www.communities.gov.uk
Arrow Valves (backflow prevention device) – www.arrowvalves.co.uk
Albion Water Heaters (cylinder manufacturer) – www.albionwaterheaters.com
Bathroom Manufacturers Association (members organisations have learning materials on plumbing systems) – www.bathroom-association.org

chapter 8

Common requirements of hot- and cold-water systems

OVERVIEW

In previous chapters, you have looked at some of the specific aspects of hot- and cold-water systems. This chapter considers the common requirements of both systems, including the range of equipment used, installation, soundness testing and maintenance:

- **Taps and valves**
  - the range available
  - their location within systems

- **System installation requirements**
  - installation of pipework
  - noise in systems

- **Soundness testing of hot- and cold-water systems**
  - procedures
  - final checks

- **maintenance and decommissioning**
  - maintenance requirements
  - decommissioning of hot- and cold-water systems

Taps and valves

Taps and valves are used to isolate supplies, reduce the flow rate through pipework, permit drainage from systems, and provide an outlet to an appliance, for example sinks, baths, washing machines etc.

Water Regulations require that all fittings must be:

- suitable for their purpose
- made of corrosion-resistant materials
- sufficiently strong to resist normal and surge pressure
- capable of working at appropriate temperatures
- easily accessible to allow the renewal of seals and washers.

Taps and fittings for both hot and cold water should conform to BS 1010, Parts 1 and 2, BS 1552 and BS 5433.

Taps and valves are usually made of brass pressings or castings, and are chrome plated when appearance counts. Chrome plating also makes them easier to clean.

The range available

Stop valve

Figure 8.1 Stop valve

Stop valve

The stop valve pictured here is a typical screw-down valve. The washer and jumper fit into the threaded part of the spindle and is raised or lowered by turning the crutch head. When lowered to its maximum extent onto the seating of the valve, the incoming supply is isolated.

The stop valve is usually located on high-pressure pipelines – for example between the incoming mains and rising main – and is therefore used as a main supply isolation valve.

Stop valves used below ground must be made of a corrosion-resistant material such as gunmetal or bronze to avoid **dezincification**, and marked CR in the UK or DRA in Europe.

Gate valve

Gate valves are sometimes referred to as full-way gate valves, because when they are open there is no restriction through the valve. The gate valve is fitted with a wheel head attached to the spindle. When the head is turned anti-clockwise, the threaded part of the spindle is screwed into the wedge shaped gate, raising it towards the head.

Figure 8.2 Gate valve

Gate valves are usually located in low-pressure pipelines, for example the cold feed from the cold-water storage cistern (CWSC) to the hot-water storage cylinder. You might also find them used on supplies to shower valves.

Quarter-turn valves (spherical plug valves)

The 'ball valve'-type quarter-turn valve (spherical plug valve) can be used as a service valve to a cistern, in which case it will have a compression nut and olive on each side of the valve. It can also be used on the hot and cold supplies to a washing machine or dishwasher, when it will have a flexible hose connection. These valves are used to isolate the supply for the service, repair or maintenance of components and appliances within the hot and cold system.

Did you know?

A gate valve is classed as a service valve – it is provided to allow servicing of equipment

Definition

Dezincification – the breakdown of zinc in brass fittings by the electrolytic corrosion between zinc and copper. This occurs in some soil and water conditions, especially in soft-water areas

Safety tip

When closing gate valves for servicing, beware of closing the valve too tightly: a common fault is that on reopening the spindle breaks and the gate stays wedged firmly in the closed position

Find out

Why should a quarter-turn valve be closed slowly on high-pressure pipework?

Figure 8.3 Lever operated spherical plug value

The valves work by the quarter-turn operation of a square head (or slotted head) located at the top of the valve, which aligns the hole in the valve with the hole in the pipe. When the head is in line with the direction of the pipe, the valve is open. The quarter-turn head could be made from plastic and operated by hand – as used on washing-machine supplies – or it could have a slotted head, operated by a screwdriver, as fitted on supplies to cisterns. It can be used on low or relatively high-pressure installations

Ball-o-fix

for servicing purposes. Care should be taken with the use of spherical plug valves on low pressure installations. The valve should be of the full bore type rather the reduced bore type shown in Figure 8.3. A reduced bore in the valve control mechanism will greatly increase the water pressure loss through the valve body which will have an impact on the water flow rate available.

Draw-off taps

Draw-off taps, such as bib taps and pillar taps, work in a similar way to screw-down valves. The more modern type do not work as screw-down valves, but allow two polished ceramic discs to turn and align with two port holes through which the water can pass (see Figure 8.7).

The bib tap

The tap shown in Figure 8.4 includes a hose union attachment used for garden hosepipes.

You are unlikely to find the bib tap shown in Figure 8.5 ('supatap') in most modern domestic dwellings, but you may come across it when carrying out maintenance work on other properties.

Figure 8.5 Bib tap: supatap

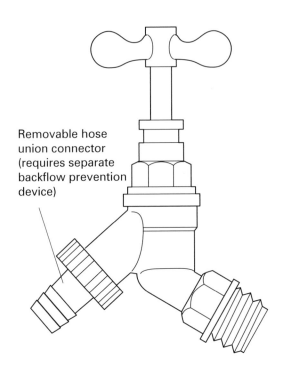

Figure 8.4 Hose union bib tap

The pillar tap

The pillar tap has a long vertical thread which passes through pre-drilled holes in the appliance. It is held in position with a backnut, and the supply pipe is attached to the thread via a tap connector. Its method of operation is very similar to the stop valve we saw earlier. The tap shown in figure 8.6 is a pillar tap with non-rising spindle. Here the tap spindle is fixed in position by means of a circlip and does not rise when the tap is opened.

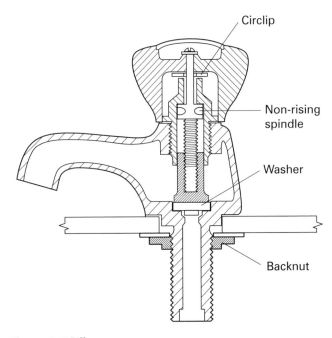

Figure 8.6 Pillar tap

The tap shown in figure 8.7 uses quarter-turn ceramic discs. This type of tap is in common use. However, they often have maximum water pressures within which they operate: if you install outside the requirements laid down by the manufacturer, the discs can shatter under operating conditions.

Globe taps

You may come across globe taps when carrying out maintenance work on an old bath installation. Instead of having a male thread connection to the supply, like a bib tap, they use a female thread. The male threaded outlet from the water supply is connected through prefabricated holes in the bath, and the globe tap is then screwed on to the fitting; they can pose a real threat to water-supply contamination, as they discharge below the spillover level of the bath. Their method of operation is similar to the stop valve.

Safety tip

Check the maximum water pressure carefully before installing a tap

Lever operation

Top disc turns

Fixed disc

Black (open)

White (closed)

Figure 8.7 Ceramic disc tap

Ceramic disc

Mixer taps

Outlet nozzle

Hot-water control

Cold-water control

Figure 8.8 Mixer tap

There are a number of design variations in mixer taps, but they all work on the principle of allowing the water from the hot and cold supply to flow from one outlet.

There are two types of mixer tap:

- single-flow outlet
- twin-flow outlet.

In the case of the twin-flow outlet nozzle, as shown in the illustration, the water is not allowed to mix in the tap swivel outlet. In a single-outlet unit, mixing can take place and cross-contamination of water can occur between the hot and cold supplies.

Figure 8.10 Mixer tap pipework

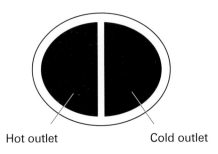

Figure 8.9 End view of outlet: twin-flow

The drain-off tap

Figure 8.11 Drain-off tap

Did you know?

The twin-flow separation of supply is achieved by a specially designed valve body and swivel outlet, which includes a complete division in the casting so that the hot and cold water do not mix until they leave the swivel outlet

Conforming to BS 2879, the drain-off tap is located at the lowest point of any system and has a 'ribbed' outlet to enable a good grip for a hosepipe connection. The tap is usually fitted with a 'lockshield head', which means it can only be operated by a purpose-made key. Combined stop valves with drain-off taps are also available.

Figure 8.12 Combined stop valve with drain-off tap

System installation requirements

Installation of pipework

Find out

The requirements of pipework installed in non-accessible locations in the Water Regulations and British Standards

You covered some of the pipework installation requirements in Chapter 5 *Common plumbing processes*. This section looks at some of the basic installation aspects of system pipework and the reasons why you need to think carefully about how you route pipework.

Accessibility

Water Regulations require that water pipes and fittings are easy to get at so that they can be inspected and repaired. However, customers and installers are not keen to see exposed pipework, so the tendency is to hide it under floors or behind wall surfaces.

BS6700 gives recommendations about the accessibility of pipes and water fittings.

Water fittings that are difficult to access are commonly named as concealed fittings. Examples include:

- a fitting installed below ground
- a fitting passing through or under any wall, footing or foundation
- a fitting enclosed in any chase or duct
- a fitting in any position which is inaccessible or renders access difficult.

The following shows acceptable methods of placing pipework in concealed locations.

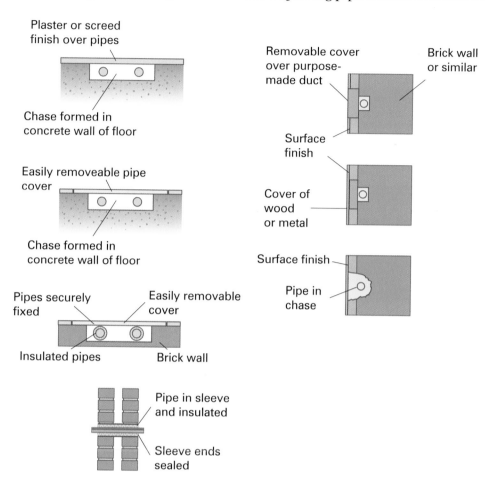

Plaster or screed
finish over pipes

Chase formed in
concrete wall of floor

Easily removeable pipe
cover

Chase formed in
concrete wall of floor

Removable cover
over purpose-
made duct

Brick wall
or similar

Surface
finish

Cover of
wood
or metal

Surface finish

Pipe in
chase

Figure 8.13 Requirements for pipes in solid walls and floors

Pipes securely
fixed

Easily removable
cover

Insulated pipes

Brick wall

Pipe in sleeve
and insulated

Sleeve ends
sealed

Pipes securely
fixed

Plasterboard

Timber stud partition

Plasterboard

Figure 8.14 Requirements for pipes in timber walls and floors

Screwed down
access panel

Insulated pipes below
boards

Clipped and supported
pipes

Pipes in notches
or holes

Air lock in hot and cold water pipework

One of the most common causes of problems in low pressure hot and cold water pipework water systems is air locks. Air locks are usually caused when pipework from cold-water storage cisterns, hot-water storage vessels and boilers is not installed correctly. The illustration shows two installation details that highlight the problem.

Figure 8.15 Air locks

Pipework running horizontally should be level or to an appropriate fall allowing air to escape from the system. Horizontal runs of gravity primary circulation pipework should also be installed to the appropriate falls.

Noise in systems

Noise in systems is usually caused by vibration. Not only is vibration a source of annoyance to the occupier of the building, but in severe cases it can also cause damage to pipework and fittings, eventually causing leaks.

Noise in systems is categorised as:

- water hammer
- flow noise
- expansion noise.

Water hammer

This is probably the most common cause of complaint from customers. When a valve is closed suddenly, shock waves are transmitted along the pipework, making a loud hammering noise. The problem is made worse where cold-water supply pipework is not adequately clipped.

Defective float valves and tap washers on cold taps can cause water hammer, so it can be cured through regular maintenance of the valve or taps. The velocity of the cold-water supply will further affect the problem; reducing the flow rate by the installation of a pressure reducing valve can help to reduce the possibility of water hammer occurring.

Flow noise

Pipework noise becomes significant at velocities over 3 m/s, so the system should be designed to operate below 3 m/s even if this means increasing the diameter of the pipe or installing a pressure-reducing valve.

Flow noise is also sometimes heard from cisterns. This can be caused by splashing noises as the incoming water hits the water surface as it fills.

Silencer tubes on float valves were once used to cure this, but are no longer allowed, except for the collapsible type.

Expansion noise

This usually occurs in hot-water pipework. As the system expands and contracts, it causes creaks and cracking sounds. The use of relevant pipe clips, brackets or pads between pipes, fittings and pipework surfaces should help to deal with expansion and contraction.

Find out

What are the main causes of air locks in a hot-water system? How can these be avoided?

Knowledge check 1

1. What is the part that holds the rubber washer in position in a stop tap known as?

2. Is a fullway gate valve particularly suited to high- or low-pressure applications?

3. What is a mixer tap that does not permit mixing of hot and cold water in the valve body known as?

4. A new hot-water storage cylinder that will not fill when all the valves are open in the pipeline and the cold-water storage cistern is full of water is most likely to be affected by what?

5. Creaking noises from hot-water pipework when water is drawn off is most likely to be caused by what?

Soundness testing of hot- and cold-water systems

Soundness testing on completed hot- and cold-water systems is essential to ensure that there are no leaks. Even the most competent plumber can make the occasional mistake, so testing is always required.

BS6700 provides the standard for soundness testing on hot- and cold-water systems.

Procedures for carrying out soundness testing

Soundness testing of hot- and cold-water systems includes:

- visual inspections
- testing for leaks
- pressure testing
- final checks.

Visual inspection

This includes making sure all pipework and fittings are thoroughly inspected to ensure that:

- they are fully supported, including cisterns and hot-water cylinders
- they are free from jointing compound and flux
- all connections are tight
- terminal valves (sink taps etc.) are closed
- in-line valves are closed to allow stage filling
- the storage cistern is clean and free from **swarf**.

It is useful at this stage to advise the customer or other site workers that soundness testing is about to commence.

Testing for leaks

When testing for leaks you should follow this checklist:

Testing for leaks checklist

1 Slowly turn on the stop tap to the rising main

2 Slowly fill, in stages, to the various service valves, and inspect for leaks on each section of pipework, including fittings

3 Open service valves to appliances, fill the appliance and again visually test for leaks

4 Make sure the cistern water levels are correct

5 Make sure the system is vented to remove any air pockets prior to pressure testing.

Pressure testing

Pressure testing of installations within buildings is done using hydraulic pressure-testing equipment. BS 6700 has separate procedures for testing rigid pipes and plastic pipes.

The procedure for testing rigid pipes (e.g. copper)

- Make sure any open-ended pipes are sealed, e.g. vent pipes

- Once the system has been filled, it should be allowed to stand for 30 minutes to allow the water temperature to stabilise.

- The system should be pressurised using the hydraulic testing equipment to a pressure of $1\frac{1}{2}$ times the system maximum operating pressure.

- Leave to stand for 1 hour.

- Check for visible leakage and loss of pressure. If sound, the test has been satisfactory.

- If not sound, repeat the test after locating and repairing any leaks.

Hydraulic pressure tester

The procedure for testing plastic pipes

BS 6700 has two test procedures for plastic pipes – procedures A and B. See Water Regulations for more details.

Test A procedure:

- Apply test pressure ($1\frac{1}{2}$ times maximum working pressure), which is maintained by pumping for a period of 30 minutes, and visually inspect for leakage.

- Reduce pressure by bleeding water from the system to 0.33 times maximum working pressure. Close the bleed valve.

- Visually check and monitor for 90 minutes. If the pressure remains at or above 0.33 times working pressure, the system can be regarded as satisfactory.

Test B procedure:

- Apply test pressure and maintain by pumping for a period of 30 minutes.

- Note the pressure and inspect visually for leakage.

- Note the pressure after a further 30 minutes. If the pressure drop is less than 60 KPa (0.6 bar), the system can be considered to have no obvious leakage. Check and monitor for 120 minutes. If the pressure drop is less than 20 KPa (0.2 bar), the system can be regarded as satisfactory.

Using a hydraulic pressure tester

Final system checks

After the system tests have been completed, the system should be thoroughly flushed out to remove any debris or swarf before carrying out a final visual check for leaks. Advise the customer and/or other site workers that testing is complete. At this point in the test procedure a performance test of the system is conducted, plus a hand-over, both of which will be covered at Level 3.

Maintenance and decommissioning

Hot- and cold-water system maintenance at Level 2 requires you to understand the principles of maintaining some of the key components. Maintenance is important as it ensures that the system components operate as they were designed to. It also reduces the chance of wasting water and energy by preventing dripping taps, float-operated valves and showers.

Maintenance work usually includes turning off hot- and cold-water supplies and then turning them on again. You need to ensure the customer is notified when you are doing this, so that they can fill kettles and turn off washers and dishwashers. Don't forget that co-workers on site might also need to store water for the period that the supply is going to be turned off – a plasterer with no water is going to be pretty irate as he/she might not be able to do his/her job properly. If a cold-water system needs to be isolated for a prolonged period of time it might be necessary to establish a temporary water supply.

The principles of maintenance/servicing are quite simple:

- find out what the fault is with the component to be serviced
- isolate the supply
- strip the components
- repair or replace the defective part
- reassemble the components
- turn on the supply and test the component for correct operation.

Maintenance requirements

Here are maintenance check lists for the following:

- stop tap
- hot- and cold-water tap
- float-operated valve
- shower mixer valve.

Stop tap

Common faults

Tap reseating kit

- The stop tap won't turn off the supply properly. This is usually due to a worn-out washer or a defective seating.

- The stop tap is stiff and hard to turn. This is due to infrequent use and lack of regular maintenance. The spindle and packing-gland nut oxidise over time which has the effect of 'bonding' the two together.

- The stop tap is leaking. If the stop tap is a compression type, the nut and 'O' ring joint could be leaking. Similarly, with a capillary fitting, the solder joint could be leaking. It is more likely, however, that the packing-gland nut will be leaking. Another possibility is that condensation has been mistaken for a leak. The incoming main is very cold, and the stop tap is usually located in a kitchen, where moisture levels are high. If the pipe is not insulated, the moisture will condense on the cold surface of the pipe and trickle down the pipe to the stop tap, giving the impression of a leak.

On the job: The wrong taps

Ben and Zoë are both apprentices working for a large firm on a maintenance contract for a housing estate. They are carrying out routine servicing of sink mixer taps that are fed directly from the incoming cold-water supply. Zoë discovers that the type of sink mixer tap is not permitted by Water Regulations and suggests to Ben that they should inform the supervisor. Ben says 'Don't bother, it will only cause a problem as the firm is already behind schedule, and anyway, it is nothing to do with us.'

1   Who is right?

2   What should be done?

Checklist

1 **Isolate the supply**
   - Make sure you have a stop-tap key in readiness.
   - Locate the external stop tap. It should be at the boundary of the property (footpath at the front of the house). On other property types, it may not be easy to locate. In some cases you may need assistance from the Water Company to isolate the supply.
   - Once you have located the external stop tap, lift the cover and locate the external stop-tap head.

- Be prepared for the stop-tap box to be filled with silt or other rubbish. You may have to clear that first to find the head.
- Once you have located the head, use the stop-tap key to turn off the supply (clockwise). If it is stiff, DO NOT FORCE IT – you may snap the head. Try twisting it both ways to get some movement. Again, if it is impossible to move it, you may have to call the Water Company.
- Return to the property and drain down the cold water from the lowest point (usually the kitchen sink). Hopefully there should be a drain-off tap close to the stop tap. On older properties, drain taps are not always fitted.

2 Strip the component

- You will need to remove the headgear from the body. This is not always easy, as it may not have been stripped for years. It often helps to apply heat to the joint between the headgear and body. Remove the headgear using adjustable grips/spanners. If a drain-off tap was not fitted, get ready for a rush of water from the pipework above the stop tap.

3 Repair the component

To fully repair the stop tap:

- Strip the headgear by removing the **crutch head** and **packing-gland nut**.
- Clean the rising spindle and the inside of the packing-gland nut using fine emery cloth (or a file in some cases).
- Replace the defective washer.
- Check the seating for defects. If the seating is severely pitted, it might be worth replacing the stop tap. Alternatively, you could re-seat the stop tap using a re-seating tool which 'grinds' the seating flush again.

4 Reassemble the component

- Fit all the component parts together, making sure that the spindle and inside of the packing-gland nut are lubricated.
- Repack the gland nut using PTFE tape.
- Refit the crutch head.
- Before refitting the headgear, make sure the washer between the headgear and body is in place and intact.
- Refit the headgear and tighten.
- If not already fitted, install a drain-off tap using adjustable grips/spanners.

5 Turn on and test

- Make sure the internal stop tap is turned off, as well as the drain tap.
- Turn on the external stop tap.
- Return to the building, and make sure the cold tap on the sink is open.
- Slowly open the stop tap and allow water to flow from the sink tap. This will help in removing any bits of debris that may have entered the pipework.
- Open the stop tap fully and assess the pressure at the sink tap.
- Turn off the sink tap and allow the supply to charge the system.
- Check that all the appliances fed by the supply are operating properly; check the cistern water levels.

Hot- and cold-water tap – kitchen sink

The servicing requirements of hot and cold taps are generally the same, but the isolation of the supply and turning it on, differs.

Common faults

Depending on the type of taps fitted, faults tend to be:

- dripping taps
- leaking taps (glands)
- tap heads that are difficult to turn
- noise when the cold tap is turned on.

On some properties, servicing valves may have been fitted to the hot and cold supply under the sink. This is good practice but is not a requirement of the Water Regulations. If servicing valves are fitted, isolation is achieved by turning them off.

The checklist below assumes servicing valves are not fitted.

Checklist

1 **Isolate the supply**
   - On a direct cold-water system, turn off the cold-water supply to the system and drain through the sink tap.
   - On an indirect cold-water system, turn off the service valve to the cold-water distribution pipe and drain through the sink tap.
   - Turn off the gate valve to the domestic hot-water cylinder.
   - Drain down the system pipework through the sink tap.

2 **Repair or replace the defective part**
   This procedure will depend on the type of taps fitted. Most taps fitted to kitchen sinks are chrome-plated brass, but some may be plastic.
   - Remove the cover from the tap head assembly; these are usually screwed through the top of the tap cover and concealed behind a chrome or plastic cap. Lift the cap with a small flat-headed screwdriver; the cover screw (which is usually cross-headed) can then be removed.
     You are then left with a headgear similar to the stop tap dealt with above.
   - The process for re-washering the tap is the same.

3 **Reassemble the component**
   - This is the same procedure as for the stop tap, but you will also need to replace the cover.

4 **Turn on and test**
   - Make sure the hot and cold taps to the sink are turned off.
   - Turn on the stop tap if it is a direct system.
   - Turn on the service valve on the cold-water distribution pipe if it is an indirect system.

Did you know?

You may come across sink taps using ceramic discs rather than the more traditional washer and seating methods of control. The ceramic discs do wear and allow leakage. These discs can be replaced, but you may have difficulty locating the parts. It may be more cost effective to replace the headgear completely

Remember

If the cover is plastic, don't overtighten the fixing screw, or you could split the plastic

- Turn on the gate valve to the hot-water storage cylinder.
- Return to the sink taps and make sure the supply pressure to the cold-water tap is satisfactory (direct) and that the cold-water (indirect) and hot-water supply is flowing smoothly.
- Test the operation of the taps.

Float-operated valve

There are three types of float-operated valve:

- diaphragm ball valve

- Portsmouth ball valve

- equilibrium valves:
   Portsmouth type
   diaphragm type

The trend on new installations is towards the use of plastic diaphragm ball valves both on CWSC and WC cisterns, but on maintenance work you could come across a range of different types.

Diaphragm ball valve

It is a Water Regulation requirement that servicing valves are fitted as closely as possible to the cistern in order to isolate the float-operated valve for maintenance work.

Common faults

- Probably the most common fault is a running warning or overflow pipe. This is caused by the float-operated valve not closing off the supply properly.

- No water in the cistern, toilet will not flush, no hot water. This is usually because the hole in the seating is blocked with tiny bits of debris that got into the system when it was installed.

- Excessive noise in the system, particularly when the WC cistern is flushed or hot/cold water is drained off from the cold-water supply cistern.

Checklist

1 **Isolate the supply**
   - If a service valve is fitted then you can simply turn this off.
   - If a service valve is not fitted, you will have to turn off the supply at the stop tap.
   - If you are working on a CWSC, run the level down to enable you to work easily on the float-operated valve. If it is a WC cistern, pull the flush to allow the float to drop. You might think this wastes water, but you need to check the correct operation of the float valve by watching it working as it fills up again.

2 **Repair or replace the defective part** (when dealing with overflows)
   - If the float valve is plastic, then it is either the washer in a Portsmouth valve that is defective, or the diaphragm in the diaphragm ball valve.

Figure 8.16 Section through a Portsmouth type float valve

Figure 8.17 A diaphragm type float valve

Brass Portsmouth ball valve

You may come across this type of ball valve on maintenance work.

Common faults

Most problems are caused by defective washers. Others are caused by oxidation or a build-up of scale between the lever arm and piston.

Checklist

Once you've split the body from the body nut, follow this sequence:

1 Remove the end cap.

2 Remove the split pin from the lever arm, this frees the piston.

3 Remove the piston cap and take out the defective washer and replace.

4 Reassemble the components:
  - Refit the cap.
  - Make sure the piston, and the end of the lever arm that locates the piston, are cleaned thoroughly.
  - Replace the piston, relocate the arm and split pin.
  - Replace end cap.
  - If a service valve is fitted, turn it on; if not, turn on at the stop tap.
  - For both the WC cistern and CWSC, allow the cistern to fill naturally. Make any final adjustments using the float adjustment.

5 Service valves should be fitted in installations where these do not exist.

Other maintenance points

WCs that are not flushing or hot-tap outlets with no hot water can often be caused by a float-operated valve problem. This is usually because small deposits, such as solder or grit, have got into the system and eventually become lodged in the throat of the seating, thus preventing the water from getting through. This is rectified by cleaning out the seating.

System noise known as 'water hammer' (see earlier) can also be a problem with float-operated valves, particularly when fed directly from the mains. You need to check the inlet pressure is not too high by adjusting it with the stop tap. Fitting an equilibrium valve, which distributes pressure equally on both sides of the washer, will also help to solve the problem.

Shower mixer valve

There are a number of shower mixer valves available. It would be impossible here to give detailed maintenance requirements for all the units on the market. We will, therefore, cover the general maintenance requirements here.

Common faults

- Constantly dripping showerhead – this is caused by the deterioration of seals within the main body of the shower on the outlet side of the showerhead.

- Erratic temperature control, particularly on thermostatically controlled showers – this is due to failure of the thermostatic control mechanism or the entry of dirt or grit into the shower unit.

Remember

Poor supply could also be caused by poor maintenance of the shower head itself. Strip down the shower head or rose first, clean and replace

If servicing valves have not already been fitted, they should be installed during the maintenance work

- Manually controlled showers rely on the hot- and cold-water supplies being at the same pressure. If one becomes partially blocked (sometimes pieces of debris find their way into systems), flattened or raised above the other, this could affect the temperature.

Checklist

1 Isolating the supply. Hopefully, the hot and cold supply to the shower from the CWSC and HWSC can be isolated using service valves. Otherwise you will have to drain down 230 litres of water. Some shower manufacturers include isolating check valves in the actual shower unit, which improves the ease of servicing. Once the supply is turned off, maintenance can begin.

2 Strip the component. Again, much will depend on the manufacturer, but it means gaining access either to the thermostatic or to the manual cartridge. This is usually done by removing the on/off control cap. Behind this should be screws holding the temperature-control housing.

3 Repair or replace components. This could involve replacing a set of 'O' ring washers (for leaks) or replacing a thermostat assembly for control (erratic temperature). It could be as simple as cleaning out the internal ports.

4 Turn on and test. Once the mixer has been reassembled, the supply is turned on and the mixer shower tested.

Decommissioning of hot- and cold-water systems

Decommissioning domestic systems means turning off the supply to the system, removing system pipework and components, and making sure that the hot or cold supplies are sealed or left so that they can not be turned back on. The correct term for this is called permanent decommissioning.

This might be necessary:

- where an old system is to be completely stripped out of a domestic property and replaced with a new, or alternative system, for example a direct hot-water system being replaced with an indirect hot-water system

- where a system is to be stripped out permanently, for example prior to the demolition of a building.

Temporary decommissioning is the process of taking a system out of action temporarily in order that work can be undertaken on the system, such as the maintenance activities discussed earlier in this chapter.

Did you know?

The Water Regulations require that when a system or pipework length is permanently decommissioned no dead legs should be left in the pipework system – this means removing any branched tee connections from an existing pipe run

FAQs

How is it that quarter-turn taps are associated with water hammer?

The sudden closing from full-bore flow to closed in a fraction of a second can cause severe shock waves and even cause damage. It is necessary to close these types of valves slowly so as not to cause the flow to come to an abrupt stop.

The cold-water supply has been turned off by the water authority for about an hour and my hot taps have now run dry. Is it safe to leave my immersion heater switched on?

Yes. Although the hot taps have run dry, the hot-water cylinder always remains full in short-term situations like this, so it is perfectly safe.

Why do I sometimes get an air lock in my hot-water system?

It is simply poor installation of pipework, where the pipe is not laid with sufficient fall to allow air to escape to the highest point. The most common problems are found in cold feeds and gravity circuits, where air becomes trapped. Forcing the air out with a hose is only a temporary measure as air will accumulate over and over again. The only real solution is to alter the pipework to give correct falls and allow air to escape naturally.

Knowledge check 2

1   What procedure is carried out immediately before carrying out a soundness test on a new cold-water system?

2   What is the recommended test pressure for a metallic pipework system operating at a maximum pressure of approximately 4 bar?

3   What is the pressure test period for a metallic pipework system?

4   What is the simplest method of dealing with a defective tap seat?

5   Water will not discharge from the outlet of a float-operated valve when the water supply is turned on. What is the most likely cause of the problem?

Additional information sources

Sources of additional information on this chapter include:

Water Regulations Advisory Scheme (WRAS) – www.wras.co.uk

Pegler Ltd – www.pegler.co.uk

Department of the Environment, Food and Rural Affairs (DEFRA) – website includes a guide to the Water Regulations – www.defra.gov.uk

Arrow valves (backflow prevention device) – www.arrowvalves.co.uk

Bathroom Manufacturers' Association (members' organisations have learning materials on plumbing systems) – www.bathroom-association.org

chapter 9

Central-heating systems

OVERVIEW

The Level 2 Certificate is limited to pipework systems and solid fuel boilers only. This nevertheless allows us to look at the various controls and components used on central-heating systems, and to think about the other types of oil and gas boilers on the market.
This chapter will cover:

- Types of central-heating system

- Building Regulations and central-heating

- Central-heating system components

- Heat emitters

- Central-heating system controls

- How the controls work together

- Soundness testing of central-heating pipework systems

- Maintenance and decommissioning of central-heating systems

Although we have discussed cold- and hot-water systems (both direct and indirect) separately, in reality, in a new domestic dwelling, they would be installed together.

In this chapter, we are going to add central-heating systems, which are closely linked to the hot-water supply as one boiler is usually used to heat both the hot water and the central-heating system.

Types of central-heating system

The various types of central-heating system can be summarised with a chart.

Figure 9.1 Types of central-heating system

Sealed systems or combination boilers will be fully covered at Level 3.

Full gravity heating systems

These are no longer installed, but you might come across an old system during maintenance work in buildings such as old village halls. They were sometimes referred to in domestic properties as background heating: one radiator, for example in the bathroom, which was fed off the primary flow-and-return pipe from the boiler. As with any gravity system, it worked on the principle of having the pipework at the correct fall to aid circulation. Flow-and-return to the radiator was usually 22 mm. The key feature of these systems was large-diameter pipes and no pump in the system for either heating or hot water.

Pumped heating and gravity hot-water systems (semi-gravity)

These can be either one-pipe or two-pipe systems.

One-pipe system

Again, you are unlikely to install these from new, but you may come across them.

Figure 9.2 One-pipe semi-gravity system

This system works on the basis of a one-pipe run, with a flow and return from the boiler. The disadvantages with the one-pipe system tend to outweigh the advantages, so they are no longer installed in domestic properties.

Advantages	Disadvantages
Lower installation cost compared to a two-pipe system	The heat emitters on the system pass cooler water back into the circuit. This means that the heat emitters at the end of the system are cooler
Quicker to install	The pump only forces water around the main circuit and not directly through the radiators. This means it is important to select radiators that allow minimum resistance to the flow of water
Lower maintenance costs	The 'flow' side to the radiator is usually installed at high level to improve circulation, creating additional unsightly pipework

Figure 9.3 Advantages and disadvantages of a one-pipe system

Careful balancing of the radiators is a must and – for a given heat requirement in a room – the radiators at the end of the circuit will need to be larger than those at the beginning.

To ensure that radiators operate correctly they must be sited as close to the pipework ring as possible. If you try to cut a couple of connections 100 mm apart into the pipework and run it for 5 m to the radiator, it will not work. The full ring has to be extended to run under the radiator with short tail connections on to it.

Two-pipe system (semi-gravity system)

Did you know?

The key to recognising whether a system is a one-pipe system is that there will be a complete ring of pipework from the flow on the boiler back to the return

Figure 9.4 Two-pipe semi-gravity system

This was a popular choice of system particularly in the 1970s. They are no longer permitted on new properties (other than with solid-fuel boilers) without additional controls being installed. Nor do they meet the requirements of the Building Regulations for extension or boiler replacements to existing oil- or gas-fired systems.

In a two-pipe system, water is pumped around both the circuit and the radiators. This improves the speed with which radiators heat up. The system can easily be balanced by adjusting the lock shield valve on each radiator.

Building Regulations and central-heating

The Building Regulations 2000 deals with the conservation of fuel and power in an approved document: *Approved Document L1: Conservation of fuel and power*.

This document sets out requirements for:

- space-heating systems controls
  - zone control
  - timing controls
  - boiler control interlocks
- hot-water systems
- alternative approaches for space heating and hot-water systems
- commissioning of heating and hot-water systems
- operating and maintenance instructions for heating and hot-water system controls.

The legislation sets out methods for improving energy conservation in dwellings by improving the control of a system using temperature control to individual rooms, permitting only the most energy-efficient boilers to be installed, and prohibiting the installation of gravity circulation systems in new properties. It is covered in more detail in Chapter 13 *Environmental Awareness*.

Meeting the minimum requirement of the Building Regulations

The next system type can be used to upgrade the central-heating controls system but only when a cylinder is being replaced in a gas or oil fired system. In the case of a boiler replacement in a gas or oil fired system, the system must now be upgraded to the fully pumped type.

Pumped heating and gravity hot water (semi-gravity)

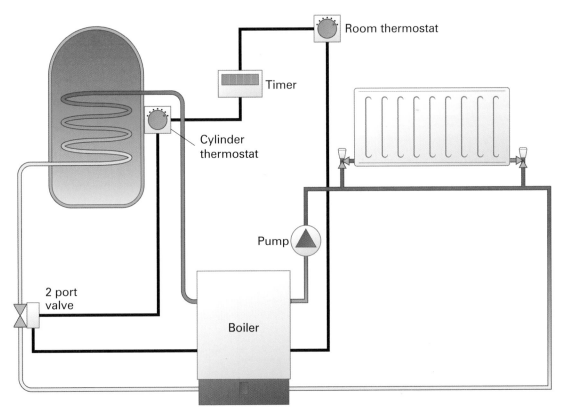

Figure 9.5 Semi-gravity system with two-port valve

This semi-gravity system uses a two-port valve and provides independent temperature control of both the heating and hot-water circuits in a pumped heating and gravity domestic hot-water heating system. The pump and the boiler are switched off when space and hot-water temperature requirements are met. Time control can be with either a time switch or a programmer. Thermostatic radiator valves (TRVs) could also be fitted to provide overriding temperature controls in individual rooms. The pump and circulation to the cylinder can be turned off independently by the thermostats, and there is an interlock to ensure that when the final circuit turns off so do the pump and boiler, thereby avoiding wasting energy.

Fully pumped systems

This type of system is often used on small and micro-bore heating systems, which are covered in detail at Level 3.

In this system, the hot water and the heating circuits are operated completely by the pump. Therefore, because there is no requirement for gravity circulation, the boiler can be sited above the height of the cylinder, giving more design options.

Installations are controlled by motorised valves. There are a number of system designs incorporating two-port zone valves or three-port valves (two-position and mid-position) that meet the requirements of the Building Regulations *Approved Document L1*.

Figure 9.6 The fully pumped system

Figure 9.7 Fully pumped system with two-position diverter valve

Fully pumped system using a two-position three-port diverter valve

This was one of the first fully pumped systems to be installed in domestic properties. However, it is no longer widely used.

This system is designed to provide independent temperature control of the heating and hot-water circuit in fully pumped heating systems. The design, when used with a programmer, satisfies the Building Regulations. The two-position three-port diverter valve is usually installed to give priority to the domestic hot-water circuit – that is, it can only feed either the hot-water or the central-heating system at any one time. Because the system is a priority control system – to produce domestic hot water – it should not be used where there is likely to be a high hot-water demand during the heating season: the heating temperature could drop below comfort level when the demand for hot water is high.

Because the system feeds only one circuit at a time, you will find that this system tends not to be widely installed.

Fully pumped system using three-port mid-position valve

This type of system is commonly installed in new domestic properties.

The system is designed to provide separate time and temperature control of the heating and domestic hot-water circuits. To fully meet the requirements of the Building Regulations, time control must be via a programmer and **TRVs**, and an automatic bypass valve must be fitted (where required).

The mid-position valve permits both hot-water and heating circuits to operate together.

On the job: Installing central-heating

Ray qualified as a plumber last year. He now has his own apprentice and needs to explain to her the installation requirements and system operation for a fully pumped system using a two-position diverter valve.

List all the points Ray needs to cover.

Fully pumped system using 2 × 2 port valves

This type of system is commonly installed in new domestic properties, particularly larger ones.

This system is recommended for use in dwellings with a floor area greater than 150 m². The main reason for this is the limited capacity of a three-port valve installation to satisfy the heat demands of a larger system. The use of the 2 × 2 port valves also gives greater flexibility in system design, with additional valves being added to the system to zone separate parts of the building. The system provides separate temperature control for both heating and hot-water circuits. Again, the features are similar to the other systems shown earlier.

Definition

TRVs – thermostatic radiator valves

Remember

Be careful when working on existing systems: the three-port diverter valve and mid-position valve system look similar. However, the three-port valve and timing device are completely different for each system

This system is not suitable for dwellings with a floor area greater than 150 m²

Figure 9.8 Fully pumped system with 2 × 2 port valves

FAQ

Can I use a small boiler with a diverter valve for individual control of hot water and heating?

Yes, but it is not widely done because while the cylinder is being heated up (HW Priority) the heating system is off. The smaller boiler is calculated on the max load of the heating system only. Once the cylinder is hot all heat is diverted to the heating system.

Knowledge check 1

1　What is the main feature of a one-pipe heating circuit?

2　What type of circulation takes place in the hot-water circuit in a semi-gravity system?

3　What are the two types of system that include three-port motorised valves?

4　Which document provides technical detail on the minimum legal requirements for energy efficiency and domestic heating systems?

5　Which type of fully pumped system is particularly suitable for properties with over 150m² of floor area?

Central-heating system components

Feed and expansion cistern

As you have seen from the diagrams in this chapter, the feed and expansion (F&E) cistern is used on all open-vented central-heating systems. While the cistern allows the system to be filled up, its main purpose is to allow water in the system to expand. The water level should therefore be set low in the cistern when filling the system. The cold feed to the system in an average domestic property is usually 15 mm minimum, and this pipe should not include any valves. This is to ensure that, in the event of overheating, there is a constant supply of cooler water to the system to prevent the dangerous condition of boiling. The servicing valve to the system should be located on the cold-water inlet pipework to the cistern.

The F&E cistern is located at the highest point in the system, and it must not be affected by the position and head of the circulating pump. To avoid any problems with gravity primaries, a minimum height can be obtained by dividing the maximum head developed by the pump by three. In fully pumped systems, the level of water in the F&E cistern should be a minimum of one metre above the pumped primary to the direct hot-water storage cylinder. If a valve were fitted in the cold feed it may be inadvertently closed. This could have disastrous consequences if the open vent also became blocked.

Space for expansion of water

The cistern and float valve must also be capable of resisting a temperature of 100°C. The system volume expands by about 4 per cent when heated, so a system containing 100 litres would expand by 4 litres. Space must be allowed in the F&E cistern to take up the additional volume when heated.

Primary open safety vent

In a fully pumped system, this should usually rise to a minimum height of 450 mm above the water level in the F&E cistern to allow for any pressure-surge effects created by the pump.

Did you know?

An 18 litre (4 gallon) feed and expansion feed and expansion cistern is normally used with an open vented central-heating system in a domestic property

Primary open vent rises to 450 mm above cistern – preventing surge effects

Water level in F&E cistern below CWSC

Figure 9.9 Open safety vent to F&E cistern in a fully pumped system

The purpose of the primary open safety vent is to:

- provide a safety outlet should the system overheat due to component failure
- ensure that the system is kept safely at atmospheric pressure.

The minimum diameter of the safety vent is 22 mm, and the pipe should never be valved.

Air separators

The purpose of the air separator is to enable the cold feed and vent pipe to be joined closely together into a correct layout to serve the system. The grouping of the connections causes turbulence of water flow in the separator, which in turn removes air from the system. This reduces noise in the system and lowers the risk of corrosion.

The diagram shows a fully pumped system containing an air separator. This is also called a close coupled method of feed and vent pipe connection, featuring only two pipes connected to the boiler.

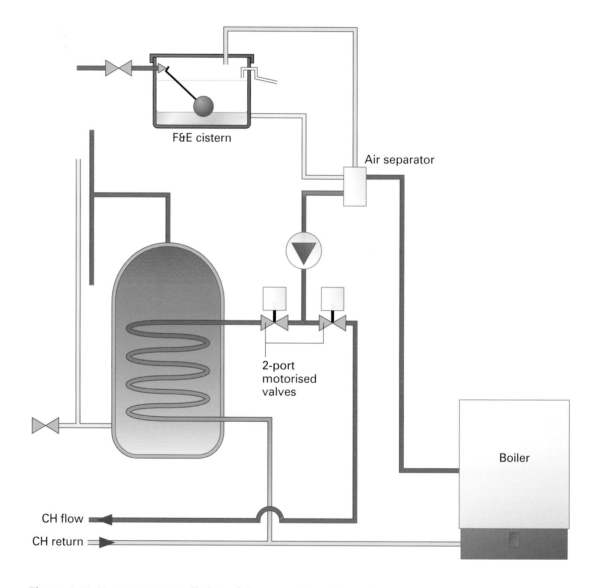

Figure 9.10 Air separator installed in a fully pumped heating system

Domestic circulating pumps

Domestic circulating pump

Ball-type pump valve

Gate-type pump valve

Pump stripped down to show the impellor

Once installed, domestic circulating pumps are fitted with isolation valves to permit service and maintenance.

Pumps are fairly simple. They consist of an electric motor which drives a circular fluted wheel called an impeller; this 'accelerates' the flow of water by centrifugal force.

It is the pump's job to circulate water around the central-heating system, ensuring that the water is delivered at the desired quantities throughout the system components. Most pumps have three settings, and pump manufacturers provide performance data for each, which shows flow rate in litres per second and pressure in kPa and m head.

Flow rate should not exceed 1 litre per second for small-bore systems and 1.5 litres per second for micro-bore systems; anything higher can create noise in the system. Most pumps deliver 5 or 6 m head. This is usually enough to overcome the flow resistance of the whole heating circuit.

It is good practice to position the pump so that it gives a positive pressure within the circuit. This ensures that air is not drawn into the system through microscopic leaks.

The pump position is even more critical in a fully pumped system because of its position in relation to the cold feed and vent pipe.

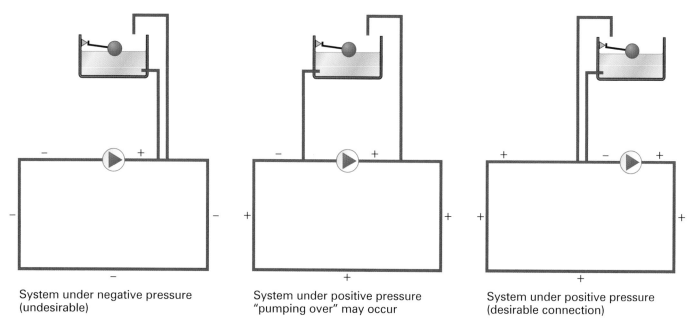

System under negative pressure (undesirable)

System under positive pressure "pumping over" may occur

System under positive pressure (desirable connection)

Figure 9.11 Pump position in relation to cold feed and vent pipe

Heat emitters

We use the term 'heat emitter' because it describes all types of devices used to heat the rooms that we live in. These include:

- cast-iron column radiators
- skirting heaters
- fan-assisted convector heaters
- panel radiators.

Cast-iron column radiators

Sometimes called hospital radiators, these are mostly found on older installations in such buildings as schools or village halls. However, some are now being installed in domestic properties as part of a 'designer' décor.

Cast-iron column radiator

Skirting heaters

Figure 9.12 Skirting heater

These work on the principle of natural convection; the fins provide a large surface area for heat output. The radiator is heated up by conduction from the heating pipe, cool air then passes through it and as the air is heated it rises and passes from the panel via louvres at the top. Skirting heating is not now widely used in domestic properties owing to the restrictions placed on output from the heater. You may still however come across it installed in some older properties.

Fan-assisted convector heaters

These work on the basis of forcing cooler air through the heating fins (heat exchanger) using an electrically controlled fan, and they therefore require connection to the central-heating control system.

Figure 9.13 Fan-assisted convector heater

Panel radiators

Despite the name 'radiator', about 85 per cent of the heat is given off by convection. The heat output of a standard panel radiator can be further improved by the addition of fins welded onto the back. These have the effect of increasing the surface area of the radiator, as they become part of its heated surface. The design of the fins will also help convection currents to flow.

Typical panel radiator

Types of panel radiator

Radiator designs have developed dramatically over the last 60 years. Manufacturers aim to provide radiators that are efficient and also offer the maximum choice of styles. The most common types of steel panel radiators are shown in the illustration.

| Single panel | Single panel with fins | Double panel | Double panel with fins |

Figure 9.14 Types of panel radiator

Manufacturers will provide a range of at least four height options, from 300 mm to 700 mm. Width measurements are from 400 mm, with increments of 100–200 mm, to a maximum of around 3 m. The recommended height from the floor to the base of the radiator is 150 mm (depending on the height of the skirting board). This allows adequate clearance for heat circulation and valve installation.

Outputs will vary depending on design. You must ensure that the output will be sufficient to heat the room the radiator is going to be in. Radiator sizing will be covered in Level 3.

Radiator styles

Seamed top panel radiator

This is currently the market leader and is the most commonly fitted radiator domestic installations. Top grilles are also available for this radiator.

in

Seamed top panel radiator

Compact radiator

Combined radiator and towel rail

Tubular towel rail

Compact radiators

These have all the benefits of steel panel radiators, with the addition of 'factory fitted' top grilles and side panels, making them more attractive to the consumer.

Rolled-top radiator

As the popularity of the compact has increased, the market for rolled-top radiators has declined. The method of manufacture means that some of the production seams can be seen following installation, making these radiators less attractive to the customer.

Combined radiator and towel rail

This product combines a towel rail and radiator in one unit. It allows towels to be warmed without affecting the convection current from the radiator. These are generally only installed in bathrooms.

Tubular towel rail

Often referred to as designer towel rails, these are available in a range of different designs and colours. They can also be supplied with an electrical element option, for use when the heating system is not required. They tend to be mounted vertically on the wall.

Low surface-temperature radiators (LSTs)

These were originally designed to conform to Health Authority requirements, where surface temperature of radiators was not allowed to exceed 43°C when the system was running at max. LSTs are now becoming popular in children's nurseries, bedrooms and playrooms, and in domestic properties where occupants are disabled.

Rolled-top radiator

Safety tip

You should not wedge the back of radiators over skirting boards or tightly under window sills, as this will prevent air circulation across the back panel reducing the output. Radiator shelves have a similar effect

Panel radiator accessories

Tappings

Radiator plug and manual air vent

Radiators are usually provided with $\frac{1}{2}$" BSP tappings, into which you attach the radiator valves and a plug and air-bleed valve. Domestic installations have the valves placed at the bottom of the radiator at opposite ends. Alternatively the connections could be placed top and bottom at opposite ends where a person may be elderly or disabled, making access easier.

Radiator brackets

A number of brackets and fixing styles are available, and are usually provided with the radiator. The bracket shown has deep hanging slots and corresponding lug positions. These provide greater stability. Plastic inserts are used to seat the radiator precisely and help to minimise expansion and contraction noises.

Typical radiator bracket

Radiator valves

There is a wide selection of valves available from manufacturers. Building Regulations require thermostatic radiator valves (TRVs) to be installed on new systems or for there to be some other means of controlling an individual room's temperature. A thermostat and motorised valve may be used, but these are usually too expensive to include in all rooms. You may be required to install manual radiator valves on repair jobs.

Wheel-head radiator valves

These enable the occupier of the building to control the temperature of the radiator manually by turning it on or off.

Rotating the plastic 'wheel head' anti-clockwise will raise the spindle through the body of the valve, lifting the valve and opening the flow to the radiator.

Wheel-head radiator valve

Lock-shield radiator valves

These valves are intended to be operated not by the occupier of the building but by the plumber. The plastic cap conceals a lock-shield head which can only be operated with a special key or pliers. The plumber will use this valve either to isolate the supply if removing the radiator, or to balance the system when commissioning. The valve illustrated shows an added feature – an in-built drain-off facility.

Lock-shield radiator valve with drain-off facility

Thermostatic radiator valve

Thermostatic radiator valves (TRVs)

These control the heat output from the radiator by controlling the rate of water flowing through it. The user adjusts the valve to maintain a desired temperature in the room; the valve then works automatically to maintain that set temperature.

TRVs are fitted with a built-in sensor which opens and closes the valve in response to room temperature. Liquid, wax or

Gland nut

Packing

Figure 9.15 Section through a wheel-head radiator valve

> ### Remember
>
> The valve head of a TRV needs to have good air circulation around it in order for it to sense the room temperature properly. If the head is going to be covered by, for example, a curtain, then a special TRV with a remote sensing element needs to be installed

> ### Safety tip
>
> TRVs should not be fitted to all radiators in a system. The radiator in the room in which the room thermostat is sited should be fitted with a wheel-head valve

Temperature adjusting head

Heat sensor

Bellows chamber

Pressure pin

Union tail to radiator

Valve

Figure 9.16 Section through a thermostatic radiator valve

gas expands into the bellows chamber as the sensor heats up. As the bellows expand they push the pin down, closing the valve. There are a number of settings on the head of the valve to enable a range of room temperatures to be selected.

You need to install a bypass valve to prevent the boiler and pump working against a closed system should all the TRVs close down and the water flow rate in the system fall to a low level.

Knowledge check 2

1   What is the recommended height that the vent pipe should rise above the water level in the F&E cistern in an open vented heating system?

2   What component can be used in an open vented heating system to closely group the open vent and cold feed connections in the system?

3   In relation to TRVs, what type of radiator valves should be fitted in a room that contains a room thermostat?

4   Most of the heat from a panel radiator is distributed by what means?

5   What component in a circulating pump is driven by an electric motor causing water to flow around the system?

Solid-fuel boilers

The only boiler you cover in the Level 2 Certificate in Plumbing is the solid-fuel boiler, although we will look briefly at the basics of other boilers.

Types of solid-fuel boiler

In domestic installations you could come across:

- solid-fuel room heater with back boiler
- independent free-standing boiler.

Solid-fuel room heater with back boiler

This type of appliance is available in a number of ratings, usually 10 kW and 13 kW. The 10 kW version will heat the room in which it is located, plus the domestic hot water, and will provide a total radiator surface of up to 13.8 m²; the 13 kW can serve up to 20.5 m² radiator surface. It fits a standard fireplace opening and is thermostatically controlled. This boiler requires a flue of 175 mm. It has two flow tappings and two returns for the domestic hot water and heating, all 1" BSP.

Convector-type solid-fuel room heater

Independent free-standing solid-fuel boiler

The model shown in the illustration is a gravity-fed boiler.

This type of boiler is available in a range of outputs from 13.5 kW to 29 kW. The model shown has a range of 13.5 kW, 17.6 kW, 23.5 kW and 29.3 kW. There is a built-in fan to boost heat output on demand. The hopper is easy to fill from the top, and a typical unit will hold enough fuel to burn continuously for 14 hours.

These boilers are designed for installation with a class 1 flue system.

Thermostat cuts fan when temperature reached

Fuel door

Flue

Thermostat control

Fuel store

Fan

Air supply

Firebed

Ashpit

Figure 9.17 Section through an independent gravity-fed solid-fuel boiler

FAQ

Surely solid-fuel burners are not used much these days?

No, they are not, but some remote areas still use solid fuel because gas is not available.

Oil-fired boilers

The following ranges of oil-fired boilers are available:

- combination
- condensing
- traditional
  - wall-mounted internal – room sealed or open flues
  - wall-mounted externally sited
  - floor-mounted internal – room sealed or open flues
  - floor-mounted external.

Outputs vary, but are typically as shown in the table.

Floor-mounted oil-fired boiler

Type of boiler	Capacity	Comment
Combination	Available in a range of capacities	
Wall-mounted internal	20 kW, 22 mm flow and return connection	Suitable only for fully pumped systems
Wall-mounted external	14.6 kW and 19 kW, 22 mm flow and return connections	
Floor-mounted internal	Outputs up to 70 kW (240,000 Btu/h)	
Floor-mounted external	Outputs range from 14.7 kW to 33.7 kW (50,000 to 115,000 Btu/h)	

Figure 9.18 Outputs of oil-fired boilers

Gas boilers

Gas boilers are covered in detail at Level 3. There are three basic types of gas boiler:

- traditional wall mounted or floorstanding
- combination boilers
- condensing.

In all cases, new boilers have to meet minimum efficiency standards. This is set out as a **SEDBUK rating**. There's more about this in Chapter 13 *Environmental awareness*.

Definition

SEDBUK rating – a measure of the seasonal efficiency of a boiler installed in typical domestic conditions in the UK

Flue
gases out

Air
supply

Heat
exchanger

Gas
burner

Gas
supply

Figure 9.19 Section
through a traditional
wall-mounted boiler

Traditional or regular boilers

These include:

- boilers with cast-iron, high-alloy steel, copper or aluminium heat exchangers

- system boilers.

All of the above can be floor-standing or wall-mounted, and are usually designed to fit in with kitchen unit installations. Gas-fired back boilers are also available; these are concealed in a chimney opening at the back of a gas fire.

Traditional or regular boilers supply hot water indirectly via a domestic hot-water storage cylinder, which is usually sited at first-floor level.

Cast-iron heat exchangers

The water in this type of boiler is heated by hot gases passing though the heat exchanger. The heat exchanger on a cast-iron boiler is a close network of waterways or, for the alloy type, a tube that passes through a series of fins. Boiler outputs for general domestic installations range from 9 kW to 29 kW.

Traditional gas or oil fired boilers are largely not now permitted to be installed in new and existing properties where boilers should now be of the high efficiency (condensing type).

Condensing boilers

Under the Building Regulations, all boilers, whether being used in new systems or as replacements, are required to be of the condensing type. Key features of the condensing boiler are:

- It is capable of extracting normally wasted heat from the flue gases to the point at which water condenses in the boiler and needs to be run safely to waste

- The condensate can be discharged to
  - a soil pipe
  - a waste pipe
  - a gulley
  - a purpose made soakaway (absorption point), used only as a last resort

- The condensate pipeline should normally be trapped, and the trap is often sited in the boiler.

Figure 9.20 A condensing boiler

System boilers

System boilers supply domestic hot water indirectly. The key system controls – pre-wired, pre-plumbed and pre-tested – are all built-in to the boiler unit, including the expansion vessel, the pump, and the system bypass.

Combination boilers

This type of boiler is designed to heat up a cold-water supply instantaneously for domestic hot water, so it is classed as direct hot-water supply. When required, it will also supply hot water for the central-heating system. Combination boilers are covered in greater detail in Level 3.

Gas-fired combination boiler

Central-heating system controls

Everyone has a moral obligation to do what they can to conserve energy. Energy conservation is also a high priority of g J7233 overnment. For this reason, Building Regulations have been amended to improve energy efficiency in buildings.

The revised Regulations have had the effect of enforcing fixed specifications for system design and, in particular, its control. We will therefore now look at mechanical and electrical controls used on central-heating systems, beginning with the basic system controls.

Time switch

A time switch is an electrical switch operated by a clock to control either space heating or hot water, or both together.

Full programmer

This can be used on gravity primary and fully pumped systems. It allows the time settings for space heating and hot water to be fully independent. More sophisticated programmers can include separate time controls for multiple space heating circuits.

Central-heating programmer

Room thermostat

This device measures the air temperature within the building or room where it is sited and controls the operation of the heating circuit (depending on whether the timing device is calling for heat). It can be used directly to switch on a circulating pump or boiler, or to operate a motorised valve.

Room thermostat

Programmable room thermostat

This is a combined time switch and room thermostat. It enables the user to set target temperatures for different periods for the central heating. This can be done on a daily or weekly cycle: for example, a higher temperature may be required for a certain period in the evening or at weekends.

Programmable room thermostat

Cylinder thermostat

The cylinder thermostat measures the temperature of the water stored in the hot-water storage vessel and, when it reaches a pre-set temperature, switches off the hot-water circuit to the vessel. This is done by connecting to the circulating pump or boiler, or by the operation of a motorised valve.

Frost thermostat

These give automatic frost protection to boilers and pipework that have to be located in areas at risk from severe cold. External boilers are a typical example.

Frost thermostat

Automatic bypass valves

These are mechanical devices used to make sure water can flow through the boiler to maintain a minimum water-flow rate should a system using zone valves or TRVs become closed. Once set, the valve opens automatically as the TRVs or zone valves close and the system pressure goes over the pre-set limit.

Use of the bypass valve reduces system noise and increases pump life by preventing it working against a 'dead head'.

Automatic bypass valve

Automatic air vents

These are used on central-heating systems in order to remove air from the system automatically. A vacuum break on the bottom of the valve prevents an air lock forming and encourages air to be released from the water.

Vents should always be installed on the positive side of the system and positioned where air is likely to get trapped.

Automatic air vent

Motorised valves

The type of valve used will depend on the system design, but the following are available:

- three-port diverter valve
- three-port mid-position valve
- two-port motorised zone valve.

Three-port diverter valve

This valve has been designed to control the flow of water on fully pumped central-heating and hot-water systems, on a selective priority basis, which is normally for the domestic hot water.

Three-port diverter valve

Three-port mid-position valve

This valve looks very similar to the diverter valve. It is used on fully pumped hot-water and central-heating systems, in conjunction with a room and cylinder thermostat, to provide full temperature control of both the hot water and heating circuits, which can operate independently of each other or both at the same time.

Two-port motorised zone valve

A single zone valve is used on gravity domestic hot water and pumped central-heating systems to enable separate temperature control of both the heating and hot-water circuits. Motorised valves are also used in fully pumped systems to provide separate control of both heating and hot-water circuits. They can be used to zone different parts of a building, for example upstairs and downstairs.

Two-port motorised zone valve

Junction box or wiring centre

These provide the connections between the electrical system components and the mains electricity supply.

Most manufacturers supply their controls in packs, and these can include a wiring centre which is designed to simplify the wiring of a particular system pack. Packs include control valves, programmers and thermostats. All the terminal connections are clearly marked, and full instructions are included with the wiring centre.

Wiring centre

Boiler control interlocks

This is a term used in Building Regulations. It is not an actual control device but an interconnection of controls – such as room and cylinder thermostats, programmers, zone valves etc. – designed to ensure that the boiler does not fire up when there is no demand for heat.

In a system with a traditional boiler, this would require the correct wiring of the room **stat**, cylinder stat and motorised valve(s). It may also be achieved by more advanced controls such as a boiler energy manager.

How the controls work together

So far, we have looked at system layouts, components and controls. We now need to think about how they work together. We will focus on the following system types, as they are the ones commonly installed.

For fully pumped systems (the only option for new installations and the preferred option for any existing system modifications or upgrades):

- fully pumped system using three-port mid-position valve
- fully pumped system using 2 × 2 port zone valves.

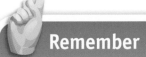

Remember

Building Regulations *Approved Document L1* is the Government publication that deals with the conservation of fuel and power in domestic premises and sets out the Regulations for the installation of domestic heating systems

Definition

Stat – a widely used abbreviation for 'thermostat'

Fully pumped system using three-port mid-position valve

Figure 9.21 Components of a fully pumped system using a three-port mid-position valve

Installation

- Three-port valve
  - water flow must be into port AB
  - port A is connected to the heating circuit
  - port B is connected to the hot-water storage vessel primary circuit

- Room and cylinder stat as for the previous system

- Time control
  - must be provided by a programmer that permits the selection of heating without hot water

- Bypass
  - an automatic bypass should be installed immediately after the pump, between the flow and return; pipework should be 22 mm, and flow should be in the direction of the arrow marked on the valve body.

How this installation works

- **Heating only.** When the room stat calls for heat, the valve is activated so that the central-heating port only is opened and the pump and boiler are switched on.

- **Hot water only.** When the cylinder stat calls for heat, the valve remains open to the hot-water storage vessel; the pump and boiler are switched on.

- **Heating and hot water**. When both stats call for heat, the valve plug is positioned so that both ports are opened. The pump and boiler are switched on; they're switched off once the stats are satisfied. The valve remains in the last position while the programmer is in the 'on' position.

Fully pumped system using 2 × 2 port zone valves

Figure 9.22 Components of a fully pumped system using 2 × 2 port valves

Installation

- Two-port valves
  - water flow must be in the direction of the arrow on the valve body.

- All other components are the same as the previous system. A programmer must be used to satisfy Building Regulations.

This system provides greater flexibility than the others. Additional zone valves and thermostats can be added so that the heating circuit can be separated into several zones. This improves temperature control and efficiency.

How this installation works

When either stat calls for heat, the respective motorised valve will be opened. Just before it is fully opened, the auxiliary switch sited in the motorised valve head will be closed and will switch on both pump and boiler. Once both stats are satisfied, the valves are closed and the pump and boiler switched off.

Knowledge check 3

1   What type of solid-fuel boiler includes a fan?

2   What type of gas- or oil-fired boiler can supply hot water and central-heating all from the same unit?

3   The central-heating control device that can provide independent timing of both heating and hot-water circuits is known as _____.

4   Which port on a mid-position valve normally takes the water flow from the pump and boiler?

5   Which type of relatively obsolete fully pumped heating system can only work on either a priority-to-heating or priority-to-hot-water basis?

Soundness testing of central-heating pipework systems

You have already looked at the testing requirements for hot and cold systems; these aspects apply equally to central-heating systems. In this section, we will also look at flushing central-heating system pipework and using additives to protect against corrosion.

Soundness testing procedures

Visual inspection

In the same way as for hot and cold water pipework, testing includes making sure that all pipework and fittings are thoroughly inspected to ensure:

- they are fully supported, including F&E cisterns and hot-water storage cylinders

- they are free from jointing compound and flux

- that all connections are tight

- that in-line valves and radiator valves have been properly tightened and are closed to allow stage filling

- that the inside of the F&E cistern is clean

- that all the air vents are closed.

Before filling, it is a good idea to remove the pump and replace it with a section of pipe. This will prevent any system debris entering the pump's workings.

Remember

Advise the customer or other site workers that soundness testing is about to commence

Testing for leaks

- Turn on the stop tap if the installation was for a complete cold-water, domestic hot-water and central-heating system, or the service valve to the F&E cistern if only the central-heating circuit has been installed.

- Allow the system to fill.

- Turn on the radiator valves fully and bleed each radiator.

- Visually check all the joints for signs of leaks.

- Drain down the system, flushing out all debris, wire wool, flux etc.

- Reinstall the pump.

- Refill and test for leaks again.

- Make sure the water level in the F&E cistern is at a level that allows for expansion of water in the system when hot.

- Heat system and flush while hot.

Pressure testing

On larger jobs, testing could be part of the contract specification. It is usual to test the system to $1\frac{1}{2}$ times the normal working pressure or 3 bar, whichever is the greater, for a period of one hour; the test pressure is achieved by using hydraulic test equipment.

Hydraulic test equipment

A safe mains connection (one that will not contaminate the supply) or a hydraulic test kit can be used to pressure test the system. All open ends in the system must be temporarily sealed prior to the test and a test point provided to connect the hydraulic kit or mains supply. The system is then increased to $1\frac{1}{2}$ times its normal working pressure and left to stand for one hour. Larger systems must be tested section by section, e.g. first floor, second floor, or by zones.

System cleaning

This part of the process is vital as it prevents corrosion in the heating system and the build-up of deposits that could lead to system failure.

The cold flush

Once you've checked and tested for leaks, it is time for the cold flush. You will need to drain the system from the drain off points as quickly as possible. First turn off the water supply, then drain off with a hose pipe, which allows air into the system at the various bleed points.

Hydraulic test kit

The cleansing process

BS 7593 details the requirements for the treatment of water in domestic heating systems. These guidelines state that all systems, as part of the commissioning process, should be treated with a proprietary system cleaner to ensure the removal of foreign matter and bodies that may cause corrosion, such as flux residues.

The cleansing is done by introducing a system cleaner into the system. The amount of cleaner required will be detailed by the manufacturer based on the system's volume in litres. The system cleaner must be added via the F&E cistern as it is re-filling. Adding the system cleaner into the cistern after it has refilled will not allow it to work properly.

Once the cleaner has been introduced the system can be re-filled and vented.

The hot flush

Most manufacturers recommend a 2 hour period for the cleaning operation, after which the system cleaner can be emptied out of the system. Turn off the system electrics so that the controls are isolated. Open the drain valves full bore and empty the system contents, being sure to fully open any bleed valves and motorised valves. Once the system contents have been drained, close all the valves and introduce the required amount of corrosion inhibitor. This is done prior to refilling at the same fill point as you introduced the cleaner. Refill the system and vent the air. Run the system back up to temperature, venting air as required and ensuring all circuits are operating correctly.

Final system checks

During the process of filling and testing the heating system, the system should be balanced and the boiler fully tested and adjusted. These procedures are fully detailed at Level 3.

Maintenance and decommissioning of central-heating systems

Fault diagnosis will be fully covered at Level 3. In Level 2 you are required to understand the principles of maintaining the components within the system (excluding the boiler). It also requires you to be able to change a defective radiator valve and central-heating pump, so we will focus on these two components.

By now, you should have realised how important energy efficiency is. Maintenance plays an important role here: it ensures that the system components continue to operate as they are designed to. For example, a pump that is not performing to its specification could make the boiler work harder, thus wasting energy.

Maintenance procedures consist of:

- finding out what the fault is
- isolating the supply
- stripping the component (if possible)
- repairing or replacing the defective part
- reassembling components (if applicable)
- turning on supplies and testing the component for correct operation.

Replacing a defective radiator valve

Valves could be one of three types:

- TRVs
- lock shield
- manual radiator valve.

Replacement of any of these valves will follow the same procedure. There are four ways to do this job:

1 Create a vacuum in the system to hold the water while replacing the valve – this is usually the preferred option as it is quicker.

2 Drain down the central-heating system – this is usually a last resort.

3 Use pipe-freezing equipment to isolate the supply to the radiator valve.

4 Some manufacturers have produced specialist servicing tools and valve inserts for both manual valves and TRVs. These enable the valve to be serviced without draining down.

The first two methods are the most commonly used, so we will focus on them.

Creating a vacuum in the system

This is a quick and simple solution to changing a component such as a radiator valve as it saves draining down the system. It works by completely plugging and closing all outlets to the system such as vent pipes, automatic air vents etc. It only works where you are working on one open pipe end (you cannot use this procedure for cutting into pipework) and it can only be used for vertical sections of pipework – making it ideal for radiator-valve replacement.

Safety tip

When creating a vacuum to access a radiator valve, there must be no existing leaks in the system

Replacing a defective radiator valve

1 Isolate the radiator valve

2 Drain the radiator

3 Relieve system pressure

4 Saw through compression ring (optional)

5 Replace valve

6 Finish the job

Safety tip

Be careful to fully support the valve when opening the union so as not to strain the pipework

Isolate the system

Make sure the electrical supply to the heating system is isolated. Turn off the boiler. Remove the fuse from the spur outlet to the wiring centre or junction box for the controls. Advise the customer of what you are doing and not to touch the controls. Only work on the system when the water has cooled. If you speak to the customer beforehand to arrange the job, it is a good idea to ask them to turn off the system so that it is cold when you get there.

Close all outlets

Turn off the service valve to the F&E cistern; close any automatic air vents. The open vent pipe and cold feed now need plugging – manufacturer devices are available, to carry this out.

Drain down the radiator

Before progressing any further, the radiator that is serviced by the valve to be replaced needs to be drained down. This is carried out by isolating both radiator valves and the loosening of the union nut on the radiator valve to be replaced. Catch the water from the radiator in a small bowl, and have a bucket handy to empty the bowl from time to time. Air needs to be let in to the radiator through its air vent to ensure that the water in the radiator fully drains out.

Relieving the system pressure

The radiator valve to be replaced should now be 'cracked open' at its control mechanism. A small amount of water should be discharged into the bowl until the pressure in the system reduces to atmospheric pressure. After a short period, the water flow should stop. If it does not, or the flow is excessive, this suggests that you have not closed all of the outlets or there is a leak in the system. If there is a leak, there is no option but to try another method of isolation, such as draining down. When the flow stops, you can move on to replacing the valve.

Replacing the valve

Check that the new valve body will fit the existing valve tail screwed into the radiator. If it does not fit, change the tail. If it is the type with an external nut, this can be done with an adjustable spanner. If not, then you will need a radiator spanner/**allen key** which goes inside the tail to remove it. When fitting the new tail, wrap PTFE tape around the threads five or six times in the direction of the thread.

It is at this point that you will need to establish whether modifications to the existing pipe are required. If you are fitting a valve by the same manufacturer, it will probably be a straight swap; otherwise, if there is some lift in the pipe, you may be able to remove the old compression ring and nut by sawing and lifting the pipe to the required height of the new valve. It is also possible that the length of pipe may have to be extended. Once the pipe has been prepared, the valve should be installed and all the nuts tightened.

Finishing the job

You are now ready to put the system back into operation. Remove the plugs from the cold feed and vent pipe, and open any automatic air vents in the system. Open the radiator valves and refill the radiator, checking for any leaks to the replacement valve. Replace the fuse and operate the system. Advise the customer that the work has been completed.

Draining down the system

This turns the relatively simple task of changing a defective radiator valve into a long job. It is not the draining down but the refilling and making sure all the system is working correctly that takes time.

Isolate the system

Make sure the electrical supply to the heating system is isolated. Turn off the boiler. Remove the fuse from the spur outlet to the wiring centre or junction box for the controls. Advise the customer of what you are doing and not to touch the controls. Only work on the system when the water has cooled. If you speak to the customer beforehand to arrange the job, it is a good idea to ask them to turn off the system so that it is cold when you get there.

Did you know?

Allen key – a hexagonal tool designed to fit into hexagonally recessed nuts and allow them to be turned

Safety tip

Always use the electrical safe isolation procedure of removing the fuse

Drain down the radiator

Locate the drain valve at the lowest point below the radiator valve to be changed. If you are working on an upstairs radiator drain only the first-floor level, not the whole system. You can do this by slightly loosening the connection from the valve to the radiator (have a bowl and cloth ready) and checking it until the water has stopped dripping. Drain-off is done by connecting a hose pipe to the drain valve and running the other end to an outside drain.

Fitting the valve

Once the system has been drained, one of two situations will exist:

1 The valve can be fitted without altering the pipework. This is because the valves are the same size, or there is sufficient 'play' in the pipe to move it up or down to fit.

2 The valve dimensions are different, so the pipework has to be altered.

If the valve can be changed without altering the pipework, remove the valve, leaving the existing nut and compression ring on the pipe.

Check that the new valve body will fit the existing valve tail screwed into the radiator. If it does not fit, change the tail. If it is the type with an external nut, this can be done with an adjustable spanner. If not, then you will need a radiator spanner/allen key which goes inside the tail to remove it. When fitting the new tail, wrap PTFE tape around the threads five or six times in the direction of the thread.

Remove the nut and compression ring from the new valve and place the new radiator valve body in position. If it is a TRV, remove the thermostatic head first. Because the compression ring has been used, some plumbers wrap PTFE tape around the joint between the compression ring and the valve body.

If you have to alter the pipework – either shortening by cutting, or extending by use of a fitting – complete the process, fit the new nut and compression ring to the pipe, and follow the next steps.

Once the valve has been secured to the pipe, fit the valve body to the radiator tail and tighten using adjustable spanners/grips. Make sure you use opposite force on the valve body to stop the valve twisting and possibly affecting the nut and compression ring joint.

Finishing the job

Turn off the drain tap, turn on the supply to the F&E cistern and begin the air-release process. This will involve:

- bleeding all the radiators, so do not forget your air key
- bleeding the pump
- bleeding the air valve (if manual type) on the pumped primary on fully pumped systems.

Using pipe-freezing equipment

There are a number of kits on the market. They vary in specification and cost, and you will have to decide what is the best kit to suit your needs and your pocket. The kits illustrated can generally be used on copper and plastic, and all are suitable for use on central-heating systems.

- Make sure you can use your kit competently.
- Only carry out this process on cold pipework with no water flow.
- If you try to use a pipe-freezing kit on warm pipes, you cannot predict how long the plug of ice will last and so you may get wet and the house may be flooded.
- Make sure the supply to the opposite end of the radiator is turned off.

Pipe freezing

- Remove the floorboards from above the pipe run to allow plenty of access to the pipe to be frozen.
- Following the manufacturer's instructions, place the freezer head or jacket at the required distance from the valve.
- Most kits will freeze the supply within 2 minutes and will last for 30 to 45 minutes.
- Removing and replacing the valve is the same as the draining-down job, but remember that here you will have to drain the excess water out of the radiator. When you're doing this, loosen the air vent to allow the water to drain more quickly. Have a bowl and cloth available to deal with any excess water.
- Once the process of replacing is complete, fill the radiator and test.

Replacing a defective heating pump

In the same way as valve replacement, advise the customer what you are doing, both in advance and during the actual job. On maintenance jobs, a pump replacement could be on any type of central-heating system, and the pump could be located almost anywhere in the building. However, it is more than likely that it will be located close to the domestic hot-water storage vessel. On solid-fuel systems or gravity primaries, pumps can be located near the boiler; on system boilers, pumps are located inside the boiler casing.

Remember

Always read the manufacturer's instructions properly

You may have to alter pipework, which will involve heat. Allow it to cool before freezing

Safety tip

You have only got 30–45 minutes, so get moving!

Safety tip

Always follow the safety instructions that come with the equipment

Safety tip

Always use a temporary continuity bond before breaking any pipework connections

Carry out some preliminary checks:

- First, make sure that the pump really is faulty – the fault could be in the electrical control, or something as simple as a blown fuse to the fused spur to the pump.

- Check to see whether the replacement pump will fit without any pipework alterations. If it doesn't, go and get a pump that will! Check that the pump valves are in good condition (which they rarely are!). If the answer is yes to both these checks, you will not have to drain the system down. If the answer is no to either, then you will.

Isolate the supply

Depending on the type and the age of the system, there may be a fused spur to the pump, or a fused isolating switch. The pump could also be part of a system control package, in which case you would have to isolate the fused spur to the wiring centre or junction box.

In all cases, remove the fuse to isolate the supply. If a wiring centre or junction box has not been used (i.e. older systems), turn off the boiler as well. If you need to alter the pipework, you will have to drain down the system in the same way as for replacing a radiator valve.

Remember

Because the electrical system is involved in this job, the safe isolation procedure must be followed at all times

Don't forget: some manufacturers produce pump extensions which will allow you to avoid draining the system, so check for these first

1 Disconnect the cable and close the valves

2 Undo union connections

3 Remove old pump

4 Clean the union surfaces

Replacing a pump

5 Fit the new pump

6 Final electrical connections

Disconnect the pump cable

Disconnect the electrical supply. First remove the plastic cover. This will reveal wiring that is the same as on an ordinary plug.

Remove the plastic cable retainer and disconnect the cables, making sure there is enough bare wire ready for reconnection to the pump fittings, and that the wires are in good condition. Cut back the wire and re-strip if not, or replace the cable, and wire the new pump before it is put in position. This makes the job easier.

Remove the old pump

Assuming that you do not have to drain down, turn off the isolating valves and remove the pump. If it has been in place for a long time, this may be difficult. You may have to apply heat to the pump valve nuts to loosen them. You will also need large **stillsons** or grips on both valve nuts to stop the pump rotating. Be careful when doing this not to disturb the joint between the pipe and the pump valve.

Cleaning the union surfaces

Once you have got the pump out, make sure the pump valve faces are free of any left-over valve washers.

Fitting the new pump

Fit the new pump in accordance with the manufacturer's instructions and making sure the new washers are fitted. If you altered the pipework to fit the new pump, ensure the pump and the new valves are fitted. Turn the water supply back on. If you drained the system, refill it as you did for the radiator valve replacement. If not, turn the pump valves back on. Bleed the pump through the slotted head in the body of the pump or pump bleed screw.

Final electrical connections

Reconnect the pump cable to the cable terminals, carry out any electrical checks to the electrical supply and test the pump. Advise the customer that the system is back in service and complete any commissioning records that may be required.

General maintenance requirement of systems

Prevention is better than cure, and more and more plumbing firms offer their customers planned maintenance packages for the full domestic system, not just heating. This will include:

- servicing of boiler and other heating appliances at least once a year (a legal requirement for gas appliances in tenanted properties)
- regular servicing of taps and valves
- cleaning and flushing of central-heating systems.

Remember

Remember your plug wiring: blue to neutral, yellow and green to earth and brown to live. On older systems the wiring could be red to live, black to neutral, green to earth

Maintenance contractors report that the majority of maintenance and repair jobs in domestic properties are caused by:

- failure of central-heating controls, boiler parts, motorised valves and thermostats

- leaks or burst pipes

- leaks caused by householders themselves, a typical example being the removal and replacement of radiators for decorating.

Replacement two-port motorised valve motor

What is of concern is that contractors receive calls from consumers who think their system is faulty, when in fact they do not understand how to work it properly. This is why it is vitally important to make sure that your customer fully understands everything about the system you install.

Motorised valves are usually very reliable, but will eventually come to the end of their working life. At one time this meant draining the system. However, it is now possible to replace defective motors or the actual power head.

On older systems, it may be necessary to flush pipework out when carrying out maintenance work. A power flushing kit is designed for this purpose.

Approved cleaning agents should be used with this equipment, which comes complete with hoses and valved connections. Always follow manufacturer's instructions when power flushing, as procedures vary from manufacturer to manufacturer.

Remember

Again, remember to include a system additive when refilling

Decommissioning central-heating systems

As with the decommissioning of hot- and cold-water systems covered earlier, this could include the following situations:

- An old system is going to be completely stripped out of a domestic property and replaced with a new system, for example a gravity solid-fuel system being replaced with a fully pumped condensing boiler system.

- A system is to be stripped out permanently, for example prior to the demolition of a building.

Power flushing kit

Make sure you:

- keep customers or other site workers informed of what work is being carried out
- make sure all supplies are isolated and capped off (where necessary) and that electrical supplies are effectively isolated
- use notices on taps or valves, for example 'not in use', 'do not use' etc.

FAQs

Why do I need to know about "one pipe" systems, hardly anyone installs them these days?

You will no doubt end up working on a one pipe system one day so it is wise to understand a little about the principles of operation.

Is it good practice to use an inhibitor in heating systems

Yes, tests have proven that an inhibitor correctly added to a system will prolong the life of boiler, radiators and other components.

Are correct settings of thermostats really that important? What difference does a few degrees make?

Well, quite a lot over time in terms of wasted fuel and money. By setting thermostats wisely we can save on fuel bills and help the environment too.

Knowledge check 4

1   Why is the visual inspection important before soundness testing a heating system?

2   What is the normal test pressure for a central-heating system?

3   What chemical is used in a new heating system prior to the addition of corrosion inhibitor?

4   When replacing a radiator valve by the vacuum method, what needs to be done to the cold feed and open vent pipe in the system?

5   What action needs to be taken before removing the electrical connections to a circulating pump that needs to be replaced?

Additional information sources

Sources of additional information on this chapter include:

Communities and Local Government (Domestic Heating Compliance Guide and Building Regulations Approved Documents) – www.communities.gov.uk

Honeywell Controls – www.content.honeywell.com/uk/homes

Pegler Controls – www.pegler.co.uk

Danfoss Randall – www.danfoss-randall.co.uk

Energy Savings Trust Best Practice Guide CE30 (Gas Systems) – www.energysavingtrust.org.uk/uploads/documents/housingbuildings/ce30.pdf

Energy Savings Trust Best Practice Guide CE29 (Oil Systems) – www.energysavingtrust.org.uk/uploads/documents/housingbuildings/ce29.pdf

Energy Savings Trust Best Practice Guide (CE47 Solid Fuel Systems) – www.energysavingtrust.org.uk/uploads/documents/housingbuildings/ce47.pdf

chapter 10

Electrical systems

OVERVIEW

At Plumbing Level 2, electrical work is restricted to basic practical skills on components that are not connected to the supply – that is, non-live work. On completion of Level 2 you *must not* regard yourself as competent to work on electrical systems. You will cover electrical work in much more detail at Level 3.

The job of a plumber involves installing and maintaining a wide range of appliances and components, many of which are powered by electricity. You need to be able to work safely and competently on the electrical supply to items such as heating controls, immersion heaters, showers etc., including being able to inspect and test electrical systems, and connect wiring from an outlet to the appliance or component. This chapter will provide the underpinning knowledge for this by covering:

- **The electricity supply to the property**
  - generation of electricity
  - supplying electricity
  - basic domestic circuits
  - earth continuity
  - bonding and temporary bonding

- **Electricity supply systems installation (Part 1)**
  - safety requirements when working with electricity
  - safe isolation
  - testing

- **Electricity supply systems installation (Part 2)**
  - types of cable and cords
  - installing pvc cables

- **Electricity supply systems installation (Part 3)**
  - installing fixed electrical appliances

- **Equipotential bonding**
- **Testing and decommissioning (knowledge only)**
  - different tests for electrical supply and earth continuity systems installations, including polarity, insulation, resistance
  - decommissioning electrical appliances

The electrical supply to the property

Here we will build on the electrical theory principles covered in Chapter 4 *Key Plumbing Principles*. We will cover the electrical supply into the property from the power station and then move on to the electrical installation of plumbing components inside the property.

Generation of electricity

Almost all electricity used in domestic and industrial premises is generated at power stations. There are two types of power station, but they all utilise the same principle of a turbine turning a three-phase alternator, which is the actual machine that produces the electricity.

Power stations are either 'thermal' or 'kinetic'; the name refers to the fuel or process used to turn the turbine that turns the alternator and so makes electricity.

Most commercial power stations in use today are of the thermal type, where a fuel such as gas, oil, coal, nuclear fusion etc. is used to heat water, producing high-pressure steam which drives the turbine.

Alternative energy – a wind farm

In recent years, as fossil fuels such as oil and gas have begun to run out, there has been increased research into the use of kinetic power stations. One specific type of kinetic power station which can be seen more and more frequently is the wind turbine, where the wind is harnessed to turn a giant propeller blade – which then turns an alternator to produce electricity. Wind turbines utilise a renewable energy source: the earth's fossil fuels will be used up eventually, but wind is unlikely to stop blowing in the foreseeable future!

Did you know?

The term 'generator' is sometimes used instead of alternator. Generator is an old term and refers to a machine that produces 'direct current'. All modern power stations produce 'alternating current' and use alternators.

Definition

Thermal – heat

Definition

Kinetic – movement

A nuclear power station

Another type of kinetic power station is water driven (known as hydroelectric stations). They harness the power of water when it falls from a great height, as down a mountain or from a dam, but some experimental stations are now using wave power from the sea to turn the alternator.

The basic components of a thermal power station are shown in Figure 10.1.

| Boiler | Steam turbine | Alternator | Transformer & pylons |

Figure 10.1 Simple thermal power-station layout

Supplying electricity

The alternator in a power station produces electricity at high voltage which is then fed through devices called transformers. These increase the voltage to even higher levels before passing the electricity into the National Grid. Electricity is transported through the National Grid at voltages of between 130 and 400 kV. The voltage is raised to these levels so that losses are reduced when the electricity is distributed.

Electricity is 'taken' from the National Grid through a series of appropriately located distribution stations. These transform the grid supply back down to 11 kV and distribute electricity at this level to a series of local sub-stations.

At sub-stations the 11 kV supply is transformed down to 400 V and then distributed through a network of underground radial circuits to customers. In rural areas, this distribution sometimes takes place using overhead lines. The electricity is generated and distributed in the form of a 'three-phase' supply, which is used without modification by some large factories and industrial sites.

The supply that we are familiar with in our homes, however, is a single-phase supply, derived from one of these three phases, and it comes into the main intake point of a home as a 230 V 50 Hz supply with phase and neutral conductors and a main earth point provided.

Did you know?

The National Grid is a network of nearly 8,000 km (5,000 miles) of overhead and underground power lines that link power stations together throughout the country

Sub-stations can be seen dotted throughout our cities. They are normally connected together on a ring-circuit basis

Generally at a domestic property main intake point you will find the following:

- a sealed overcurrent device (fuse or circuit breaker) which protects the supply company's cable

- a metering system to determine the customer's (consumer's) electricity usage.

The consumer's installation must be controlled by a main switch, which should always be located as close as possible to the supply-company equipment. In the average domestic installation, this device is merged with the means of distributing and protecting the final circuits in what is known as the consumer unit (formerly known as a fuse box).

Figure 10.2 illustrates the final part of the electrical supply journey.

A domestic electricity meter

Did you know?

Under the new requirements of BS 7671, the individual circuits in the consumer unit will need to include an RCD, as most domestic circuits include embedded cables

Figure 10.2 Electrical supply to a domestic property showing the meter installation and consumer unit

Basic domestic circuits

As a domestic plumber you will usually come into contact with three types of wiring circuits.

Lighting circuit

This is a **radial circuit** which feeds each overhead light or wall light in turn. To allow the lights to be turned off and on, the live or phase wire is passed through a wall-mounted switch. Two-way switches are also used (at the top and bottom of stairs or in long passageways), and these require special switch controls.

The lighting circuit is usually fed by a 1.5 mm twin and earth PVC-insulated cable and is protected by a 6 amp circuit protection at the consumer unit. You will often find that the lighting in domestic houses is split into two separate circuits, one for upstairs and one for downstairs.

Ring main circuit – 13 amp socket outlets

The sockets used in domestic properties feeding televisions and stereos etc. will normally be 13 amp sockets fed from a continuous **ring circuit**. As with the lighting circuit, cables circulate from the consumer unit round each socket, but the difference is that they return to the consumer unit forming a continuous loop – hence the term ring main. The ring main permits the cables to be kept to an optimum size as electricity is permitted to flow in two directions to reach the socket.

The ring-main circuit is fed using a 2.5 mm twin and earthed PVC cable and is protected by a 32 amp circuit protection device.

Spur outlets

Spur outlets are usually used to connect into a ring-main circuit on an existing system (you would not usually encounter spurs on new installations) where it is inconvenient to place a socket from the ring main using the conventional two cables. The spur is connected to the ring main through a junction box, or is wired directly from the back of an existing socket. Spurs can be either fused or non-fused.

Figure 10.3 overleaf shows examples of all the features described above.

Definition

Radial circuit – starts at the consumer unit and connects to every point on the circuit, terminating at the last one

Definition

Ring circuit – connects the consumer unit and every point on the circuit, and returns to the consumer unit, creating a ring

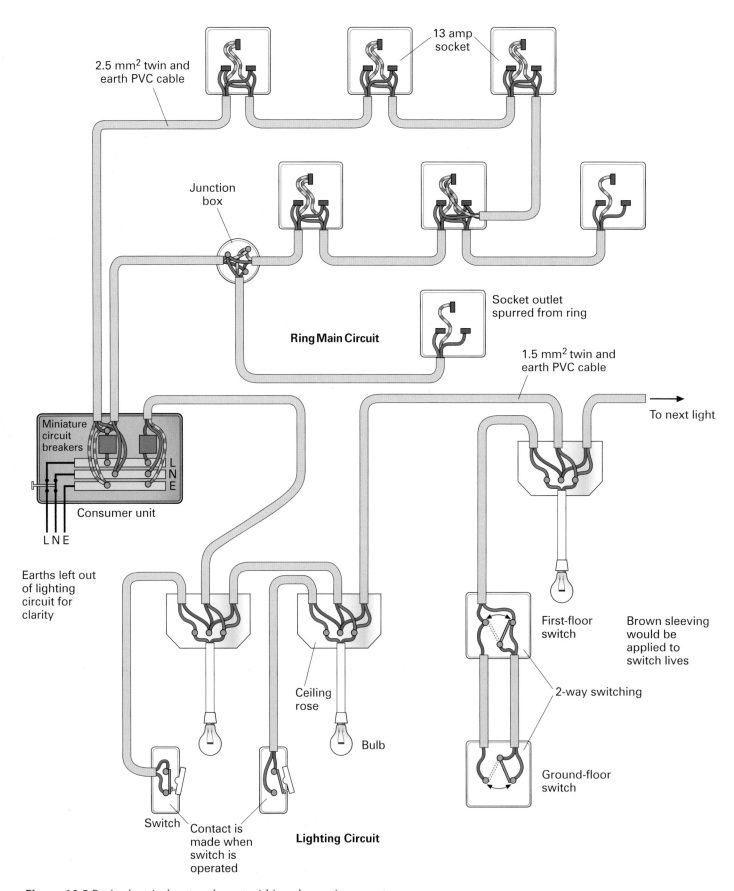

Figure 10.3 Basic electrical system layout within a domestic property

Earth continuity

In order for electricity to flow, we know that we must have a complete circuit of conductors, a voltage source to create a potential difference across the ends of the circuit and, for the circuit to be of practical use, a known value of circuit resistance to limit the amount of current flowing.

In the case of a short circuit, when the known circuit resistance is by-passed or breaks down, very high currents can flow, causing serious fire risks due to the heating effect of the current and danger to life from the risk of electric shock. It is for this reason – to protect people from danger – that we **earth** electrical equipment and installations.

The earth can be thought of as a very large conductor at zero potential (voltage). By connecting together all of the metalwork of an installation (other than the metalwork such as copper circuit conductors which are designed to carry current) we ensure that dangerous potential differences cannot exist between metal parts of the installation and earth (ground). If they did, persons touching the 'live' metalwork could get a serious electric shock or even be killed.

One simple way to remember what earthing is all about is to think of electricity as being lazy. It is always flowing around a circuit through a load with a known resistance to control the current, and it flows from a high potential to a low potential. As it is lazy, it will always try to find the path of least resistance to get to the zero potential of earth. If it can, it will do that through you (you have a low resistance) when you touch live un-earthed metalwork, instead of fighting its way around the circuit through the high resistance.

Earthing is designed to provide a very low-resistance path to earth through the exposed installation metalwork so that, in the event of a fault, a very high fault current will flow which will almost instantaneously operate the circuit protective devices such as fuses and circuit breakers.

Safety tip

Electrical installations must be earthed to prevent electric shocks

Remember

Electricity is lazy – it will always find the path of least resistance to get to earth

The definition of earthing from the IEE Regulations (BS 7671 – the electricians' 'rule book') is:

> *the act of connecting the exposed conductive parts of an installation to the main earthing terminal of that installation*

To more fully understand the concept of earthing, study the illustration.

Figure 10.4 Earth fault loop path

Figure 10.5 shows a three-phase supply (the star-shaped bit at the top left of the diagram) which is supplied, by the local electricity authority, to a local sub-station. One 'phase' conductor (single 230 volt) supply is taken into the premises to the consumer's supply input, and a second 'neutral' conductor is also provided. This is connected at the sub-station to the centre of the three-phase supply, which is called the 'star point'. This star point is connected to earth at the sub-station.

You will also notice a dotted line (the earth conductor) which is taken, again from the star point, into the consumer's supply input. At the consumer's end, the earth conductor is connected to a separate terminal called the **main earthing terminal**. It is important to notice at this point that the neutral and earth conductors are connected together at the same point at the electricity authority's supply but are connected to separate terminals when they arrive inside the premises. It is the supply authority's job to maintain the value of the star point as near to zero (earth) potential as possible so that a good low-resistance earth path is always available.

The consumer's responsibility starts inside the premises, where care must be taken that all of the earthing conductors connect to the main earthing terminal to maintain the low-resistance path. The total earth path, all around the installation and back to the star point of the transformer, is known as the '**earth fault loop**' and is given the symbol Ze.

During normal operation, the current will flow through the fuse from the consumer's supply along the phase conductor R1, through the load which has a separate earth connection, and return via the neutral. We know that the neutral is connected to earth back at the sub-station, so the only high potential difference that exists is between the phase conductor R1 and earth.

Imagine now that the load (which we will assume to be an electrical appliance, such as a kettle or toaster) has not been earthed and an internal fault develops which will allow electricity to flow in the casing of the appliance. If the consumer were to come along and touch it, the result could be death!

Figure 10.5 An earthing fault taking place

Because the appliance was not earthed, the fault current flowed through the consumer to the low-resistance earth return path. (Remember: why would the lazy electricity struggle to flow through the high-resistance load of the appliance when it can get to the cosy earth through you?)

The type of installation discussed in this section is known as a **TN-S system** and it is very common in this country. As a plumber, you do not need an in-depth knowledge of different types of installation earthing arrangements, but a brief explanation may be useful to you for reference purposes.

The most common types of installation system are listed below, with a brief explanation of each.

TT system

The first T in each of the systems stands for TERRA from the Latin word for 'earth'. The second T means that the exposed metalwork is connected directly to earth by a separate earth electrode (usually a copper rod driven into the earth). This system is most common in rural areas, where the electricity is often fed by local overhead lines.

TN-S system

Again the T is for TERRA. N means that the exposed metalwork is connected to the main earthing terminal. S means that a separate earth conductor is used throughout the installation. This is a very common system, but it is gradually being overtaken by the TN-C-S system.

TN-C-S system

This is sometimes called 'Protective Multiple Earthing' or PME for short. The T is for TERRA. N means that exposed metalwork is connected to the main earthing terminal. The third letter, C, means that for some part of the system (usually in the supply section) the function of the neutral conductor and earth conductor are combined in a single common conductor. The final letter, S, means that for some parts of the system (usually in the consumer's part of it) the functions of neutral and earth are performed by separate conductors. This type of system is provided in most new housing developments and is the most commonly installed system for new-build environments. While it is very effective, there is a small number of potentially very serious risks with this type of system if it is not installed or maintained properly. For this reason, certain rules are laid down for TN-C-S systems. The main ones are:

1 The neutral conductor must be earthed at a number of points.

2 Neutral and phase conductors must be made of the same material and be of the same cross-sectional area.

3 The neutral conductor must not be fitted with any neutral link or device that can break the neutral path.

You almost certainly will not remember all of the above, and the information is provided for reference purposes only in case you need it.

Bonding and temporary bonding

We have studied earthing and why it is essential to provide a good earth on an installation. Another related term is '**bonding**' and this is something that you will almost certainly come across in your work as a plumber.

When we talk about earthing we generally refer to the designed earthing arrangement within an electrical system where earth conductors are connected to earth terminals in equipment and appliances specifically provided for the purpose.

However, remember that earthing is designed to provide a very low-resistance path to earth through the exposed installation metalwork so that, in the event of a fault, an extremely high fault current will flow, which will almost instantaneously operate the circuit protective devices such as fuses and circuit breakers.

When considering this low-resistance path to earth, we must also think about all of the metalwork inside a modern house that could conduct electricity in the event of a fault occurring. The spidergram lists a number of conductors of electricity that are not specifically designed to carry current.

Figure 10.6 Conductors of electricity in a modern house

The use of plastic and non-metallic plumbing fittings and PTFE tape makes the need to connect all of these devices together with low-resistance copper wire even greater, as we cannot safely assume that the total metallic pipework system provides a low-resistance return path for fault currents. The process of connecting these devices together and into the earthing system is called bonding.

Consider, for example, a hot-water radiator central heating system in a house. We could be forgiven for assuming that as the whole system is metal and a good conductor of electricity, then the total resistance must be low.

Now think about it more deeply. Each of the valves connected into the system is likely to have PTFE tape around the thread to improve the water seal. PTFE tape is an excellent insulator. Also it is possible that the system may contain one or more plastic fittings on the cold-water feed side – and again, plastic is a good insulator.

To prevent any build-up of potentially dangerous voltages should a fault occur on the electrical system, all the metalwork is connected together using conductors and 'earth clamps', or clips, in a process known as bonding. It is sometimes called '**equipotential bonding**' because all the metalwork is kept at equal potential so that dangerous potentials cannot exist.

It is a requirement of the electrical wiring regulations BS7671 that all exposed metalwork in a building must be bonded together and connected to the earthing block within the consumer unit. The following bullet points briefly describe the steps that must be taken to ensure the safe earthing of all metallic materials within domestic properties.

- Gas, oil and water pipes can provide a path for stray electrical current. This could lead to corrosion of the pipework and also the potential of electric shock for anyone touching or removing a section of the pipework.

- The bonding of all exposed metal components in a dwelling that are not part of the electrical installation is known as equipotential bonding. The equipotential bonding conductor should be found close to the consumer unit.

- In certain areas of domestic property supplementary bonding may be required. Supplementary bonding is required to link sections of central heating or cold- or hot-water pipework where the metal pipework has been separated by a plastic fitting or length of pipe. This will ensure earth continuity throughout the property.

- When maintenance processes are being undertaken and it is necessary to remove a length of metal pipework, it is essential that the earth continuity be maintained. This is achieved by 'bridging' the gap exposed by the removed section of pipe with a temporary bonding wire. It is vital that the temporary bond is securely fixed in place before the length of pipe is removed.

- Earth clips should be used when connecting bonding wire to pipework. These are designed to indicate clearly the importance of the connection and to show that it ensures a safe electrical connection.

The following illustration shows a typical bonding arrangement for domestic premises. All of the incoming services to the property, such as gas, water and electricity, are bonded together close to the point of entry to the building and are connected to the main earthing terminal.

Consumer unit

Extraneous
metallic pipework
such as CH
systems

Gas
pipework

Max of 600 mm
from consumer's
side of meter

kWh

Main
electrical
supply

Main earth bar

Water supply
pipework

Oil supply
pipework

Figure 10.7 Example of the main bonding system in a domestic property

Knowledge check 1

1 Electricity is normally supplied to a domestic property at what voltage?

2 The component that contains the circuit protective devices in a domestic property is known as a _____.

3 What is the normal size of cable used in a domestic ring main circuit?

4 Is a lighting circuit an example of a circuit for fixed or portable equipment?

5 What is supplementary bonding?

6 What is the maximum distance of an earth bonding clamp from the outlet of a domestic gas meter?

Electricity supply systems installation (Part 1)

Plumbers carry out work on electrical systems, which is mainly restricted to connecting supplies to appliances and controls. You are unlikely to be required to install electrical circuits in either new or existing dwellings, but it is important that you have a basic understanding of how circuits work.

Anyone working on electricity supplies *must be competent*.

Safety requirements when working with electricity

The two most important factors are:

- circuit protection
- safe isolation (avoidance of electric shock).

Circuit protection

The Electricity at Work Regulations 1989 require that an electrical current is disconnected automatically if a greater current flows through a conductor (cable) than the circuit was designed for. This should happen within 0.4 seconds for anything with a socket outlet and 5 seconds for fixed appliances, e.g. an immersion heater.

Faults in circuits are caused by:

- overloading
- short circuits
- earth faults.

These faults are usually down to badly designed circuits, faulty appliances and/or poor installation. Circuits are protected against overload and short circuits by either fuses or circuit breakers. Circuit protection against earth faults includes **fuses**, **circuit breakers** or a **residual current device** (RCD) – this provides the greatest level of protection. Don't forget that many electrical circuits in domestic properties will now need to include protection by a Residual Current Device (RCD) as they undoubtedly include cables embedded in walls which require this level of circuit protection.

Electric shock

There are two ways in which a person can receive an electric shock:

- by direct contact with a live supply
- by indirect contact with a live supply.

Safety tip

Never be tempted to try any electrical work for which you have not been fully trained

Definition

Fuse – a small unit containing a strip of wire that will melt when the current exceeds a certain level, thereby breaking the circuit.

Circuit breaker – a similar device to a fuse that can be set to interrupt an electric current in certain situations

Residual current device – a highly sensitive unit that measures changes in the electrical current between different electrical conductors in a system and automatically disconnects the circuit if a small change occurs

Direct contact

There are ways of preventing direct contact. For example, the conductors in a cable or flex are protected with a plastic coating that insulates the live parts of the system that carry the current. Components such as pumps and valves using electric motors are also insulated internally.

Another method is to ensure that, under normal circumstances, live components are placed within an enclosure, such as a junction box, which means the exposed cabling cannot be touched.

Indirect contact

Indirect contact is prevented by a method called **earth equipotential bonding**, which ensures that all exposed metalwork in a building is bonded together and connected to the main earthing terminal. Should any exposed metal become live due to an electrical fault, the current will be discharged to earth. The fault current will also be detected by the protective device, which will automatically cut off the supply.

Safe isolation

Before working on any electrical supply you must make sure that it is completely dead and cannot be switched on accidentally without you knowing. Not only is this a requirement of the Electricity at Work Regulations 1989, but it is *essential* for your personal safety and that of your customer or co-workers.

The proper way to test if a circuit is live is to use an approved voltage-indicating device, similar to the one shown in the photograph below.

Voltage-indicating devices may use either an illuminated lamp to indicate the presence of a voltage or a meter scale. Test lamps are normally fitted with a 15 watt lamp and must be so constructed that they are not dangerous if the lamp is broken. They must also be fitted with protection against excess current, either by a fuse not exceeding 500 mA or a current-limiting resistor and a fuse. The test leads should be held captive and sealed into the body of the voltage detector. Test lamps and voltage indicators should be clearly marked with the maximum voltage that may be tested by the device and the maximum voltage that the device will withstand.

An approved voltage indicating device

Testing

Most fatal accidents involving electricity occur at the isolation stage. This is when you must be really careful and be aware of what you are doing, as you may have no idea of the type of supply you are confronted with. Do not take any risks, and if you are not sure, seek assistance. We will look at how to:

- identify sources of supply
- isolate
- secure isolation
- test that the equipment or system is dead
- begin work

Standardised procedures have been drawn up by the Electrical Contracting Industry and these are also used as the standard for safe working in the plumbing industry.

Identify sources of supply

It is important that you identify both the *type* of supply and the *source* of supply. In domestic dwellings, this will be a single-phase 230 volt supply to the circuits.

In most cases there are two types of circuit:

- 32 amp ring main for 13 amp socket outlets
- 6 amp lighting circuit.

In addition to these, you may also encounter:

- 16 amp supplies to fixed components such as immersion heaters
- supplies to other fixed components such as cookers and central heating systems and showers.

As a plumber you are unlikely to work on lighting circuits, but you need to be able to recognise the circuit layout and cable sizes. In older properties you may also find a 16 amp radial circuit. These supply the socket outlets but, unlike the ring main, the circuit terminates at the last socket and is not returned to the consumer unit. They are not as popular as the ring main, because the number of sockets that can be installed is limited.

To identify the source of supply:

1 Locate the plug socket nearest to the point where you intend to work on the component or appliance.

2 Make sure the socket is live by testing it – this can be done by plugging in a power tool to check that it works.

3 If it is sound, a check at the consumer unit should clearly indicate which circuit breaker (or fuse on an older property) will isolate the supply.

Isolate

Now that the type and source of supply have been identified, it needs to be isolated. Regulations require that a means of isolation must be provided to enable skilled persons to carry out work on or near parts that would otherwise normally be energised (live).

Isolating devices (fuses, miniature circuit breakers, RCDs) must comply with British Standards, and the isolating distance between the contacts must comply with the requirements of BS EN 60947-3 for an isolator. The position of the contacts must either be externally visible or be clearly, positively and reliably indicated.

Secure isolation

To prevent the supply being turned on accidentally by the customer or other co-workers, the fuse or circuit breaker should be removed and kept in your pocket, or the isolator locked off, with a special locking-off device. As an extra precaution, a sign saying 'work in progress and system switched off' must be left at the consumer unit or the area in which you are working.

Test equipment and checking system is dead

Any circuit you work on *must* be tested to ensure it is dead. Test equipment must be regularly checked to make sure it is in good and safe working order. Your test equipment must have a current calibration certificate, indicating that the instrument is working properly and providing accurate readings. If it is not calibrated, test results could be inaccurate. Before starting work:

- Check the equipment for any damage – look to see if the case is cracked or broken, indicating a recent impact, which could result in false readings.

- Check that the insulation on the leads and the probes is not damaged, and that it is complete and secure.

If you have any doubts about an instrument or its accuracy ask for assistance. These instruments are very expensive and any unnecessary damage caused by ignorance should be avoided.

Test the voltage indicator on a proven supply before you start; this will confirm that the kit is working. The best piece of equipment for doing this is a **proving unit**.

Only now can you use the voltage-indicating device to establish that the circuit you are to work on is dead. You should check phase (live)-to-neutral conductors, phase-to-earth conductors and neutral-to-earth conductors to make sure all connections are dead.

You are not quite ready to begin work – you should again check the test equipment on a known supply to make sure it is working correctly and has not become damaged during the testing procedure.

Only if all the above procedures have been followed correctly should you consider the circuit to be dead and safe to work on.

Begin work

Make use of warning notices: 'Plumber at work'. It may be helpful to put your name and contact number on the notice, so that if you have to leave the job while the customer is out, they can contact you to find out why the power has been turned off.

The flow chart below shows the procedure for isolating an individual circuit or item of fixed equipment.

Figure 10.8 Isolating a circuit

Knowledge check 2

1   Which piece of legislation lays down procedures for safe working on electrical systems?

2   What are the two main ways of receiving an electric shock?

3   What is the procedure used to confirm that an electrical circuit is safe to work on?

4   What piece of test equipment is used to determine that a circuit has been turned off in order to work on it safely?

5   What piece of equipment is normally used to check that the test equipment is working properly when making sure that an electrical circuit is safe to work on?

Electricity supply systems installation (Part 2)

By the end of this section you should be able to describe the practical requirements for installing appliances and controls on electrical systems.

Before we look at a specific installation, we will consider some aspects of installation good practice.

Types of cable and cords

PVC (Polyvinylchloride) insulated and sheathed cables are the most popular type of cables in current use and are employed extensively for electrical installations in domestic dwellings. PVC is versatile, tough, cheap and easy to work with and install, and is generally the most economical material for this type of work. However, its level of insulation is limited in conditions of excessive heat and cold, although this is not a problem in domestic properties. It can also suffer mechanical damage unless you apply additional mechanical protection in certain situations.

PVC insulated and sheathed flat-wiring cables

This cable is used for domestic and industrial wiring. It is suitable for service wiring where there is little risk of mechanical damage. It is available as a two- or three-core. Two or three plain copper, solid or stranded conductors are individually insulated with PVC and sheathed overall with

Flat-wiring cable

PVC. An uninsulated plain copper circuit protective conductor (CPC) lies between the cores.

The core colours for two-core cables are brown and blue plus earth. For three-core cables, they are brown, black and grey plus earth. The sheath colours are normally grey or white. The construction of three-core cables is exactly as mentioned above with the inclusion of a CPC between the coloured cores.

Multi-strand PVC-insulated and sheathed cable

This cable is suitable for surface wiring where there is little risk of mechanical damage. It is normally used for 'meter tails' to connect the consumer unit/ distribution board to the PES (Public Electricity Supply) meter, and as single core for conduit and trunking runs where conditions are difficult. The construction of this cable is PVC-insulated and PVC sheathed solid or stranded plain copper conductors. The old core colours are normally black and red, new colours are blue and brown. Sheath colours are normally black, red or grey, although other colours are available.

Multi-strand cable

Single core PVC-insulated and sheathed cable

This cable is used for domestic and general wiring. It comprises a PVC-insulated copper conductor with a PVC sheathed overall. The old core colours are red or black, new colours are brown or blue and the sheath colour is normally white or grey.

Single-core cable – old colour

Heat-resisting PVC-insulated and sheathed flexible cords

These flexible cords are suitable for use in ambient temperatures up to 85°C. They are not suitable for use with heating appliances. Their construction comprises plain copper flexible conductors insulated with heat-resisting (HR) PVC and (HR) PVC sheathed.

The core colours for single- and twin-core are brown or blue; for three-core, brown, blue and green/yellow; and for four-core, blue, brown, black and green/yellow. The sheath is white.

Installing PVC cables

Cables are fixed at intervals using plastic clips, which incorporate a masonry-type nail.

Where PVC cables are installed on the surface, the cable should be run directly into the electrical accessory, ensuring that the outer sheathing of the cable is taken inside the accessory to a minimum of 10 mm. If the cable is to be concealed, a flush box is usually provided at each control or outlet position.

Cable-fixing clips

Installing and clipping PVC cable

- In order to ensure a neat appearance, PVC cable should be pressed flat against the surface between the cable clips

- The cable can be straightened by running the thumb or fingers over it before clipping. The palm of the hand can be used for bigger cables.

Flattening cable before clipping

This sequence of forming the cable should be carried out after inserting the previous cable clip and before fixing the next cable clip.

Fix the first clip

Fix the last clip

Fix the intermediate clips

- When a PVC cable is to be taken around a corner or changes direction, the bend should be formed using the thumb and fingers as shown.

- Care must be taken to ensure that the bend does not cause damage to the cable or conductors and that cable clips are spaced at appropriate intervals.

Clipping round a curve

The finished result

Cable runs

Cable runs should be planned so as to avoid cables having to cross one another, which would result in an unsightly and unprofessional finish. When installed in cement or plaster, they should be protected against damage. This is done by covering the cable with a plastic or metal channel or by installing them in an oval PVC **conduit**.

PVC-sheathed cables should not come into contact with polystyrene insulation, as a chemical reaction takes place between the PVC sheath and the polystyrene resulting in the migration of polymers from the cable, known as **marring**.

Terminating cables and flexible cords

The entry of the cable end into a fitting is known as a **termination**. In the case of a stranded conductor, the strands should be twisted together with pliers before terminating. Care must be taken not to damage the wires. BS7671 requires that a cable termination of any kind should securely anchor all the wires of the conductor. This is to prevent any undue stress on the terminal or socket, causing it to overheat.

The cable, under working conditions, will expand and contract due to heating and cooling, again putting pressure on the terminal connection. This may result in increased resistance and probably overheating. So, when terminating the flex, it should be gripped with a flex clamp, tightened down onto the protective sheathing.

Stripping PVC cables

Definition

Conduit – a preformed PVC channel that covers cable runs and protects cables against damage

Pulling cable apart

Removing the sheathing

Stripping the insulation

- Nick the cable at the end with a knife and pull apart.

- When the required length has being stripped, remove the surplus sheathing with a knife or snips.

- The insulation can also be stripped from the conductors with a knife or purpose-made strippers. Examine the conductor insulation for damage.

- An alternative method of stripping the insulation from the conductors is with a pair of purpose-made strippers.

Connecting to terminals

There is a wide variety of conductor terminals. Typical methods of securing conductors in fittings are:

- pillar terminals
- screw heads
- nuts and washers
- claw washers
- strip connectors.

Pillar terminals

A pillar terminal has a hole through its side into which the conductor is inserted and secured by a setscrew. If the conductor is small in relation to the hole it should be doubled back. When two or more conductors are to go into the same terminal they should be twisted together. Care should be taken not to damage the conductor by excessive tightening.

Figure 10.9 Pillar terminal

Screw-head, nut and washer terminals

When fastening conductors under screw heads or nuts, it is best to form the end of the conductor into an eye, using round-nosed pliers; the eye should be slightly larger than the screw shank but smaller than the outside diameter of the screw head, nut or washer.

The eye should be placed in such a way that rotation of the screw head or nut tends to close the joint in the eye. If the eye is put on the opposite way round, the motion of the screw or nut will tend to untwist the eye and will probably result in a bad contact.

Figure 10.10 Screw head and nut-washer terminals

Claw washers

In order to get a better connection, claw washers can be used. Lay the looped conductor in the pressing. Place a plain washer on top of the loop and squeeze the metal points flat using the correct tool. The diagram illustrates this method.

Figure 10.11 Claw washer

Strip connectors

The conductors are clamped by means of grub screws in connectors, which are usually made of brass and mounted in a moulded insulated block. The conductors should be inserted as far as possible into the connector so that the pinch screw clamps the conductor. A good, clean, tight termination is essential in order to avoid high resistance contacts resulting in overheating of the joint.

Figure 10.12 Strip connector

Electricity supply systems installation (Part 3)

Installing fixed electrical appliances

In this section we will look at how the electrical appliances and components are connected into a domestic ring-main circuit. This will also cover basic practical installation requirements for:

- immersion heaters
- dishwashers
- washing machines
- macerator-type WCs
- waste-disposal units.

Immersion heaters

Immersion heaters are not wired directly from the ring main but from the consumer unit in 2.5 mm twin and earth PVC cable, and have an overcurrent protection device rating of 16 amps. They are fed to a double pole switch, which has to be located a maximum of 1 m from the connection to the immersion heater. The cable between the pole switch and the immersion heater is 1.5 mm heat-resistant flex, usually butyl. A double-pole switch

Immersion heater flex to double pole switch

Remember

We discussed the installation of pumps, electric showers (or water heaters) and central-heating controls in Chapters 8 and 9.

Find out

What should you do to ensure that a section of pipework can be safely removed without any risk of electric shock?

is safer than a single-pole version, because when it is turned off both the live and neutral are isolated. It is most likely that you would be required to disconnect and reconnect the supply when replacing a defective immersion heater.

Dishwashers and washing machines

These are normally connected via a three-pin plug into a 13 amp socket outlet. While most appliances come with pre-moulded plug heads, you will occasionally need to change a plug. This is best learnt by a practical demonstration from a qualified person.

Macerator-type WCs and waste-disposal units

Like central heating control units, these are connected to the system using a spur outlet. A spur outlet is a bit like a hidden plug connection; it has the advantage that it cannot be unplugged accidentally thus turning off the supply to a control or appliance. It is also less likely to get damaged.

You will probably wire a component or an appliance from the fused outlet. On some occasions, however, you may be required to connect into the ring main.

Installing a fused spur outlet

Before you start, make sure the cables and fittings are correct for the job. Check for any damage to the cable, and that it is the right size and type, and that the fittings have the correct amp rating. *Remember the procedure for isolating the supply*. The spur outlet should be located close to the appliance or component it is intended to serve. Check the manufacturer's instructions for any specific requirements.

You need to connect to the ring main as close as possible to the spur outlet. This can be done by locating the nearest socket outlet, lifting a floorboard and then tracing the wire until it is at the closest point to the spur.

The cable is cut. It should be tested to make sure that it is the ring-main circuit and that the cable is sound (see the next section on testing). The joint on the ring main will be made using a junction box.

Old and new junction-box designs

Running the cable to the spur outlet

The spur outlet can be surface-mounted or flush-fitting. If using a flush fitting, the plaster and block work has to be **chased out** to take the box; keep this as neat as possible to reduce the mess and the amount of making good. A chase is cut from the hole for the box to the floor level; make sure this is wide enough to receive plastic capping, which you will have to fix to protect the cable before making good. In some cases, you may be able to thread the cable under the floor. Much will depend on the position of the joists, but in some cases you may have to drill a hole in the floor.

Fit the spur box by drilling and screwing to the wall. If it is a metal box, you must include a rubber grommet in the hole that the cable will pass through. Push the cable through the hole in the spur and under the floor, and pull the cable to the junction box.

If you are running the cable through the joists, the hole should be drilled in the centre to prevent any weakening. Alternatively it can be clipped to the side or the underside of the joists.

The ends of the ring-main cables and the cable to the junction box are prepared, and connected live-to-live, neutral-to-neutral and earth-to-earth.

The cable from the junction box to the spur, and the outlet from the spur, are connected to the back of the spur fitting. Again, if it is a metal box, you need to take an earth wire from the box to the fused spur. The cable is now ready to connect to the component or appliance.

When you have completed the work, carry out a visual inspection to make sure all the connections are sound, the cable routes and fixings are still where they should be, and the cable has not been damaged during installation.

The installation is tested while still switched off (we will look at testing in the next section), and, assuming everything is in order, the supply can be turned on.

Equipotential bonding

This is a very important area of electrical wiring, as most of the systems we install require some form of bonding protection, so you should be able to tell if the earthing and bonding have been installed correctly. You may also have to carry out work on these systems, but should do so only if you are competent.

Main equipotential bonding

The method explained here of protecting against indirect contact using equipotential bonding and automatic disconnection of supply meets the requirements of the Wiring Regulations (BS 7671). Metal pipework can provide a route for stray electric current to go to earth. This could cause an electric shock for someone touching the live pipework unless the metal is properly earthed. It can also cause the pipework to corrode.

Definition

Chasing out – cutting a recess into plaster or block work for the socket or cable

Remember

Under the new requirements of BS 7671, when you add a spur to an existing system incorporating embedded cables, you may need to upgrade the circuit to include RCD protection

Safety tip

Never carry out work of this nature if you are not sure. If this is the case, contact an approved electrician

On the job: A bonding problem

Callum, a final-year apprentice, is required to carry out a repair job on the copper pipework of an existing heating system, and this involves removing a section of pipework. When he checks the earthing/bonding arrangements, he discovers that none exist. His supervisor has gone to a boiler breakdown several miles away and is not due back for several hours.

1   What should Callum do now?

2   What would you do in this situation?

3   What should you check to make sure that equipotential and supplementary bonding was correctly installed, and how?

Figure 10.13 Earth-bonding conductors

In Figure 10.13, note the size of the bonding conductors, and the need for the clamp to be labelled. The bonding conductor from the main terminal to the earthing clamp is 10 mm. The bonding to gas, water or other services should be as close as possible to the point of entry, and for the gas supply within 600 mm of the meter.

Bonding the pipework as shown in the diagram will provide a safe route to earth and is described as main bonding.

Earth bonding to pipe

Supplementary bonding

You need to be able to tell if a plumbing installation is correctly bonded. You can see from the previous diagram that the equipotential bonding connects only to one point of the pipework.

There are other exposed metal parts within the domestic hot-water, cold-water and central heating systems, which may not be protected because they have been isolated from the earth by plastic fittings, cisterns etc. used in the system.

It is important that, on completion of a job, the services as shown in the illustration – including any supplementary bonding requirement – are carried out. This work will normally be done by a qualified electrician and your company should have advised the customer that the work is necessary or should have made arrangements to carry out the electrical work at the same time as you are completing your plumbing systems installation.

Temporary bonding

If you are required to work on a repair or maintenance job, it is important that you check that the pipework you are going to work on is bonded correctly. If you are going to remove a section of metal pipework or fittings, it is essential that earth continuity is maintained before any cutting or disconnection takes place.

A typical piece of kit consists of crocodile clips and a 10 mm conductor, 250 V rating minimum. It is suitable for up to 28 mm metal pipe.

The clips should be placed in a position to bridge the gap of the pipe or fitting that is going to be removed. Work can then safely take place. Only remove the clamps once the job is complete.

Testing and decommissioning

At Level 2 you need to understand the basics of testing so that you know what must be carried out when completing electrical work. Practical testing of electrical systems is covered at Level 3.

In this section we will look at other tests that confirm the safety of an electrical installation. As a plumber you need to test for earth continuity and the polarity of wires at an appliance or a component. In addition to these, an insulation-resistance test should be carried out which will tell you if the cabling used is in good condition.

Remember

Plumbers are in constant contact with metal pipework and metal surfaces when carrying out their job – customers can be too!

Safety tip

Remember: by removing the section of pipework, you have removed a section of the system from the main bonding conductor and may, therefore, have exposed yourself to real danger – this one can kill!

Electrical tests

The systems we deal with include:

- continuity of earthing conductors
- polarity
- insulation resistance.

Whatever the test, you must use the correct type of test equipment. For electrical work associated with plumbing, the testing and the proper equipment required includes:

- continuity – low-reading ohmmeter
- polarity – low-reading ohmmeter or insulation and continuity tester
- insulation resistance – insulation-resistance tester.

Your training will include showing you examples of the above test equipment.

Earth continuity

Earth continuity is making sure that should there be an electrical fault, all exposed metalwork in a building is bonded together and connected to the earthing block in the consumer unit, leaking the current to earth and automatically disconnecting the supply. An earth continuity test will verify that exposed metalwork in a building is bonded together and connected to the earthing block in the consumer unit. The ohmmeter leads are connected between the points being tested, between simultaneously accessible extraneous conductive parts – pipework, sinks etc. – or between simultaneously accessible extraneous conductive parts and exposed conductive parts (metal parts of the installation). This test will verify that the conductor is sound. To check this, move the probe to the metalwork to be protected. This method is also used to test the main equipotential bonding conductors. There should be a low resistance reading on the ohmmeter.

Figure 10.14 Testing for continuity

Supplementary bonding conductor

Low reading ohmmeter

Polarity

Testing for polarity makes sure that phased conductors are not crossed somewhere – neutral from mains connected to live and vice versa (reversed polarity). In this situation, the system might still function as expected; however, when isolated from a switch, the system would be in dangerous mode.

Testing across the conductors, as detailed above, is carried out to make sure that no wires have been crossed.

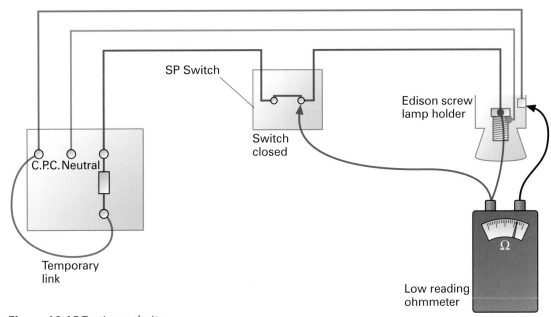

Figure 10.15 Testing polarity

Insulation resistance

Insulation-resistance tests make sure that the insulation of conductors, electrical appliances and components is satisfactory, that electrical conductors and protective conductors are not short-circuited, or do not show a low insulation resistance (which would indicate a defective insulation).

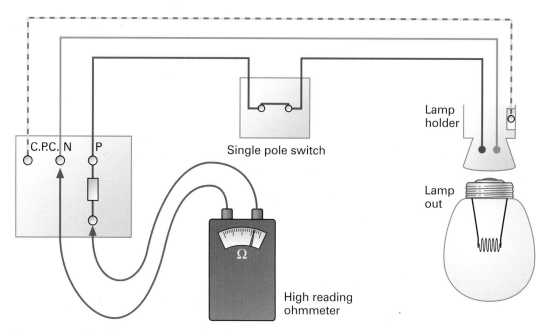

Figure 10.16 Insulation-resistance check

Before testing, ensure that:

1 Pilot or indicator lamps and capacitors are disconnected from circuits to avoid an inaccurate test value being obtained.

2 Voltage-sensitive electronic equipment such as dimmer switches, delay timers, power controllers, electronic starters for fluorescent lamps, emergency lighting, Residual Current Devices etc. are disconnected so that they are not subjected to the test voltage.

3 There is no electrical connection between any phase or neutral conductor (e.g. lamps left in).

An insulation resistance of no less than 0.5 M ohms should be achieved.

Decommission electrical appliances or components, including associated wiring

Your job will require you to take out components and appliances, and that will include the electrical supply. Generally speaking, the work will be a repeat of what you've covered earlier, in terms of safety, isolation and testing, because you're working on an electrical supply.

In some cases you'll decommission an appliance or a component to repair or replace it, so it's disconnect, reconnect. There may be instances, however, where an appliance is going to be removed altogether, which would mean the supply would need to be made safe. This would be done by taking out the cable back to its isolation point, e.g. immersion heater isolating switch, fused spur outlet. All of this should be done following the safety procedures that we have already learned.

FAQs

Why is electricity included in a plumbing qualification?

It is not intended to make plumbers into electricians but plumbers need to know quite a bit about safe methods of working and isolation in the installation and maintenance of plumbing components with electrical connections. There is a wide range of plumbing/electrical components that have to be tested for correct operation, and plumbers often remove sections of pipework which may affect earthing arrangements.

What should I do if I am unsure about the correct action to take in order to safely isolate a component?

The simple answer is DO NOT try to do the work at all, seek help and guidance from someone who is competent in this area of work. Do not take any risks, You are a long time dead!

Surely locking an mcb in the off position while working on a circuit is an overreaction. Is it really necessary?

Yes, it is absolutely necessary, how else can you be sure that the mcb will not be switched on by someone while you are working.

Do I have to be qualified in electrical work in order to carry out any testing of a system controls circuit for correct operation?

No, but you do have to be competent. The rules on this may change in the future, however, so keep up to date.

Knowledge check 3

1   What type of cable is normally used for embedding in walls in domestic properties?

2   What type of cord is normally used to make an electrical connection to a circulating pump?

3   The electrical connection of a new central-heating system into an existing ring main circuit is known as a _____.

4   What is the normal size of an earth bonding conductor to a water pipe?

5   What device is used to maintain earth continuity when making a branched connection into an existing water pipe?

Additional information sources

Sources of additional information on this chapter include:

Building Regulations Approved Document P – www.planningportal.gov.uk
HSE website dealing with Electrical Safety – www.hse.gov.uk/electricity/index

chapter 11

Above ground discharge systems

OVERVIEW

Above ground discharge systems (AGDS) include sanitary appliances, sanitary pipework and fittings, and rainwater systems. This chapter will look at the various appliances on the market, how to fit them and pipe them up, and how they are tested and maintained. This chapter also covers rainwater drainage – that is, guttering and rainwater-pipe systems. AGDS are essential to keep our environment clean and hygienic.

This chapter will cover:

- **Sanitary appliances**
  - British Standards and Regulations
  - types of appliance and materials
  - working principles of sanitary appliances
  - installing appliances
  - other types of appliances

- **Traps**
  - types of trap
  - access
  - why traps lose their seal

- **Sanitary systems**
  - primary ventilated stack system
  - stub stacks
  - general discharge stack requirements

- **Pipework installation, access and materials**
  - preparing to install AGDS pipework
  - the installation and jointing of AGDS pipework
  - connections to drains
  - connecting the soil and vent pipe

- **Soundness testing**
  - soundness testing of AGDS

- **Rainwater and guttering**
  - materials
  - gutter and rainwater systems

- **Maintenance and decommissioning**
  - basic maintenance and cleaning
  - decommissioning systems

Sanitary appliances

Sanitary appliances can be divided into two main groups: those used for washing purposes and those used for the removal of human waste.

Sanitary appliances play a very important role in our lives: they mean we can keep ourselves and our eating utensils clean, and we can use the toilet in privacy and comfort.

Although we are mainly concerned here with domestic installation we will extend coverage to urinals and flushing cisterns which you may come across when working in public toilets or larger buildings.

British Standards and Regulations

British Standards

British Standard 8000 is the Code of Practice covering workmanship on building sites. Part 13 covers above ground drainage and sanitary appliances, including:

- materials handling
- site storage of components
- the preparation of work, materials and components
- the installation of sanitary appliances
- inspection and testing.

BS8313 gives guidance on the holes, chases and ducts required for pipework.

Water Regulations

Water Regulations do not refer specifically to sanitary appliances, but do deal with how they perform by covering undue consumption, waste of water and protection against backflow.

The regulations governing WC cisterns, urinals and automatic flushing cisterns are limited to the capacity of water for flushing.

Backflow prevention is achieved either through mechanical methods (using valves) or non-mechanical means (using air gaps between outlet and appliance water levels) to prevent the possibility of water being drawn back into the supply pipework and into the main. This could occur if an open tap outlet was allowed to become submerged in a sink or bath, and if the main outside stop valve was turned off and water drained from the system. As the water drained down from the system, water from the sink or bath would be drawn into the pipework due to siphonic action. Water Regulations will be covered in greater detail at Level 3.

Remember

Backflow prevention has a bearing on sanitary appliances, which are categorised by the Water Regulations in terms of risk. This covers WC suites, washbasins, bidets, baths, shower trays and sinks

Siphonic action uses differences in the atmospheric pressure in two parts of a pipe to draw water through the pipe. Check back in Chapter 4 if you cannot remember how this works

Types of appliances and materials

There is a vast range of styles and designs of sanitary appliances available. However, British Standards ensure that there is standardisation in terms of dimensions, including the size of waste outlets.

Baths

Manufactured to BS 4305 (EN198), baths can be supplied in acrylic sheet, heavy-gauged enamelled steel, and vitreous enamelled cast iron. Victorian-style free-standing designs give a traditional 'period look'. Vitreous enamel provides a smooth hardwearing surface that is corrosion resistant and easy to clean.

Baths come in a huge variety of styles, and taps and waste can be either end-mounted or centre-mounted. Space-saving styles include corner baths and compact baths.

More expensive options include enamelled baths manufactured from heavy-gauged porcelain enamelled 3.5 mm thick onto heavy gauge steel. Steel sheet is pressed into shape and then coated in layers of vitreous enamel which is fired at high temperatures. The enamel and steel bond during firing to give the enamel a glass-like high gloss finish.

Did you know?

Acrylic sheet baths are vacuum-formed to the desired shape

This type of bath is particularly suitable for installation in hotels, or in housing association and local-authority housing where durability is important.

The jet-system bath is a popular choice in the 'luxury' market. It is like a standard bath but contains a number of jets (generally eight) in the side of the bath to provide a massage effect. In some cases additional outlets are sited in the floor for foot massages.

Figure 11.1 Typical jet-system bath

The water is circulated through pipework located around the outside of the bath and forced through the jet nozzles by a pump.

Great care needs to be taken with the electrical connections to this type of bath as the bath is a high risk appliance containing water. The electrical safety precautions that must be employed are significantly higher. A qualified electrician will usually deal with the connection of such appliances.

Baths are manufactured with two holes for pillar taps, or single holes for monobloc fixings. Most come with an overflow hole, and waste hole that will take a $1\frac{1}{2}''$ threaded waste. Combined overflows and waste fittings are also provided, as shown.

Water closets (WCs) and cisterns

Again, there are a lot to choose from. They are categorised as:

- back to the wall
- close-coupled
- low-level
- high-level
- concealed.

In the main, WCs are manufactured from **vitreous china** conforming to BS 3402. Vitreous china is made from a mixture of white burning clays and finely ground materials. These are fired at high temperatures, and even before glazing the material cannot be contaminated by bacteria, so it is totally hygienic.

Figure 11.2 Combined waste and overflow fitting

Vitreous china is coated with an impervious non-crazing vitreous glaze in either a white or coloured finish. The material is stain proof, burn proof, rot proof, rustfree, non-fading and is resistant to acids and alkalis.

WCs used in public places may also be manufactured in stainless steel. These are more resistant to vandalism than vitreous china.

WC cisterns

Prior to 1993 the capacity of a WC flushing cistern was 9 litres. The Water Regulations brought this down to 7.5 litres, and from January 2001 reduced it further to 6 litres. In addition, Water Regulations also permit the use of dual-flush cisterns; these deliver 6 litres for a full flush, and 4 litres for a lesser flush. The dual flush cistern is now specified more often that the single flush.

There are two types of cistern:

- siphonic
- dual flush valve (often known as a drop valve).

With the siphon type, when the lever is pressed the water in the bell of the siphon (1) is lifted by a disc (or diaphragm) up and over into the leg of the siphon (2) (see figure 11.3). This creates the siphonic affect, which continues until the water level in the cistern has dropped to a level which allows air to enter the bell.

Definition

Vitreous china
– produced from a solution called slip, or casting clay. Slip has the consistency of pouring cream and contains ball clay, china clay, sand, fixing agent and water

Dual flush button — Single flush button

Single flush – siphon type

Dual flush – valve type

Figure 11.3 Different types of cistern

The dual-flush valve-type cistern is operated by pressing either the full flush or lesser flush buttons, which have to be clearly marked on the cistern. These operate a valve which releases the water into the WC pan.

Another type of fitting (see figure 11.4), is also available for 6 litre capacity cisterns. This works on the principle of lifting a hinged flap and is suitable for full flush only.

Cisterns of 9 litres and 7.5 litres are still available. Water Regulations permit these to be installed as replacements for existing cisterns. This is because the WC pan was designed to work on those capacities.

Overflow pipe

Link to flush handle

Hinged flap valve

Flush pipe connection

Figure 11.4 Cistern fitted with hinged flap mechanism

The overflow pipe can now discharge into the WC pan using an integral overflow, eliminating the need to provide an external overflow pipe. This pattern of cistern is now supplied with most WCs.

WC pans

There are a number of designs for WC pans, but there are only two main types:

- wash-down WC pan
- siphonic pan.

The wash-down pan, which tends to be the most common type, uses the force of the water from the cistern to clear the bowl. The principle of the siphonic pan is to create a negative pressure below the trap seal.

With the single-trap pan, this is done by restricting the flow from the cistern and is achieved by the design of the pan.

Remember

WC pans and cisterns are manufactured as a single unit and therefore replacement cisterns to existing pans must always be the same capacity as the original. It is not possible to 'mix and match' as each component has been designed carefully in conjunction with the others to operate properly as a 'suite'

The double-trap close-coupled pan uses a pressure-reducing device between the cistern and the pan. As the water is released into the second trap, it has the effect of drawing air from the void between the two traps, and siphons the content from the bowl.

Figure 11.7 shows what the pressure-reducing valve looks like between the cistern and pan.

Figure 11.5 Wash-down pan

Figure 11.6 Single and double trap siphonic pan

Popular WC types include:

Concealed – where the cistern is hidden

Close-coupled – where cistern and pan are joined in one unit

Low-level – where the cistern is no more than one metre above the pan

High-level – where the cistern is more than one metre above the pan necessitating flushing by a chain

Figure 11.7 Typical pressure-reducing valve between cistern and pan

The choice of WC suite depends on a number of factors:

- cost – the customer's budget
- location – for public toilets, factors such as durability and ease of cleaning are a consideration, for example stainless steel units are often used because of their resistance to vandalism
- aesthetics (how good it looks) – Victorian versions of the high-level cistern are now considered very desirable.

On larger housing contracts, or in public buildings, the choice of WC may have already been made by the client's architect as part of the contract specification document.

Some cisterns are reversible – that is, the handles, overflow outlet and water inlets can be either left- or right-handed. In some cases, the cistern is plastic rather than vitreous china.

Kits are usually supplied for the soil outlets, and will be either a 'P' trap or an 'S' trap.

Washbasins

These fall into two main types:

- fixed to the wall
- countertop basins.

Fixed to the wall

These are secured by using brackets or are supported by a pedestal. Larger washbasins, fixed using brackets, are mostly found in non-domestic installations. The pedestal type is probably the most popular choice in homes. Appearance is the main factor as all the pipework and fittings can be seen with the bracket-mounted basin.

Wall-hung basin

Pedestal basin

Figure 11.8 Fixed to wall

Countertop basins

There are three types:

- countertop – sits proud of the work surface
- semi-countertop – used where space is tight as the front of the basin projects clear of the top
- under the countertop – has the flange surface on top of the basin.

All basins are available with either single tap holes for a monobloc mixer tap, or dual tap holes for pillar taps. Basins are also available with three holes for fittings with an independent spout.

Countertop basin

Under countertop basin

Semi countertop basin

Figure 11.9 Countertop

Countertops are installed in all sorts of buildings, from homes to pubs and restaurants. They are manufactured from either vitreous china or a high-impact plastic called acetyl.

The basin outlet is designed to take a $1\frac{1}{4}$ in slotted waste fitting. Waste fittings can be the standard slotted waste with a plug and chain or a pop-up type, which is operated by lifting and pressing a knob on the top of the basin.

Hand-rinse basins

Called hand-rinse basins due to their size, these are used in cloakrooms with WCs. They can be a corner design or traditional, either countertop, pedestal or bracket-mounted.

Figure 11.10 Pop-up waste fitting

Countertop range

There are three types:

Standard countertop – this fits into a pre-cut hole and is secured by fixing brackets that secure the underside of the basin to the countertop; the joint is sealed using waterproof sealant

Semi countertop – can be made from vitreous china or acetyl

Under countertop – this appliance is made from acetyl.

Shower trays

Shower appliances were covered in Chapters 7 and 8. Here we will concentrate on shower trays. As with other sanitary appliances, there are a number of designs. Trays are available in reinforced cast acrylic sheet, fireclay for heavy-duty applications or resin-bonded. The outlet is designed to take a $1\frac{1}{2}$ in threaded waste. Corner trays are used where space is limited.

Bidets

Made from vitreous china, there are two types:

- over-the-rim supply
- ascending spray.

Figure 11.11 Ascending spray bidet

Lowest level
of tap outlet

Type AUK2 air gap

Spill-over level
of appliance

Figure 11.12 Air gap requirement for over-the-rim bidet

The ascending-spray bidet has to be piped up correctly to avoid risk of contamination to the supply and cannot be used on mains supply, for example with combination boilers and unvented hot-water systems. This is to meet the requirements of the Water Regulations and will be dealt with at Level 3. The bidet can be supplied for use with pillar taps, monobloc fittings and pop-up waste.

An over-the-rim bidet presents less of a water contamination risk and may therefore be connected into supply pipework fed directly from the mains provided that the Type AUK2 tap gap is maintained above the spillover level of the sanitary appliance. If the over-the-rim bidet includes a flexible hose, then it should be treated as though it is an ascending spray type as the flexible hose and spray head may be dropped into the contaminated water in the bidet bowl. An ascending spray bidet or an over-the-rim bidet must be fed from storage and have its own dedicated piping system in order to avoid potentially serious contamination of the water supply.

Sinks

Belfast sinks and London sinks

Belfast sinks are used in both domestic and commercial situations. They are usually fed by bib taps, but can also be supplied by either pillar taps or monobloc taps in domestic installations.

London sinks are similar, but without an overflow. Both types are made of heavy-duty fire clay, which is manufactured in a similar way to vitreous china. They are available in white only and are designed to take a $1\frac{1}{2}$ in threaded waste.

455 mm

595 mm

255 mm

Figure 11.13 Section through a Belfast sink

Kitchen sinks

Again, a vast range of designs is available. Sinks are produced in stainless steel, plastic-coated pressed steel, fire clay and plastic in the form of acetyl. They can be single drainer, double drainer; single and basket (suitable for connection to a small waste-disposal unit) and double sink with single drainer. They are designed to take a $1\frac{1}{2}$ in slotted waste. This can be slotted for use with a sink overflow, or plain. Taps can be monobloc or pillar.

Figure 11.14 Section through a modern acetyl plastic kitchen sink

Urinals

Usually installed in non-domestic buildings, urinals are manufactured in vitreous china for urinals and stainless steel for bowls fixed to the wall, and stainless steel, plastic or fire clay for slab urinals.

Urinals are fitted with an automatic flushing cistern. Its automatic siphon operation is very simple:

Figure 11.15 A slab-type urinal

- as the cistern fills, and the water rises, the air inside the dome of the automatic siphon is compressed

- the increased pressure forces water out of the U tube, which reduces the pressure in the dome

- the reduction in pressure causes siphonic action to take place, flushing the cistern

- when the cistern has emptied, the water in the upper well is siphoned into the lower well.

Figure 11.16 Automatic flushing cistern

Water Regulations state that auto-flushing cisterns must not exceed the following maximum volumes:

- 10 litres per hour for a single urinal bowl, or stall

- 7.5 litres per hour, per urinal position, for a cistern servicing two or more urinal bowls, urinal stalls or per 700 mm of slab.

The flow rate can be achieved using urinal flush-control valves which allow a small amount of water to pass into the system.

Timed flow control valves can also be used. These have the additional advantage of being switched off when the building is not in use, for example evenings, weekends, factory shutdowns, school holidays etc. They can be set for 'one-off' hygiene flushes, say once every 24 hours. Many public buildings now control the supply of water to the automatic flushing cistern using infra-red sensors that detect motion. On detecting motion, similar to a burglar alarm that detects motion, a solenoid valve is opened permitting a flow of water to the automatic flushing cistern.

Figure 11.17 Hydraulic flush-control valve

Figure 11.18 shows a typical example of a urinal set-up showing individual wall fixed urinals (bowls). These are vitreous china. The picture also shows the cistern and flush-pipe arrangements. On slab urinals the flush pipe is often mounted on the face of the urinal and has a number of holes which are used to wash the face of the slab and the channel. These are called **sparge pipes**.

Figure 11.18 Typical two-bowl urinal layout

700 mm

320 mm

1370 mm

675 mm

700 mm

Working principles of sanitary appliances

The main working principles that you need to consider with regard to sanitary appliances apply on the whole to WC pans and cisterns.

Some WC pans clear their waste by the wash-down action of the water as it is released into the bowl. Other designs use siphonic action: siphonic WC pans, siphons in cisterns and automatic siphons. We covered the 'science' aspect of siphonage in Chapter 4 *Key plumbing principles,* and we looked at how siphonage is used earlier in this chapter.

Installing sanitary appliances

Preparation

It is important to prepare properly before you start installing any appliance. This should ensure there are no hold-ups later owing to missing fittings, or pipe chases not having been prepared as part of the specification.

Storage of materials on-site

This is particularly important on larger jobs where you might take delivery of a number of items of sanitary ware. However, some aspects also apply to 'one-off' jobs. Sanitary appliances are expensive items, so it is important that they are stored properly to prevent damage or theft.

Here is a storage checklist:

- Before a delivery takes place, make sure there is somewhere suitable to store the appliances – a lockable materials cabin, with adequate room, is ideal. The storage surface should be raised off the ground of the cabin (use pallets or similar), and it should be clean and dry.

- Before accepting delivery, check the appliance carefully for signs of damage. This does not mean removing the wrappings – these should be left on as long as possible. Production defects are rare, but keep an eye out for these as well.

- Check all traps, taps, water plugs, brackets and seats are with the appliances. There is nothing worse than starting a job and then finding that a waste fitting or bracket is missing.

- When you are happy that everything is complete, check off all the items ordered against the delivery note. If you have the authority, sign for the delivery. You will receive a customer's copy of the delivery note. If anything is not correct, note the missing items and contact the supplier immediately. If you are desperate for the delivery as it stands, sign for the items and get the person making the delivery to countersign that some items are missing.

- Take care when handling the material. Some sanitary items are heavy: remember your lifting technique from Chapter 2 *Health and Safety*.

- Store the materials in the materials cabin. As mentioned, these should be clean and dry. The appliances should also be stored away from plaster and concrete, and should also be kept away from areas where other materials could fall onto or into them (brick stacks, etc.)

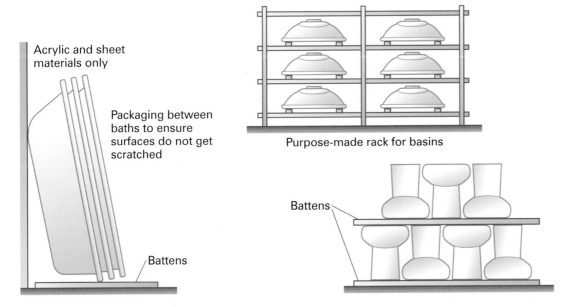

Acrylic and sheet materials only

Packaging between baths to ensure surfaces do not get scratched

Battens

Purpose-made rack for basins

Battens

Battens

Figure 11.19 Storage patterns for baths, WCs and basins

The illustrated ways of storing appliances are simple but effective. WC pans should be stacked no more than four high. Make sure the storage surfaces (battens/pallets) are clean: battens are used for all sorts of things on site, and if they are covered in grit, storing materials on them will defeat the object.

Checklist for the work site

Check access to the work area, and check that the work area itself is free from obstructions and potential hazards. This should ensure that the risk of accidents is kept to a minimum.

- Make sure you have any job details or specifications/drawings showing the fixing positions of the sanitary appliances and any specific fixing requirements. You will also need the manufacturer's fixing instructions.

- If you find any installation problems, or defects in the appliance as you are working, report them to your supervisor.

- On some jobs you will cut holes or chases or install ducts for pipework. On other jobs, this preparatory work is done by other trades in advance, so you will need to check that it has been done. If not, report back to your boss or the site supervisor.

- When you're at the stage of fixing the appliances, the **first-fix** pipework must be installed. This may have been done by you, particularly on a single-dwelling job, or it could have been done by another plumber on a large housing contract.

- Make sure servicing valves have been installed where required. Good practice would see them put on every appliance that requires regular service and maintenance.

- On first-fix jobs, open-ended pipes should have been capped. This is to prevent debris getting into the system.

- On jobs where the appliances are being fixed against tiles, baths and showers should be installed before tiling, and concealed bidets, washbasins and WCs after tiling. This makes the job easier for the tiler as they won't have to cut around the shape of the appliance.

- Edges of baths should be set into plaster or plaster board. Careful planning and liaison with other trades can ensure that plasterwork is not taken to full floor level on the bath-edge wall, and in the case of stud-partitioned walls, extra struts can be fixed to support the edges of the plaster board.

- Finally, as you are carrying out the work, avoid standing on any of the appliances. This is sometimes unavoidable with showers, or baths with a shower appliance installed above it. In each case, make sure the surface of the appliance is fully protected with dust sheets and cardboard packaging.

Installation

For the purpose of this topic, we are going to work through the installation of a typical new bathroom suite, including:

- bath
- pedestal washbasin
- close-coupled WC – wash-down pattern
- shower tray and enclosure
- bidet.

Preparing the appliances

This is often referred to by plumbers as 'dressing the suite' – it means installing the taps, wastes and, in the case of the bath, the cradle frame or feet. It will also include installing float valves, overflows, siphons or flushing valves and the handle assembly to the WC cistern.

Fitting the taps

Fitting the taps is a relatively simple process. Tap manufacturers' designs have to comply with regulations, so your installation will also satisfy the regulations. Remove just enough of the protective coating to make sure a clean joint can be achieved between the fittings and the surface of the appliance. Also, double check to make sure the surface is clean and dry. You will not have to use any other jointing compound, the washers supplied by the manufacturer are all you need. There may be slight differences in the washer kits supplied by manufacturers, but the illustrations here show typical details.

Remember

The correct tap conventions are: hot on the left, cold on the right

Pillar tap to a basin

The grip washer goes on the underside of the tap. Sometimes a thicker washer is also supplied, which goes between the back nut and the underside of the basin. Tighten the back nut using a purpose-made basin wrench. You can test if it is tight enough by checking by hand that the tap does not move. Make sure both taps are pointing outwards in parallel; some plumbers turn the taps slightly inward.

Grip washer

Backnut

Figure 11.20 Seal to basin tap

Figure 11.21 Monobloc tap connections

Monobloc taps

These are often supplied with an 'O' ring, which should be fitted between the base of the tap and the basin. The rubber washer goes underneath as shown. The assembly is completed using a metal washer held in place by a fixing nut.

Tap and spout assembly – three-hole

Obviously, these are installed in three-hole basins and ascending spray bidets.

- Remove all the parts that go above the basin
- Fit the spout with its washers
- Loosely connect the tubes between the body and side valves
- Raise the assembly into position from below, making sure the sealing washers are in place
- Fit the body seal, washer and lock nut loosely, then fit the valve flanges on top of the basin, aligning them with the top of the side valve as in Figure 11.22
- Carefully tighten all the components, and then finish by fitting the valve head gear, any shields, drive inserts and hand wheels.

Figure 11.22 Three tap-hole basin

Remember

For all tap installations make sure nuts are not cross-threaded

Safety tip

Never use linseed oil compounds on plastic wastes as they degrade the plastic causing it to fail; it can ruin the bath and shower tray as well!

Other appliances

The same principles apply to installing taps on other appliances. Remember to make sure that you use the washers supplied by the manufacturer.

Waste fittings

For washbasins, these include slotted waste fittings, moulded seals and pop-up wastes.

Slotted waste fittings

These can be sealed using non-setting plumbers' compound or silicone sealant. This is done by moulding a thin strip of compound around the flange of the waste and then placing the waste into the basin, making sure that the slots in the waste line up with the overflow slots of the basin. Clean off excess sealant or compound with a soft dry cloth.

Mould a thin strip of compound around the flange of the waste

Line up slots and apply more compound or sealant around the bottom of the basin and the thread of the waste

A plastic washer should be pressed into the compound, and the back nut is tightened to complete the joint

In the case of some pedestal basins, a bracket has to be included to fix the basin to the pedestal.

This is done by placing the metal bracket between a rubber washer and the back nut, as shown above. The basin is secured using the two adjustable clips which are tightened against the pre-cast lugs in the pedestal.

Moulded seals

These are supplied for pop-up wastes and combined wastes and overflow, as used on baths, plastic washbasins and sinks. Installation details are shown in Figure 11.23.

Standard fixing   **Plastic basin with overflow fitment**   **Pedestal basin fixing**

Figure 11.23 Moulded seals

The moulded seal fits between the waste flange and the appliance. For pedestal and plastic basins, a shaped seal and 'O' ring are used together to make the joint for the shaped metal connecting washer, or alternatively the metal pedestal-fixing bracket.

Where the appliance requires an overflow assembly, a rubber ring should be fitted between the overflow flange and the back of the appliance. When fitting this set-up:

- do not cross-thread the overflow grill and the fitting to the rear of the appliance

- do not over-tighten the fitting

- make sure you do not damage the flexible overflow tube when tightening the overflow fitting.

'Pop-up' wastes

Pop-up wastes are available for washbasins and baths. On a bath, the plug unit is raised by twisting the 'circular knob' anti-clockwise, which serves as both an overflow outlet and pop-up waste control. The plug unit is lowered by twisting the control knob clockwise, which allows the plug to drop into the waste by gravity.

Checklist for fixing a washbasin waste

- Fit the waste body into the basin outlet

- Remove cap nut and control rod

- Position the appropriate seals, then couple the pop-up body onto the waste tail thread

- Tighten with the control port correctly aligned with the back of the basin

- Slacken the lock nut on the stem of the pop-up plug and fit into the waste body

- Locate the control rod into the control port so that the end of the rod is in the hole at the end of the plug stem

- Assemble the cap nut

- Check the operation of the plug; it should be flush with the waste flange, and should then lift about 10 mm when operated

- To reduce the lift, the plug is turned clockwise, and to increase the lift anti-clockwise

- Remove the cap nut and control rod and take out the plug

- Tighten the lock nut against the strainer stem of the plug, and then refit the plug and control rod, and tighten the cap nut

- Fit the lift rod through the fitting and locate with the eyelet on the shackle

- Adjust so that the left knob is just clear of the fitting when the plug is fully open

- Tighten the screws on the shackle and check for smooth operation of the assembly

WC suite

Our installation includes a close-coupled wash-down WC pan and siphonic action cistern.

Dressing the cistern

This will include the siphon, overflow fitting and float-operated valve. The cistern will have holes for either side inlet and outlet, or bottom inlet and outlet. Most cistern components will be plastic, and should be installed using the washers provided.

The shaped rubber sealing for the siphon, and flat rubber washers for the float valve and overflow union, should go inside the cistern.

Fit the siphon first, making sure that the metal bracket for the close-coupled connection is in place, and that a rubber washer has been placed between the bracket and the underside of the cistern. The back nut completes the assembly. When tightening the back nut, make sure the bracket is kept in the correct position for the bolts and locating holes in the pan.

Fit the float valve and overflow.

Assemble the lever and linkage mechanism and check that it works correctly before installation.

Fit the cistern to the pan.

Before securing the cistern to the pan, make sure the soft foam-rubber close-coupling washer is fitted over the siphon nut. Then carefully place in position over the pan. Locate the fixing bolts into the holes in the pan, and tighten the cistern to the pan using the rubber and metal washers and the bolts supplied. Test the joints between the pan and soil pipe. Make the overflow and cold water connections.

Safety tip

When tightening any back nuts:

- do not over-tighten
- do not cross-thread.

Baths

The cradle frame or brackets and legs are usually fitted before the taps and waste, but as part of the dressing process. Bracket and feet assemblies will vary depending on the manufacture, but a typical assembly detail is shown in Figure 11.24. This is sometimes called a cradle frame.

First, place the bath upside down on a dustsheet over a clean, flat surface. Each leg is positioned in the location spigots as shown in Figure 11.24. They should be pushed in until the centre section is flush with the baseboard, making sure the sides of the brackets are plumb. Then the centre leg and bracket are located in the centre of the baseboard.

All the brackets are screwed in position using the wood screws provided.

You can then screw the brackets to the location spigots using self-tapping screws.

Figure 11.24 Bath fixings

The feet are assembled into the brackets. Stand the bath on its edge and fit the clip-on wall brackets and handles. The taps and waste are installed as described previously.

It is handy at this stage to adjust the feet while the bath is on its edge. This can be done by using a tape and getting the feet to an approximate measurement required and then 'fine tuning' the feet to the exact measurement using a straight edge and tape. Final adjustments can be carried out with the bath in place.

Other appliances

The waste requirements for shower trays are straightforward and are fitted using the washers supplied with $1\frac{1}{2}$ in waste fitting. On fire-clay trays, non-setting plumbing compound is applied in the same way as with the standard slotted washbasin waste installation.

Figure 11.25 Bidet washer and dial arrangements

The waste fitting for a bidet is a $1\frac{1}{4}$ in threaded unit, either slotted or solid depending on whether the bidet is fitted to an overflow; pop-up wastes are also available. Taps can either be pillar or monobloc for over-the-rim applications. A typical arrangement for a douche-type (under-rim) bidet is illustrated in Figure 11.25.

The installation instructions are similar to those of the tap-and-spout assembly of the washbasin outlined previously.

The installation can now begin

There are no set rules for the sequence used to install the appliances. Plumbers will work out the best sequence based on the position of the appliances and the size and layout of the bathroom. They will base their decision on what they think will be the easiest and quickest way.

On jobs where you are replacing a suite and the dwelling has only one WC, the first appliance you install will be the WC suite, to keep any inconvenience to a minimum.

It is not very often that you will install a bath and shower tray in the same room, unless it is a large property. Most baths have shower units installed over the appliance, so a shower tray may be used in an en-suite room, particularly where space is tight, or the occupier has difficulty in using a bath due to age or illness.

Generally speaking it is preferable to install the bath first.

Installing the bath

- Place the bath on its feet and in position.
- Check the measurement from the top edge of the bath and or bath panel to the floor. It should be the manufacturer's recommended distance.
- Check the bath is level across its length and width. Adjustment to level and height can be made via the feet.
- Once the bath is level and at the correct height, tighten the locking nut on each foot.
- Mark the position of the fixing holes for the feet and the wall fixings. If the wall is plastered, and you have been instructed to let the bath into the plaster, mark it off for the full length of the bath. Take out the bath.
- If the floor is wood, drill pilot holes to receive the wood screws for the feet. This will make fixing much easier. If the floor is solid, drill and plug the floor. Drill holes for the wall fixings.
- Chase out the plaster for the wall brackets and, if letting the bath into the plaster, chase out the plaster where you marked the line of the bath: this enables the bath to be let into the wall and provides a really good watertight seal once tiled.
- Refit the bath and screw the wall and floor brackets into position. If fixing to timber floors, make sure the length of the screws will not penetrate the underside of the floor.

- Service and waste connections are then made. It is best to pre-fabricate the tap connections if using soldered fittings, as this avoids using the blowtorch under the bath. The trap is connected and waste pipe extended to the outside (if there is an external waste system).

- If the bath is fixed flush to the plaster, fill the bath to one-third full, so that the weight causes the bath to settle, and then seal the joint between bath and wall with silicone sealant – after removing the protective film from the bath's surface.

The method of fixing bath panels varies depending on the manufacturer, so their fixing instructions should be followed, but generally speaking the following should serve as a guide.

- Check the measurement from the underside of the roll edge of the bath and the floor. Check for any adjustments needed in the event of the floor being out of level.

- If the panel has to be cut, firmly support the panel and cut it using a fine-tooth saw, and finish off with a file.

- In some cases, panel support frames or bath leg clips are supplied, and these should be fitted.

- Use a level or plumb line to mark out for a fixing batten at floor level; once you have made an allowance for any 'kicking space' on the panel, mark the position for the batten. If self-adhesive pads are to be used, an allowance for their thickness should also be made for the position of the batten (four pads should be sufficient).

- Alternatively screws can be used. Mirror screws provide an attractive finish, and some manufacturers provide screw-head caps that match the colour of the bath panel. Pre-drill the panel to take the screws. The panel can then be positioned by inserting the top edge of the panel between the bath roll and the timber frame of the bath, and secured at the base with either the self-adhesive pads or the screws.

Installing the washbasin

Our sample bathroom specification includes a pedestal basin. The taps and waste are in, and during dressing, the metal bracket that will hold the basin to the pedestal has also been fitted.

Figure 11.26 Basin fixing bracket

- Loosely attach the clamps using the pins and washers. Do not tighten them at this stage.

- Run sealing compound along the top edge of the pedestal; this will provide a better finish to the front.

- Place the basin on the pedestal. Adjust the clamps so that they rest against the pre-cast shoulders of the pedestal.

- Check the basin for level, make any final adjustments; carefully tighten the pins and then clean off any excess sealing compound.

The basin in this example has moulded screw holes at the back edges of the bowl for fixing it to the wall (Figure 11.27).

- Offer the basin up to the wall; make sure it is level and in the correct position (you might have to pack the pedestal to achieve level). Make a mark on the wall through the holes.

- Remove the basin and pedestal and place them where they will not get damaged, preferably laying them down on a dust sheet.

- The moulded holes in the basin are angled. You need to estimate this angle and drill the hole accordingly. If you drill it level, as you screw the basin to the wall the edge of the basin will crack off adjacent to the hole!

- Reposition the basin and pedestal and screw back to the wall using brass or alloy screws and soft washers. This will make the basin easier to remove for any maintenance work or replacement.

- Some pedestals are supplied with fixing holes in the base. If so, the floor should be marked at the same time as the basin. On wooden floors, pilot holes should be drilled before refitting. On solid floors, plugs are placed in the drill holes prior to repositioning the pedestal.

Figure 11.27 Screw fitting of basin

- Some basins are supplied without fixing holes at the back of the bowl, but have moulding holes on the underside edge of the bowl.

- Fit the bracket to the basin and tighten so that it holds firmly in the hole.

- Offer the basin to the wall; then adjust the brackets so that they are flush with the wall. Tighten the fixing bolt into the basin, and then mark the position of the bracket holes.

- The pipework installation can be prefabricated so that it fits neatly behind the basin and the pedestal. Make sure you have fitted fibre washers before tightening up tap connections for connection to pillar taps. Monobloc taps are usually supplied with compression fittings. Mark off the position of the joints between the pipework stabs and prefabricated pipe at the same time as you make the fixing holes.

Remember

Do not fully tighten at this stage because you will need to adjust the bracket

Installing the WC suite

All the components have been installed in the cistern during dressing, and the cistern has been fixed to the WC pan. The installation process is very similar to the washbasin:

- Offer the cistern to the wall, check the level of the cistern and pan and mark the holes through the back of the cistern. At the same time, mark the fixing holes through the base of the pan. If a soil pipe is already in, either remove the pan connector or ease the pan outlet into the pan connector.

Remember

Don't forget to use pilot holes on wooden floors

- Carefully take out the close-coupled suite, drill and plug the holes.

- Refit and screw back the cistern; screw down the WC pan. Again, with the cistern use soft washers, and brass or alloy screws. Some manufacturers supply screws with caps that match the colour of the suite.

- Some plumbers, once they have positioned the pan, drill through the fixing holes rather than removing the close-coupled suite to do the drilling. This is acceptable, but great care should be taken not to damage the suite with the drill, and to make sure the plug is fully located in the fixing holes.

- We have assumed that the hole for the overflow has already been prepared. If not, this should be marked out and drilled before the suite is fixed.

- Most float valves are plastic, so, when connecting the pipework with capillary fittings, do not solder the joint when it is connected to the plastic thread because it will melt and distort.

- Make sure a fibre washer is in place before tightening the tap connector.

- Fit the overflow pipe, which is usually made from plastic.

Shower tray and enclosures

The waste is fitted to the tray during dressing. It will help with future cleaning and maintenance if a removable waste is fitted. The waste trap on a shower tray should always be accessible for maintenance purposes. It is a common mistake made by plumbers not to leave access which can cause real problems and potential embarrassment a few years down the line!

Figure 11.28 Typical shower tray trap

The most important thing to ensure when fitting the tray is that it is level in all directions. If it is not, you will face big problems when fitting the enclosure.

Fixing to timber floors

- First, check whether the trap and waste pipework can be accessed under the floor. If not, you may have to raise the tray in a similar way as for an installation on a solid floor.

- If the pipework can be accessed, make sure the floor is clean and dry and coat the underside of the tray with a recommended sealant; alternatively the tray can be bedded on the tile cement.

- Position the tray and press firmly into place. Make sure the tray is level in all directions.

- Once the bed is dry or cured, the tray edges can be sealed to the adjoining walls.

Fixing to solid floors

On a solid-floor installation, it is likely that you will have to raise the shower tray to gain access to the trap and waste pipe. This is done by building a subframe out of external grade plywood and timber battens. The thickness of the battens will be governed by the amount of clearance you need, but they should be placed at 250 mm intervals. Once you have fixed the subframe, install the tray as you would for a timber floor.

Fitting enclosures

Shower enclosures come in all shapes and sizes, and the best advice here is to follow the manufacturer's instructions carefully. However, here are a few pointers:

- Check the walls are plumb. Most manufacturers' products make allowances for the walls being slightly out of plumb. On older properties, if the wall is seriously out of plumb, some plastering work may be required before you start.

- It is almost certain that the frame of the enclosure will be fixed to a tiled surface, and this will involve drilling the tile.

- Using the manufacturer's measurements for the frame, mark out for the fixing holes. The best way is to mark the wall the full length of the frame, making sure it is plumb. Then offer the frame to the line and mark out for the fixing holes.

- The remainder of the process will involve fitting the panels and the door. Make sure the doors open and close correctly. The installation should be completed by fitting any sealing strips supplied with the kit, and applying a recommended sealant as per the manufacturer's instructions.

Installing the bidet

There is little that is different about the installation of the bidet, other than that it is screwed to the floor using brass or alloy screws. Marking and connection to supply pipework follows a similar pattern to the other appliances.

That completes the installation of our suite.

Did you know?

Acrylic trays often have panels to allow access to the underneath; they do not need to be raised

Safety tip

To avoid the drill sliding over the smooth surface of the tile and causing damage (and possible injury to you), stick a piece of clear tape on the hole markings. This will provide a key for the drill bit as you start to drill the hole. You should manually rotate the chuck to break through the glaze and start the drill without hammer action

Other types of appliances

Installing a washbasin using brackets

There is a range of brackets used for fixing washbasins – mainly in industrial or commercial situations.

Figure 11.29 Selection of wall-mounted basin-fixing devices

When selecting a bracket for the job, you should:

- make sure the bracket is the correct pattern for the basin
- be confident that it will support the weight of the basin.

The normal fixing height of a basin is 800 mm from the floor to the front rim.

It is a good idea when marking out for the brackets to do the following:

- First, mark the centre line on the wall where the basin is going to be fixed.
- Place the basin on the floor with its rim 800 mm from a wall.
- Put the brackets in position on the basin and measure the distance to the fixing holes from the wall, and at the same time, the centres of the fixing holes between the two brackets. Halve that distance, and that will be the measurement from the centre line marked on the wall to the centre of each bracket.
- Transfer the measurements to the wall where the brackets are to be fixed. Position the brackets on the centre lines and mark the holes.
- Drill and fix the brackets, then place the basin in position and check height and level.

Alternatively, if you are working with a mate, mark out the centre line for the basin and the rim height. Then, one person holds the basin in position, checking that it is in line and level. The other person puts the brackets in position and marks the holes.

Countertop basins

Usually, countertop basins are supplied with a cutting template. You must follow this carefully when cutting out the hole. Use a jig-saw blade with fine teeth and a downwards cutting action, as this will reduce any damage to the finish of the worktop.

Once you have cut the hole, seal the exposed edges with a waterproof varnish to prevent any water/moisture damage to the chipboard.

The joint between the lips of the bowl must be sealed using a recommended sealant, and the bowl is then secured using the brackets supplied with the basin.

The edge of the hole for an under-countertop basin is finished off with a strip of laminate in the same material and colour as the countertop.

Wall urinals

These are usually installed at a height of 610 mm from floor level to the front lip of the bowl. Quite often, when a number are installed, one or possibly two will be installed at a height of about 510 mm for children or smaller adults. Dress the urinal bowls first, fitting the inlet spreader and waste.

Urinals are fixed on the type of bracket shown in the illustration, and the measurements can be set out in a similar way to those of washbasin brackets.

Figure 11.30 Urinal bowl fixing brackets

Slab urinal installation

Slab urinals are found mostly in public buildings. In the main they are manufactured in fire clay, and can be supplied to any length on one or more walls, either in one piece or smaller slabs. The floor channel is manufactured so that the internal surface is laid to a fall. This means the actual channel block can be installed level.

Specialist sanitary appliances

WC macerator unit

The macerator unit is installed behind the WC pan. It collects waste from the WC and reduces it to a 'liquid' state so that it can be pumped via a 22 mm internal bore pipe into a discharge stack. There are a few things to remember about macerators.

Figure 11.31 WC macerator unit

- A drain-off valve should be fitted at the base of a vertical rise.

- The macerator can only be fitted when there is another conventional WC in the dwelling.

- The electrical supply to the unit must be via a fused spur connection.

- The unit is capable of discharging its contents to a maximum of 4 m vertically upwards and 50 m horizontally but not at the same time.

- Horizontal pipework must be laid to a fall of 10 mm per metre minimum.
- Waste pipework is typically of 22 mm diameter for short pipe runs. With larger pipe runs, the diameter of pipework will need to be increased in accord with manufacturer requirements. Simple charts are included in the manufacturer installation instructions to make the pipe sizing easy based on the overall length of pipe run.

Food waste-disposal unit

These units are usually installed under kitchen sinks to dispose of refuse such as vegetable peelings, tea bags, left-over food, etc. They require a larger waste outlet than normal, and manufacturers produce sink tops to suit this. The unit should:

- be connected to the electrical supply via a fused spur outlet
- be connected to a waste trap – this must be a tubular P trap as for all kitchen sinks (see page 370)
- be supplied with a special tool to free the grinding blades should they become blocked (keep those fingers clear!)
- have some form of cut-out device that will turn the unit off should it jam.

WC macerator and food waste disposal units will be covered in much more detail at Level 3.

Waste pipe   Waste plug

Jamming key

Reset button

Unswitched fused electrical supply

Tubular trap

Blades

Figure 11.32 Waste-disposal unit

Knowledge check 1

1. What type of fireclay sink does not include an integral overflow?

2. What are the maximum flush capacities for a new dual flushing drop valve in a WC flushing cistern?

3. What are the two main types of wash basin?

4. Name an acceptable sealant material for use with a plastic slotted waste fitting.

5. Which is the most common type of WC pan?

Traps

Traps are used on above ground discharge system pipework and appliances. They are mainly manufactured in plastic (polypropylene to BS EN 274), although they are also available in brass for use on copper pipework, where a more robust installation is required, and can be chromium plated to provide a pleasing appearance.

The purpose of a trap is to retain a 'plug' of water to prevent foul air from the sanitation and drainage pipework entering the room. The depth of this 'plug' of water will depend on where it is to be used. Part H of the Building Regulations includes a table detailing the minimum trap diameter and hence connecting pipe diameter for a range of sanitary appliances. It also includes the requirements for minimum depth of seal.

A selection of plastic traps

Appliance	Diameter of trap (mm)	Depth of seal (mm of water or equivalent)
Washbasin[1] Bidet	32	75
Bath[2] Shower[2]	40	50
Food waste disposal unit Urinal bowl Sink Washing machine[3] Dishwashing machine[3]	40	75
WC pan - outlet <80mm WC pan - outlet >80mm	75 100	50 50

[1]The depth of seal may be reduced to 50 mm only with flush grated wastes without plugs on spray tap basins
[2]Where these appliances discharge directly to a gully, the depth of seal may be reduced to not less than 38 mm
[3]Traps used on appliances with flat bottom (trailing waste discharge) and discharging to a gully with a grating may have a reduced water seal of not less than 38 mm

Figure 11.33 Minimum trap sizes and seal depths

Find out

Why do you think it is necessary to install a trap?

It is vital that you ensure that the minimum trap seal depths are maintained. If you don't follow these requirements then noxious smells and insanitary conditions may be created in the building!

The depth of the trap seal is measured as shown in Figure 11.34:

Figure 11.34 Depth of seal

P and S traps

You can see why they are called P or S traps from their shape (note that they are turned here to show how they fit into the appliance). They can be obtained in tubular design, or with a joint connection which allows a few more options when installing pipework and fittings. The traps can also include a cleaning eye for cleaning and maintenance.

Figure 11.35 P and S traps

Figure 11.36 Swivel elbow

P traps are often used where the waste pipe is installed directly through a wall from the appliance and into a drain or directly into a stack. The S trap is used where the pipe has to go vertically from the trap through a floor or into another horizontal waste pipe from another appliance. P traps (and bottle traps) can be converted to S traps using swivel elbows like the one illustrated. S traps can be a real problem as the fall of pipe from the basin is generally too steep, which can lead to trap seal loss and obnoxious smells entering the building.

Bottle traps

Figure 11.37 Bottle trap

Figure 11.38 Cross-section of bottle trap

Bottle traps are often used because of their neat appearance and they are easier to install in tight situations. Figure 11.38 shows what the trap looks like inside and how the depth of seal is measured. They should be avoided on sinks, and to a lesser extent baths and shower trays where they can be prone to blockage due to the accumulation of food or soap deposits.

Tubular swivel traps

These are particularly useful on appliance replacement jobs, as they allow more options when connecting to an existing waste pipe without using extra fittings or altering the pipework. On new jobs they are often used on sinks with multiple bowls, again because of their multi-position options.

Figure 11.39 Tubular swivel trap

Low-level bath traps

These are designed so that they can fit into tight spaces under baths and shower trays. This type of trap is available with a 50 mm seal depth making it suitable for connection into a soil and vent stack. Be careful – the 38 mm seal depth version is only suitable for waste connection into a ground floor gulley.

Figure 11.40 Low-level bath taps

Figure 11.42 Straight-through trap

Straight-through trap

These are used as an alternative to an S trap where space is limited. They are also easier to hide behind pedestal basins. The main problem with this design is the two tight bends which slow down the flow of water.

The self-sealing valve

The self-sealing valve can be used as an alternative to a trap. It works on the simple principle of using an internal membrane as a seal. The membrane allows water to flow through it when the water is released, then closes to prevent foul air from entering the building. The valve can be used on systems meeting BSEN 12056 – Part 2. It is ideal for fitting behind pedestals and under baths and showers, and is supplied with a range of adaptors so that it can be used in various situations. The valve has the potential to revolutionise the installation of above ground systems, the requirements of which will be covered in the next section. System-design procedures for the straight-through trap will be covered fully at Level 3.

Figure 11.41 Self-sealing valve

Figure 11.43 Running traps

Running traps

You might see these used in public toilets where one running trap is used for a range of untrapped washbasins. On domestic installations it could be used where a P or S trap arrangement is not possible. Running traps are sometimes used with a washing machine waste outlet or for dishwashers, although specialist traps are available for these appliances.

Figure 11.44 Combined sink trap fittings

Figure 11.45 Combined sink/washing-machine trap

Resealing and anti-siphon traps

If an above ground discharge system is designed and installed correctly, loss of trap seal should be prevented. We will look at some of the reasons for loss of trap seals next, but these particular traps are designed to prevent seal loss due to the effects of siphonage (the Hepworth valve provides a very similar function although it is less complicated). These types of traps could be specified or fitted in situations where normal installation requirements cannot be met.

Anti-siphon trap	Resealing trap (sections)
This uses an anti-vacuum valve located on the top of the trap. If the pressure drops inside the pipework, the valve is activated, allowing air to enter the system, to equalise the pressure	This type has a by-pass within the body of the bottle. The dip pipe allows air to enter the trap via the by-pass arrangement. As the seal is lost due to siphonage, air is allowed into the trap, breaking the siphonic effect

Figure 11.46 Resealing and anti-siphon traps

Access

Whatever type of trap is fitted, it is important that you can get to the trap for cleaning. Some traps have cleaning eyes; others can be split at their swivel joints to enable a section of the trap to be removed.

Trap seal loss

Trap seals can be lost as a result of poor practice and by not following the regulations – the detail of which we have just covered.

There are three main causes of loss of trap seal caused by poor practice:

1 self-siphonage

2 induced siphonage

3 compression.

Self-siphonage

This is most common in washbasins, as their shape allows water to escape quickly. As the water discharges, a plug of water is formed, creating a partial vacuum (negative pressure) in the pipe between the water plug and the basin. This is enough to siphon the water out of the trap. Ensuring that the waste pipe is within the lengths allowed and to the correct fall, or that it is ventilated, should prevent self-siphonage. Resealing traps would also avoid this problem.

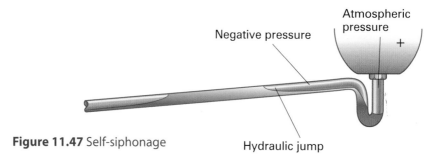

Figure 11.47 Self-siphonage

Induced siphonage

This is caused by discharge of water from an appliance that is connected to the same waste pipe as other appliances. As the water plug flows past the second appliance connection, negative pressure is created between the pipe and appliance which siphons the water out of the trap. This arrangement is not acceptable on a primary ventilated stack. In Figure 11.48, fitting a branch ventilating pipe between the two traps would solve the problem (as would fitting a resealing trap).

Figure 11.48 Induced siphonage

Compression

As the water is discharged from an appliance into the main stack (usually a WC at first-floor level), it compresses at the base of the stack, causing back pressure which can be enough to force the water out of the trap, thus losing the seal.

The use of large-radius bends and minimum 450 mm length between the invert of the drain and lowest branch pipe are defined in the regulations in order to prevent this.

Figure 11.49 Compression

Why traps lose their seal

The main reason for traps losing their seal is bad pipework design and poor installation. However, a trap may lose its seal through natural causes:

Cause	Effect
Capillary action	Only occurs in 'S' trap arrangements, and is not a regular occurrence. If a thread of material becomes lodged as shown in the illustration, water could be drawn from the trap by the effect of capillary attraction.
Momentum	The seal of a trap is removed by force of water. This can happen if you pour a bucket of water into a sink or toilet.
'Wavering'	Wind pressure across the top of the soil and vent pipe, particularly in exposed locations, causes the water in the trap to produce a wave movement and wash over the weir of the trap. This does not happen often, and can be avoided by fitting a 90° bend or a cowl to the top of the vent pipe.
Evaporation	The most common form of natural seal loss. Occurs in very warm ,dry weather when the water in the trap simply evaporates. It is unlikely to occur in traps with a 75 mm seal.

Figure 11.50 Why traps lose their seal

Sanitary systems

So far we have looked at the sanitary ware used in above ground discharge installations and the various types of traps used to prevent foul air entering the building. We now turn to the pipework systems that connect the appliances to the drainage system.

At one time – and you may still come across these on maintenance jobs – pipework systems were two-pipe: the pipework from the WC was separate from that of the washbasin, bath and sink. The wastewater only joined together once it entered the drain, as shown in Figure 11.51.

Today, all pipework systems used on domestic dwellings are one-pipe systems, of which there are four types:

- primary ventilated stack system
- ventilated discharge branch system
- secondary modified ventilated stack system
- stub stack system.

Figure 11.51 External two-pipe system

Soil and vent pipe systems can be installed either inside or outside the building. Whichever system is used, the design and installation must comply with Building Regulations Part H1. The general requirements of Part H1 are that the foul water system:

- conveys the flow of water to a foul water outfall (a foul or combined sewer, a cesspool, septic tank or settlement tank)

- minimises the risk of blockage or leakage

- prevents foul air from the drainage system from entering a building under working conditions

- is ventilated

- is accessible for cleaning blockages.

The other important document regarding sanitary pipework is BS EN 12056 Part 2. This gives guidance on the minimum standards of work and materials to be used.

The primary ventilated stack system is the most commonly used.

The main aim of the design principles we are going to cover is preventing the removal of the trap seal with the result that obnoxious smells enter the building.

Primary ventilated stack system

Sometimes called the single-stack system, this type of system is often specified for domestic dwellings because it costs less in terms of materials and installation time, as it does not need a separate ventilating pipe like other one-pipe systems. This is good for the business and the environment.

Building Regulations Part H1 sets out a number of rules about the design of this system. The rules governing pipe diameter and the minimum depth from the lowest connection above the invert

Figure 11.52 Primary ventilated stack, with full range of domestic appliances

Basin 32 mm diameter

Bidet 32 mm diameter

Bath 40 mm diameter

Cross-flow prevention arrangement

Shower 40 mm diameter

Sink 40 mm diameter

WC 75-100 mm diameter

450 mm min

Washing machine 40 mm diameter

of the drain of 450 mm are shown in
Figure 11.52.

	Pipe size	Maximum length	Minimum slope
Basin	32 mm	1.7 m	18–20 mm fall per metre run
Bath	40 mm	3.0 m	18–90 mm/m
Shower	40 mm	3.0 m	18–90 mm/m
WC	100 mm	6.0 m	18 mm/m

Figure 11.53 Limits to length and slope of pipes.

In addition, there are limits to the maximum lengths of the branch connections and their gradients. These are shown in the table.

In this system it is necessary to have the appliances grouped closely together. There is some flexibility, however: for example, if you did install a shower with a 50 mm waste it could be located up to 4 m away from the stack as opposed to 3 m if using 40 mm pipe. The size of the branch pipes should always be at least the same diameter as the trap. The maximum pipework lengths and gradients may be exceeded but only when additional ventilation is provide into the system using a device such as an anti-siphon trap or a self sealing valve.

Branch connections

The location of a branch pipe in a stack should not cause cross-flow into another branch pipe. **Cross-flow** can be prevented by working to the details shown in the illustration – again the problem we are trying to prevent is trap seal loss.

It is, however, permissible to have connections from two WCs in opposing positions.

You might find on some installations that it is easier to run the kitchen-sink waste pipe into a gully rather than pipe to the stack. This is permissible as long as the pipe end finishes between the grating or sealing plate and the top of the water seal.

Figure 11.54 Cross-flow

Where these conditions cannot be met, separate ventilation will need to be installed.

This can be done in two ways:

1   ventilating each appliance into a second stack – the ventilated discharge branch system

2   directly ventilating the waste stack – secondary ventilated stack system.

Ventilated discharge branch system

Secondary ventilated stack system

Figure 11.55 Ventilated discharge system and secondary stack system

The trap sizes and seals apply as described in the section on traps. The illustration shows what this looks like.

Branch ventilating pipes

A way of overcoming excessively long pipe lengths or steep gradients is to ventilate the pipework system using separate ventilating pipework. This does not happen on many new installations, as anti-siphon traps or self sealing valves are now commonly used as a modern alternative, reducing material costs and installation times.

There are some factors to consider:

● The branch vent pipe must not be connected to the discharge stack below the spill-over level of the highest fitting served

● The minimum size of vent pipe to a single appliance should be 25 mm. If it is longer than a 15 m run or serves more than one appliance, then it must be 32 mm minimum

Figure 11.56 Branch ventilating pipes

Figure 11.57 The stub stack system

- The main venting stack should be at least 75 mm. This also applies to the 'dry part' of the primary vented stack system.

Stub stacks

These reduce the amount of ventilating pipework on a ventilated discharge stack. They also mean that on internal stacks the need for roof weatherings is avoided. The highest waste connection allowed is 2 m above the invert of the drain, and 1.3 m to the base of the WC. The length of the branch drain from the stack is 6 m for a single appliance connection and 12 m if there is more than one. The stub stack is commonly used to connect the ground floor appliances in a property containing more than one bathroom where the high point in the drainage run includes a ventilating pipe to fully ventilate the drainage system.

These dimensions can be exceeded when an air admittance valve is used to permit air to enter the system. The valve, which operates on negative pressure (the open valve lets air in) and positive pressure (the closed valve contains the smells), can be located in the roof space or the pipe-boxing arrangement to the WC, subject to certain requirements.

General discharge stack requirements

All stacks should have access for cleaning and clearing blockages. Rodding points and access fittings should be placed to give access to any length of pipe that cannot be reached from any other part of the system. All systems pipework should be easy to get to in case of repair.

Figure 11.58 An external stack must be terminated as shown

Figure 11.59 A terminal mesh guard should be fitted to prevent the possibility of birds nesting

Figure 11.60 A vent cowl could be fitted where the stack is sited in exposed windy conditions

Knowledge check 2

1 What is the minimum trap seal depth requirement for a bath waste connected to a ground floor gulley?

2 What is the minimum waste pipe diameter for a wash hand basin?

3 An alternative to using a trap is to use a _____.

4 Which type of sanitary appliance can be particularly affected by trap seal loss when fitted with a straight through (S trap) trap?

5 What type of trap can be fitted to overcome trap seal loss due to excessive pipe gradient?

Pipework installation, access and materials

This section overlaps with some of the previous work you have covered, such as materials and methods of jointing. Here, however, we will show how the methods and materials apply to above ground drainage systems. We've also covered preparing for installation in the common plumbing process section, so this will be a bit of revision.

Preparing to install above ground drainage system pipework

You may have to drill or cut your own pipework holes for external discharge pipework. If not, make sure the bricklayer knows in advance where they should be and when they should be cut. On some larger housing jobs, the holes may be left by the bricklayer as the brick/blockwork proceeds, and then made up afterwards.

The installation and jointing of above ground drainage system pipework

Types of material

For domestic plumbing, plastic is generally the material used for above ground drainage systems. For push-fit waste systems, polypropylene to BS 5254 is used. For solvent-welded systems PVC and ABS plastics are used. These meet the requirements of BS 5255. Compression fittings are available which are suitable for both types of pipe. Waste pipe is available in 32 mm, 40 mm and 50 mm diameters, in 3 m lengths.

The most common plastic for soil and vent pipes and fittings is PVC-U to BS 4514, which uses ring-sealed push-fit joints. Pipe diameters for domestic use include 82 mm and 110 mm (this is the outside diameter of the pipe), and they are available in lengths of 2.5 m, 3 m and 4 m, either as pipe ends or single-socket pipes.

Remember

You can refer back to the generic systems knowledge section in Chapter 5. The documentation relevant to the AGD pipework is Building Regulations H1 and BSEN 12056.

Range of fittings

Waste systems

These come in three different types: push-fit, solvent welded and compression. There is an overlap with the design of the fittings, so we will just illustrate the main features here.

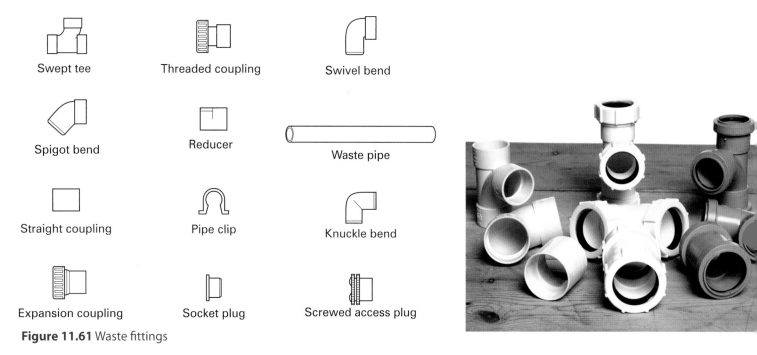

Swept tee

Threaded coupling

Swivel bend

Spigot bend

Reducer

Waste pipe

Straight coupling

Pipe clip

Knuckle bend

Expansion coupling

Socket plug

Screwed access plug

Figure 11.61 Waste fittings

Soil and vent systems

Below is a selection of soil and vent fittings.

Single socket

Double socket bend

Metal pipe clip

Plastic pipe clip

Double socket

Single socket bend

Offset bend

Socket/spigot bend

Figure 11.62 Soil pipe fittings

Access fittings

We said in the last section that providing access to clean pipework and to clear any blockages is a requirement of the regulations. Access fittings are designed to allow this.

Joints to soil and vent pipes

There are a number of options here. As you have seen from the illustrations, some soil and vent fittings are supplied with a boss access already in place, ready to install the waste pipe. Others have blocked ends in the pipe which can be drilled out and fitted with a rubber boss adaptor.

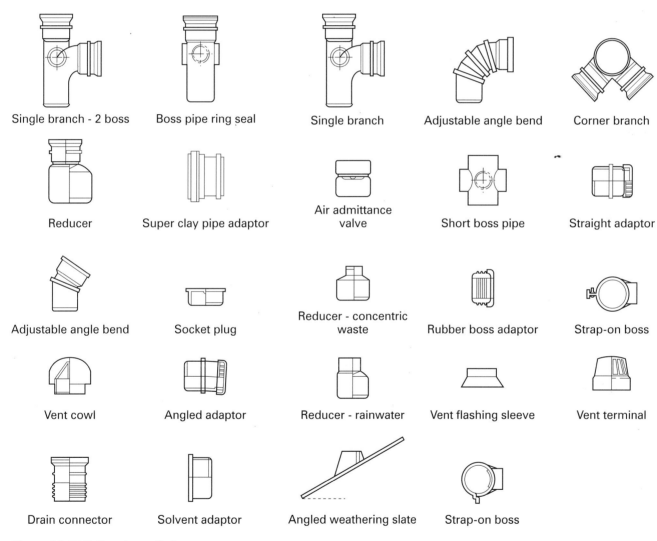

Figure 11.63 Soil and vent fittings

Boss pipe fittings

Sometimes, these fixed connections may not suit the installation requirements. In these cases you can drill your own fixing position in the soil and vent pipe and then fit a strap boss. The best way of doing this is to use a hole saw. Make sure it is the correct diameter for the strap boss.

The strap boss is fixed using gap-filling cement.

Fixing details

Plastic waste pipes should be clipped at the intervals given in the table.

Pipe size	Horizontal	Vertical
32 mm	0.8 m	1.5–1.7 m
40 mm	0.9 m	1.8 m
50 mm	1.0 m	2.1 m

Figure 11.64 Waste-pipe sizes and clipping distances

Soil and vent pipes are normally fixed vertically every 2 metres.

Connections to drains

Plumbers very rarely work on below-ground drainage systems, other than on very large projects. This tends to be a job done by ground workers on large housing contracts.

However, you need to know the basics of the systems and how to connect the soil and vent pipe to the drain, because that *is* a plumber's job. Essentially, you must ensure that **foul drainage** cannot enter the surface water system.

This is because with certain types of system, **surface water** is discharged cheaply via separate pipework to streams, etc. – we cannot afford foul-water drainage to be discharged to such pipework, as it could be a potential health hazard.

There is a simple rule here: always know what you are discharging into – then you can't go wrong.

There are three main types of drainage system:

1   combined system

2   separate system

3   partially separate system.

The combined system

In this system, the water from the sanitary appliances (foul water) and the rainwater all go into one sewer.

> **Definition**
>
> **Foul drainage** – anything discharged from a sanitary appliance – WC, bath, basin, sink, etc.

> **Definition**
>
> **Surface water** – the term used for water collected via the rainwater system

svp = soil vent pipe
rwp = rainwater pipe
fwg = foulwater gully

Figure 11.65 Combined drainage system

The separate system

Here, the foul water runs into a separate sewer, and the rainwater likewise into a sewer for surface water.

Figure 11.66 Separate drainage system

Partially separate system

This system still uses two pipes, but some of the surface water is discharged into a watercourse, soakaway or drainage ditch.

The advantages and disadvantages of each system are shown in the table below.

Figure 11.67 Partially separate system

System	Advantages	Disadvantages
Combined	• Cheap and easy to install • Gets a good flush-out during periods of heavy rain	• More costly to treat the water at the sewage works • At times of heavy rain, inadequately sized drains could overflow • All gullies have to be trapped
Separate	• No need for water treatment of surface water • No need for trapped gullies on surface water drains	• Danger of cross-connections – foul to surface water
Partially separate systems	• Greater flexibility with the system design	• Danger of cross-connection

Figure 11.68 Advantages and disadvantages of different drainage systems

Connecting the soil and vent pipe

On a new job, say on a large building contract, this information will be on the drawings and specification. When you go into the property, it is obvious because all the outlets to the drain will be in place for the soil and vent pipe or downstairs WC.

On a replacement job, the connection will be the same as the existing soil and vent pipe, so you will use that.

The soil and vent pipe are likely to be connected to a drainage pipe. Fittings are made to connect the two. Some drainpipe terminations, particularly on older systems, are finished with a collar. Drain connectors, such as the one illustrated, are inserted into the collar and a joint is made using a sealing compound. This is finished off with a sand and cement joint of a 1 to 4 mix.

Figure 11.69 Drain connector

Knowledge check 3

1   What is the most common type of sanitary pipework system?

2   What is the minimum vertical drop across 2 m of waste pipework to a bath?

3   What is the minimum distance between the lowest branch connection to a soil stack and the invert level of the drain in a two-storey dwelling?

4   The top of a vent pipe to a soil stack should be fitted with which type of device?

5   PVC-U for use in sanitary pipework systems is manufactured to which British Standard requirement?

Soundness testing

Sanitary appliances discharge large volumes of waste water, so it is important that systems are watertight. The test should be carried out in accordance with BS EN 12056, which you'll look at soon. You also need to know what to do to make sure people are aware that testing is going on, and when it is safe for them to use the appliances.

Soundness testing of above ground drainage systems

Figure 11.70 shows a test kit. Look at it and then read the testing checklist.

Hand bellows

Test nipple

Water gauge hose

Assembled plug

Hollow plug

Plug testing Y piece

Figure 11.70 Test kit

Testing checklist

- First, seal the system using hollow drain plugs. Fix a test nipple in the one to be used for the test. The bottom test plug can be inserted through the access cover at the base of the stack. If this proves awkward, a testing bag can be inserted and inflated in position, as shown in Figure 11.70.

- Fill the traps with water, and cover the test plugs to make sure they are fully airtight

- The rubber hose, bellows and hand pump are connected to one end. Air is pumped into the system to give a water head of 38 mm. Once this is reached, the plug cock is turned off and the test continued for three minutes

- Where a pressure drop is found, all the joints should be tested using leak detection fluid

- Smoke tests are also useful to locate leaks, but are not advisable because of the chance of affecting the plastic pipework and rubber seals

Water maintaining seal above test plug

Drain test plug

Manometer

38 mm pressure head

Hand pump

Traps filled with water

Hand pump

Testing bag

Access cover

Air test in action

Figure 11.71 Different tests that can be done on a system

Charging procedures

Once the testing equipment has been removed, all the appliances should be filled to their overflow levels and water released; the WC should be flushed at the same time.

Once you have done that, check that the trap seals are not less than 25 mm deep. This is done as shown in the diagram.

Figure 11.72 Performance testing trap

Figure 11.73 Depth of seal

Remember the depth of seal is measured as shown in Figure 11.73.

On the job: Testing an installation

Jenni was asked to test and charge the installation shown here.

1 Label the diagram

2 How should she go about the job? Include the equipment she should use, and how she should make sure the trap seal is adequate.

Knowledge check 4

1   During air testing, what device is used to measure the air pressure reading in the system?

2   What piece of test kit is used to raise the pressure in the system to the required level?

3   What piece of test kit can be used to plug a soil pipe where access to the internal pipework system to make a watertight seal proves awkward?

4   What is the pressure head (in millimetres) water gauge that the system is required to be tested to during a soundness test?

5   During testing for trap seal loss, what is the minimum depth of water seal that must be retained in the trap during simulation of the most adverse operating conditions?

Rainwater and guttering

Gutters are used to collect rainwater that falls on the roof of a building. It flows down the rainwater pipe and into surface-water drains, combined drainage systems or soakaways. Life without gutters and rainwater pipes would see damage to gardens from water running off the roof, staining on brickwork and water getting into houses, leading to such problems as rising damp.

Materials

Most gutters and rainwater pipework systems installed on new domestic properties tend to be plastic. It has the advantages of being:

- light
- flexible
- cheap
- low-maintenance.

It is available in white, black, brown and various shades of grey. Because PVC gutters and rainwater pipes have a smooth internal surface, they provide a better flow rate than other materials.

Plastic has a high expansion rate, so it can be affected by the variations of winter and summer temperatures. Allowance for expansion and contraction is made in the fittings, so it is important that manufacturers' fitting and fixing instructions are followed.

Did you know?

Plastic gutters and rainwater pipes do not require painting, making them very low maintenance

Other materials

You are quite likely to come across cast-iron and asbestos cement guttering on maintenance jobs.

Cast iron

This is a very strong material, but it requires constant painting to prevent rusting. Leaking joints also take longer to repair/replace than their plastic counterparts, as it takes time to remove the gutter bolt fixings which tend to be corroded and difficult to unscrew. More often than not they have to be cut off using a hacksaw.

Cast iron may still be specified on jobs where its strength is needed, or on listed buildings where original materials have to be used.

Asbestos cement

Asbestos cement gutters were quite popular before the introduction of PVC. They are no longer available because of the dangers of working with asbestos.

If you are called to a maintenance job, you must follow the procedures as outlined in Chapter 2 *Health and Safety*. Asbestos guttering should not be repaired, but removed and replaced with plastic – with all safety procedures being followed.

Gutter and rainwater systems

Gutters are specified by their shape in cross-section.

Guttering fittings

Rainwater pipes are either square or round. Guttering is supplied in 2 m and 4 m lengths, and can be obtained in a range of sizes from 75 mm to 150 mm; 112 mm is in common use for domestic dwellings, with rainwater pipes in lengths of 2.5 m, 4 m and 5.5 m lengths. For most domestic jobs, 65 mm square and 68 mm external diameters are used.

Standard half round   Square   Ogee

Figure 11.74 Types of guttering

Fitting rainwater systems

On new work, the gutter will be fixed to **fascia** boards using a fascia bracket. The fascia is the piece of wood (or plastic) that is fixed to the rafter, as shown in Figure 11.75. A fascia bracket is also shown.

On some older properties, fascia boards may not be fitted. On these jobs, metal rafter brackets like the ones shown in the following illustrations can be used for metal and plastic gutters.

Basic fascia bracket

Union bracket

Angle

Running outlet

External stop end

Internal stop end

Cast iron ring

Figure 11.76 Guttering parts

Figure 11.75 Fascia fixed to rafter

Wind blows water behind gutter if underfelt too short

Wind

Underfelt should be left long enough to lap gutter

Soffit

Fascia

Figure 11.77 Gutter brackets

Step-by-step fitting guide for plastic guttering fixed to fascia boards

- The first job is to set out the gutter brackets. Gutters can be laid level or to a slight fall. The fall should be 1 in 600, which for a length of 6 m gutter would be 10 mm – this is hardly noticeable once the gutter is fixed in position.

- Once you have worked out the fall over the total length of the installation, the first bracket is fixed at the highest level on the run, and the last bracket is fixed to give the amount of fall required.

- A string line is fixed between the two brackets. This is used to position the rest of the brackets, which are usually spaced at 1 m centres, unless the manufacturer states otherwise.

Figure 11.78 Positioning the brackets

Remember

Make sure the roof felt is positioned in the gutter

Did you know?

Don't get confused! Rainwater pipes are called down pipes or fall pipes in different parts of the country

- The gutter can then be positioned into the brackets. On the first length, depending on the roof shape, you would fit the stop end (or corner bracket) and union bracket.

- The next joint would be done in position, as shown in Figure 11.79.

- Make sure you allow for thermal movement. On a plastic gutter, a raised fixing mark is usually on the inside joint. If expansion marks are not on the gutter joints, you should allow 3 mm for every metre run.

- Once the gutter system is complete, you can start on the downpipe. The brackets are marked out using a plumb line. This is dropped from a masonry nail driven into the nearest mortar joint beneath the running outlet. The brackets are centred with the plumb line and the wall marked through the fixing holes.

Locate back edge of gutter union bracket

Pull front edge down until level wih lip of bracket

Snap the gutter under the lip of the bracket

Figure 11.79 Fixing the gutter

- You usually have to install an offset (known as a swan neck), to clear the width of the fascia and **soffit**. The easiest way to do this is to install the top section of downpipe with the first offset on, placing the second offset against the outlet of the running outlet, and taking the measurements in position.

- The fall pipe is finished off at the base with a rainwater shoe or connection to the drain.

- Clip spacings should be every 1.5 m and allowance should be made for thermal movement: 6–8 mm should be left from the end of the pipe to the inside shoulder of the fitting. Clips should also be placed where pipes are jointed.

- All screws used for fixing should be alloy or stainless steel so that they do not rust

- Cast-iron down pipe and gutters should be cut using an angle grinder.

The sizing of gutters will be covered at Level 3.

Figure 11.80 Assembled rainwater pipe

Safety tip

Always follow all safety precautions when using power tools

Maintenance and decommissioning

In previous maintenance sections, we have talked about the importance of regular maintenance and cleaning in order to make sure the system works to maximum efficiency. Above ground drainage systems are no different, and while these systems do not have as many controls as, say, central heating systems, if they are not looked after they will eventually deteriorate.

As a plumber, you may be called out to a blocked drain or discharge pipe, so we will look at that as well.

Figure 11.81 on page 394 shows some equipment for unblocking drains, but remember there is much more sophisticated equipment now available.

Basic maintenance and cleaning

In domestic dwellings this is usually restricted to the manual cleaning of pipework. Traps often accumulate hair, soap residue, toothpaste and other objects that are small enough to fall through the grid of the waste hole. Integral overflows are also prone to this, particularly on Belfast sinks. Traps should be cleaned through the access points, if fitted; if not, the trap should be broken at its joints or removed completely.

Overflows can be rodded with wire and flushed. Chemical cleaning agents are also available that can be used for both traps and overflows. When using cleaning agents always follow the manufacturer's instructions.

Access covers to soil and vent pipes should be checked to make sure they are operating properly, and a visual inspection should be made of waste traps and fittings for signs of leakage.

Flexible pipe drain cleaner

Universal

Lockfast

Black rods with metal couplings

Universal

Lockfast

Blue rods with brass couplings
(riveted to the end)

Rodding eye

Clearing wheel

Disk plunger

Double worm screw

Drop scraper

Plunger

Figure 11.81 Equipment for unblocking pipes and drains

Figure 11.82 Drain auger

Blocked pipes and drains

Blockages in discharge pipework on sinks, washbasins and baths can often be cleared using a 'force cup' or plunger.

This is done by filling the appliance with water. Make sure you have blocked the overflow with a cloth or similar, then press down repeatedly on the handle of the force cup. This creates a positive pressure on the blockage, and a partial vacuum when it is withdrawn. This is usually enough to remove the blockage. Always check the trap for signs of leakage when you have finished.

Access to blocked drains can be through a drainage inspection chamber or, on newer jobs, through a rodding eye installed for that purpose.

Flexible metal pipe and sink cleaners are also available, which work like 'mini' drain-cleaning rods.

A blocked WC can be cleaned using a drain rod with a rubber plunger. These can also be used to clear blocked drainage pipework. Sections of the rods are joined together until the location of the blockage is reached.

Additional attachments are available. A double worm screw is used for pushing through obstructions or for catching an object (e.g. a cloth) and pulling it back to the operator. A drop scraper is a tool with a hinged steel blade: it is flat when travelling forwards but the blade is lifted when the rods are being pulled back. This can be used for pulling rubbish back to the operator. Clearing wheels help by guiding rods around bends or over obstructions.

In the event of a more serious blockage a drain auger may be used. Simple drain augers may be manually operated by hand. More sophisticated devices, primarily for soil and drain runs, may be electrically powered. High pressure jetting of the system may be undertaken in cases of serious blockage. This job is usually carried out by a specialist contractor who has been fully trained in its use. The cost of the kit is expensive as well!

Gutters should be checked regularly and cleaned as required. They do collect silt residue off roof tiles, leaves and bird droppings. The silt can build up and becomes a garden for seeds carried by the wind. Plastic gutters can be stripped in lengths between joints, emptied into a bucket and wiped clean. Cast-iron gutters should be painted to prevent rusting.

Leaking joints on cast-iron stacks can be cleaned out and remade using non-setting pipe-jointing compounds. Leaking cast-iron gutter joints can be repaired by removing the old joint, thoroughly cleaning the jointing surfaces of the gutter and fitting, applying jointing compound, and fitting a new gutter bolt.

Decommissioning systems

Decommissioning above ground drainage systems will normally involve stripping out old appliances and pipework to replace with new.

When removing appliances, care should be taken not to damage them.

Removing a cast-iron bath, particularly if it has to be carried downstairs, requires careful handling. Some plumbers break the bath into four pieces for easier removal. A club hammer is the best tool for doing this, and it should be done wearing full-face protection and gloves. Once the appliance has been removed it should be stripped of any scrap metal to be taken for recycling.

On externally mounted pipework, you will have to chop out the mortar between the pipe and the masonry. Take care in doing this: it will mean less making good after you have installed the new pipework.

Taking down cast-iron soil and vent pipes can be dangerous due to their weight, so they need careful handling. It is best to try and take down short sections by partially cutting them with an angle grinder and then tapping the pipework with a hammer, which will cause it to shear; a rope should be tied to each section in turn so that they can be lowered to the floor. Make sure no one is in the area where you are working. Fixing lugs can be broken from the joint, and the nails prized out using a wrecking bar.

Once the stack is removed, make sure the joint to the drain is covered or the joint capped to prevent anyone tripping on it, prevent debris from entering the drain or prevent obnoxious smells coming from the drain – a range of blanking plugs for decommissioning work is invaluable.

Don't forget to use rubber gloves when handling old lengths of sanitary pipework and sanitary appliances, as you could develop a health problem if you came into contact with the contents. Also be careful when dealing with blocked WCs and blocked pipework as they can be a collection point for potentially dangerous items, such as hypodermic syringes deposited by drug users or incorrectly disposed-of feminine hygiene products. Don't just go exploring with the hands; make a visual inspection first!

Safety tip

Broken vitreous china has edges as sharp as broken glass, so it should be handled wearing thick gloves

Remember

Old pipework systems could be in lead, so you must take the usual precautions when handling this material

Safety tip

Historically, many soil and vent pipes were made in asbestos – always follow the handling rules for asbestos

Remember your personal protective equipment – always wear a hard hat for these jobs!

FAQs

Why is so much fuss made over correct installation of sanitation systems?

It is perhaps the most widely abused system for incorrect installation as people often believe that waste pipes are not very important anyway. In reality it is very important, as when things go wrong it causes major disruption and nuisance, not to mention the safety and hygiene problems that can arise.

Do I really need to follow all the installation guidelines and measurements given in the Building Regulations?

Yes, you do. There are regulations that MUST be followed in all cases, anyway, and it is in all of our best interests to have sanitation systems that do not become faulty, for obvious reasons.

Why does my loo fail to flush easily? It takes several sharp presses of the lever handle in quick succession to get it to flush.

It is almost certainly a worn diaphragm, the thin plastic washer in the siphon has torn or become displaced. Simply renew it for a return to normal operation.

Why is such a fuss made over allowing rainwater into a foul drain? It is relatively clean rainwater anyway so why not?

The extra quantity of rainwater will place an unnecessary burden on the sewage system as it may not be designed to take this excess, and treatment of sewage is an expensive process, which we all pay for.

Knowledge check 5

1   What are the three main types of gutter profile?

2   What type of gutter bracket is normally used to make fixings to a timber surface?

3   What is the minimum recommended vertical drop from one end of a 12 m length of gutter to the other end?

4   In a combined system of drainage, into what section of drainage pipework should the surface water discharge?

5   What type of unblocking device is normally used first to remove a blockage in a WC pan?

Additional information sources

Sources of additional information on this chapter include:

Building Regulations Approved Document H – www.planningportal.gov.uk
Range of resources on the Bathroom Manufacturers Association – www.bathroom-academy.co.uk/guides
Armitage Shanks/Ideal Standard Blue Book – www.thebluebook.co.uk
Hepworth Building Products (plastic pipe manufacturer) – www.hepworth.wavin.com
Geberit (plastic pipe manufacturer) – www.geberit.co.uk
Hunter Plastics (plastic pipe manufacturer) – www.hunterplastics.co.uk
Wavin Plastics (plastic pipe manufacturer) – www.wavin.co.uk

chapter 12

Sheet lead work

OVERVIEW

Plumbers on-site are sometimes required to carry out weatherings to chimneys, soil-vent-pipe penetrations through roofs and weatherings between building surfaces such as outbuildings. This is done by fabricating or working sheet lead, either by bossing, or by welding it together using oxy-acetylene equipment.

This chapter covers:

- **Basic principles:**
  - types of sheet lead
  - codes and sizes of sheet lead
  - characteristics of sheet lead
  - tools and equipment used for sheet lead installation.

- **Fabrication techniques:**
  - safety factors when working with lead
  - sheet lead bossing
  - sheet lead welding
  - fire safety when working with lead
  - welding equipment and safety

- **Fixings to sheet lead components**

- **Weatherings to abutments**
  - apron flashings at abutments
  - abutment flashings with soakers and step flashings
  - abutment step and cover flashings to single lap tiles
  - abutment flashing with secret gutter

- **Chimney weathering sets**
  - installation requirements of a chimney weathering set

- **Lead slates**

- **Testing sheet weatherings**

- **Maintenance**

Basic principles

The covering of complete roofs, and the weathering of building details in sheet lead have traditionally been the job of the plumber. Over recent years, much of this work has been undertaken by specialist roofing contractors. However, plumbers still carry out sheet weathering in lead on domestic dwellings. This usually takes the form of:

- simple abutment flashings
- chimney flashings
- lead slates.

Types of sheet lead

Sheet lead is categorised by the way it is manufactured. There are two types:

- Cast sheet lead – this was the original production method for sheet lead, and it is still made this way as a craft operation. Production, by specialist lead-working firms, is in relatively small amounts created by running molten lead over a bed of sand. There is no British Standard for this material, and its sheet sizes and thicknesses vary – it tends to be used on historic monuments.

- Rolled (milled) sheet lead to BS EN 12588 – formed by passing a slab of lead back and forth on a rolling mill between two closing rollers, until it has reduced to the required thickness. The sheet is then cut to a standard width ready for distribution.

Rolled sheet lead is used by plumbers for sheet weatherings. It is manufactured to British and European Standards and will not vary in thickness by more than 5 per cent at any given point. The standards also set down requirements for its chemical composition.

Codes and sizes of sheet lead

BS EN 12588 lays down the production standards for sheet lead, in particular the thickness (code of the material) and hence its approximate weight per square metre. Plumbers use the code number shown in the table on the next page to order the thickness of sheet lead required for a particular job.

Sheet lead for flashings is supplied in rolls in widths from 150 mm to 600 mm in steps of 30 mm: 150 mm, 180 mm and so on. It is usually supplied in 3 m or 6 m lengths. Larger width and lengths are available.

Generally speaking:

- Code 3 sheet lead is normally used for producing soakers.

- Code 4 sheet lead is normally used for producing lead-welded components and in some lighter lead bossing applications.

- Code 5 sheet lead is normally used for lead bossing applications.

Don't worry too much about these leadworking terms as they will be covered in greater detail later in the chapter.

Did you know?

Rolled sheet lead was first used as an alternative roofing material to cast sheet lead at the turn of the twentieth century

Did you know?

The Lead Sheet Association (LSA) produces detailed technical guidance on the use of rolled sheet lead on the outside of buildings

BS 12588 code no.	Thickness (mm)	Weight (kg/m²)	Colour code
3	1.32	14.97	Green
4	1.80	20.41	Blue
5	2.24	25.40	Red
6	2.65	30.05	Black
7	3.15	35.72	White
8	3.55	40.26	Orange

Figure 12.1 Rolled sheet lead codes

Characteristics of sheet lead

The following characteristics of sheet lead make it ideal as a weathering material:

- malleability
- fatigue and creep resistant
- thermal movement
- durability
- corrosion
- patination
- fire resistant
- 'recyclability'.

Malleability

Lead is the softest of all plumbing materials. It is malleable and easily worked, using hand tools and a process called 'bossing', into the most complicated of shapes, or 'dressed' to fit the many types and shapes of roof tiles.

Fatigue and creep resistant

Fatigue is a loss of strength in the lead due to thermal movement, which eventually leads to cracking. The strength of lead relies on the grain structure of the metal. The chemical composition of lead is governed by BS EN 12588, which effectively controls the grain structure to make the lead sheet more resistant to thermal fatigue without affecting malleability.

Creep describes the tendency of metals to stretch slowly over the course of time. Making sure you correctly size and fix individual pieces will reduce the risk of creep.

Thermal movement

Lead has a high coefficient of expansion at 0.0000297 for 1°C. It is important therefore to include regular expansion joints in lead flashings to allow for expansion and contraction due to changes in temperature. Flashings may also be secured into a wall by lead wedges, thus restricting its movement so expansion joints would be needed here as well.

Did you know?

Before metrication, lead was defined by its imperial weight per square foot, i.e. 3lb, 4lb, etc. These weights are where the present day code numbers originate

Definition

Malleability – the ability to be worked without fracture

Did you know?

Creep does not mean that a piece of lead has slipped down a pitched roof when the fixings have failed!

Durability

When specified and fitted correctly, sheet lead will provide maintenance-free weather protection for many years. It is extremely resistant to atmospheric corrosion, and will withstand severe weather conditions – hot and cold.

Corrosion resistant

Sheet lead is resistant to most forms of corrosion found in a roofing situation. However, you need to heed the following precautions:

Abutment flashing

- Mortar – unprotected lead damp proof courses and cavity trays may corrode in the presence of some mortars.

- Lichen growth – the acid run-off from lichen or moss on a roof may cause small holes to appear in lead sheet under the drip-off point from tiles or slates. A 'sacrificial' flashing may be fitted to the lead gutter. The other option is to treat the growth with a chemical fungicide.

Patination

With time, lead develops a strongly fixed patina (or sheen) which is silver-grey and won't dissolve. However, in rainy or damp conditions, new lead sheet flashings will produce an initial, uneven white carbonate on the surface. Not only does this look unsightly but the white carbonate can be washed off by rain, causing further staining of materials (e.g. brickwork) below the flashings. The use of patination oil, applied evenly with a cloth as the job progresses, will prevent this.

Fire resistant

Lead is incombustible, but melts at 327°C.

'Recyclability'

This is an important property of any material. Sheet lead is totally recyclable. In the UK scrap lead is recovered from buildings due for demolition by a national network of reclamation merchants. It is then returned to manufacturers, where it is carefully refined for re-use in the rolling mills.

Because of this well established recovery network, the sheet lead industry has a solid reputation for environmental awareness.

Definition

Abutment flashing – used where a pitched roof needs weathering against a brick wall

Did you know?

Lead can corrode very quickly when brought into contact with aluminium in seaside locations because of the high level of salt in the atmosphere

Tools and equipment used for sheet lead installation

Your Certificate in Plumbing and NVQ will require you to be assessed in fabricating and fixing sheet-lead components. The illustrations here show what these components look like.

You fabricate these components either by lead **bossing** or lead **welding**.

Definition

Bossing – the term for working sheet lead into shape using a range of mallets and dressers

Welding – the jointing of sheet lead materials by the application of heat to the two surfaces to be jointed, normally with the addition of lead filler material

Figure 12.2 Chimney weathering set detail

Figure 12.3 Lead slate detail

Lead to brick chimney

Lead slate in position

The lead bossing tools are illustrated here. Originally made of boxwood, the main tools are now available in durable plastic.

Figure 12.4 Lead bossing tools

	Name	Use
1	Dresser	To make the metal lie flat on a surface (remove any humps)
2	Bossing stick	To boss the lead into shape; used with a mallet
3	Bending stick	To boss the lead into shape; used with a mallet
4	Setting in stick	To reinforce or sharpen fold or angles
5	Mallet	To boss the lead in conjunction with the other bossing tools
6	Chase wedge	For setting-in the corner, a fold or a crease
7	Snips	To cut the lead sheet to shape (straight or curved cuts)
8	Plugging chisel	To remove mortar from existing brickwork joints; used with a club hammer
9	Club hammer	To remove mortar from existing brickwork joints; used with a plugging chisel
10	Lead knife	To cut lead sheet to the required shape (straight cuts)
11	Spirit based marker pen	To mark out the sheet lead for cutting
12	Ruler	To measure sheet lead dimensions (a folding ruler is commonly used as this can also measure the angle of components to be cut)

FAQ

Why is lead work in the qualification? Many plumbing firms don't do it.

The industry required sheet lead to be included so that all trainees become skilled in the basics of a chimney weathering and a lead slate. This means that wherever you go to work in the country, employers will know that these skills have been achieved. There will be further courses for those who then wish to expand their range of lead skills in a range of advanced work.

Knowledge check 1

1. What is the ability of sheet lead to be worked into various shapes known as?

2. What is the British Standard number for rolled sheet lead?

3. By how much will a piece of sheet lead 1.5 m long increase in length when heated by 40° C?

4. Which organisation produces detailed technical guidance on the use of rolled sheet lead?

5. Which leadworking tools are normally used to make a straight cut to a piece of sheet lead?

Fabrication techniques

In this section you will look at what is required to form sheet lead into the various shapes we refer to as flashings. It can be done in two ways: bossing and welding.

Most sheet lead work on larger sites will be done using welding techniques. There will be occasions, however, when you might not have access to welding equipment, and you will need to know how to shape the lead by bossing.

Safety factors when working with lead

Refer back to the lead safety section in Chapter 2 *Health and Safety* before continuing.

Sheet lead bossing

We will now look at the way the tools used for bossing are actually used.

Lead at ambient (ordinary surrounding) temperatures is only 300°C below its melting point: compare this to copper – at ambient temperatures it is 1056°C below its melting point. You can therefore imagine that, in many ways, lead behaves at lower temperatures in the same way as harder metals at higher temperatures.

Lead is an outstanding metal for bossing because:

- it is the softest of the common materials
- it is ductile – it will stretch quite a lot before fracturing or splitting
- it does not harden much when it has been worked (work hardening).

Bossing techniques

Bossing is not a skill that can be learnt overnight, but once you have mastered it, you will get a lot of satisfaction from the finished job. The best way to learn it is by watching an experienced lead worker, and then practising.

Did you know?

The term 'bossing' has its origins in the Middle Ages, when 'boss' meant, 'to beat out metals into a raised ornament'.

Definition

Annealing – the process of softening something by heating it. As it has such a low melting point, ambient air temperature has an annealing effect on lead

We are going to look at two techniques for bossing an external corner and an internal corner. Figures 12.5 and 12.6 show what the finished jobs look like.

Figure 12.5 An internal corner to an abutment flashing

Figure 12.6 An external corner to a chimney front apron

The main aim of bossing is to achieve the required shape without making the lead too thin or too thick.

Internal corner, step by step

Mark out internal corner

Setting in/starting

Part complete

Nearing completion

Trimming away surplus

Final job

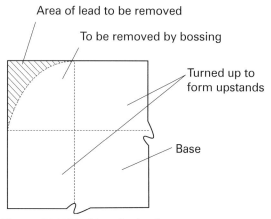

Figure 12.7 Marking the lead

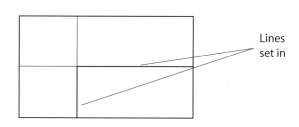

Figure 12.8 Lines to 'set in'

1 Using a spirit-based marker pen, straight edge and tape or ruler, set out the corner. Check the dimensions with a square.

2 Turn the upstands up by 90° using a timber former. A piece of timber about 600 mm long × 100 mm wide × 50 mm thick is ideal.

3 The angles are then 'set in' using the setting in stick. This is done by placing the blade of the setting in stick on the line and tapping it with a mallet. This process fixes the position of the upstands, whilst the bossing is in progress.

4 To get a good square base to the corner and to fix its position, use the mallet to raise a slight groove in the base of the sheet.

Did you know?

When forming an internal corner by bossing surplus lead is produced and this needs to be removed by cutting

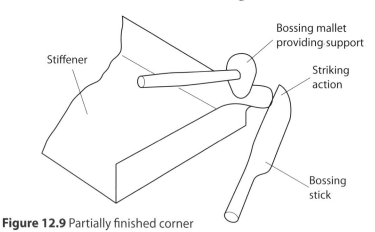

Figure 12.9 Partially finished corner

5 Lead is then worked as shown in Figure 12.9, using a bossing mallet on the inside of the corner and a bossing stick.

6 As the process goes on, surplus lead builds up as shown in Figure 12.10; this can be trimmed off if it gets in the way.

7 Keep checking that the corner is flush and square.

8 Check the dimensions of the upstands, when finishing off, mark off any excess material and trim off with the snips.

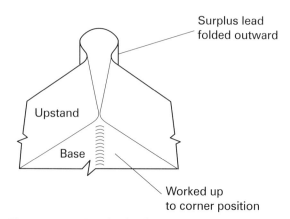

Figure 12.10 Surplus lead

External corner step-by-step

Marked out front apron

Starting bossing

One third completed front apron

Two-thirds completed front apron

Final stage front apron

Completed front apron in-situ

The principle is the same as for the external corner, that is moving surplus lead around. The marking out is slightly different, however.

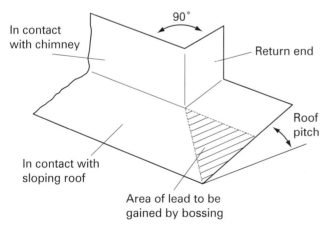
Figure 12.11 External corner – lead gain

Figure 12.12 Bossing an external corner

- Use your marker pen, tape measure, and straight edge as for the internal corner.

- For an external corner, such as the apron on a chimney breast, the lead has to be gained as shown in Figures 12.11– 12.12.

- The first job is to fold the upstands to the angle of the roof pitch.

- Trim the return end to the angle of the chimney breast (usually 90°). It should now look something like Figure 12.12.

- As shown, a hump is formed which is worked into the desired position using a dresser, bending or bossing stick.

Finally, a couple of pointers about bossing sheet lead. As the job proceeds and the surplus lead is removed, there's a tendency for it to thicken and crease. If creases are allowed to form they will cause cracking. When bossing an external corner, you have to watch that you do not stretch the lead into position. This will thin the lead, and could cause splitting.

On the job: Lead safety

Matt's company specialises in leadwork, particularly on historic buildings, and has just won a contract to replace the lead on the roof of a local church. Matt is given the job of stripping out old sheet lead prior to replacement. The lead is heavily corroded. The vicar is not keen on letting workmen inside the church so won't give Matt's company access to the vestry, which has the only source of running water on the site.

1   What items of PPE should Matt's company provide before Matt starts work?

2   Should Matt ask his employer to insist that the vicar allows the workers access to the running water supply in the vestry? If so, why?

Lead welding

Lead welding is a process of joining two pieces of lead by melting the edges of the lead together (called the parent metal) while a filler rod of lead is added. It is called **fusion welding**. The technique can be used to form an internal or external corner, as an alternative to bossing them into shape.

Figures 12.13 and 12.14 show what the finished jobs look like.

Figure 12.13 External corner to chimney front flashing

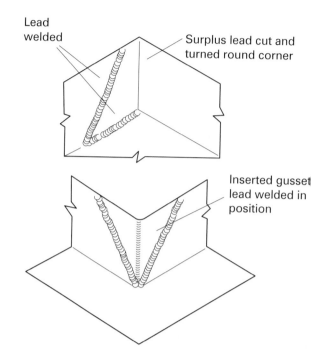

Figure 12.14 External and internal corner

Setting out the external and internal corner for welding

External corner

Marking out

Folding

Tacking

Welding

Nearly complete

Completed welded external corner

External corner

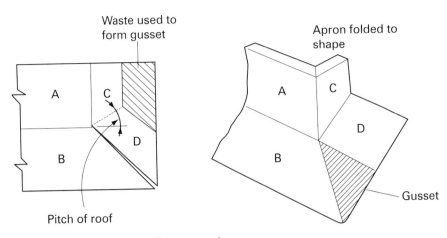

Figure 12.15 Setting out of an external corner

Figure 12.15 shows the setting out of an external corner. It will be used to form part of a front apron.

The apron will be marked out as shown and the joint cut between (B) and (D). The apron is worked into the required position and a gusset cut to fit as shown by the shaded area. This fits flush with the edges of the metal at (B) and (D) and is called a **butt joint** because the two pieces of metal are butted together.

Internal corner

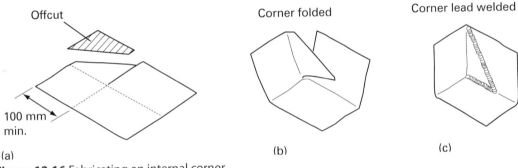

Figure 12.16 Fabricating an internal corner

- Mark out the corner in the same way as the bossed corner, but this time cut the lead as shown in Figure 12.16 (a).

- Fold the corner as in Figure 12.16 (b)

- This creates an overlap on the two upstands and it is the diagonal lap that will be welded as in Figure 12.16 (c). This type of joint is called a **lap joint** because the two pieces of metal overlap each other.

The welding process

Some plumbers still refer to it this as lead burning, a term that dates from when crude welding techniques were used for jointing lead.

Like lead bossing, this is a very skilful job that will take time to master. Once you have done so, however, it is a very rewarding aspect of the job.

There are two types of lead welded joint:

- butted seam
- lapped seam.

Butted seam

Lapped seam

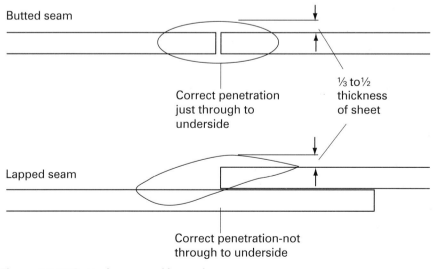

Figure 12.17 Butted seam and lapped seam

Lead welding tools and equipment

You will need:

- Oxy-acetylene welding equipment with a lead-welding torch. These torches have interchangeable heads numbered 1 to 5. In general, nozzle sizes 2 and 3 are used for lead sheet codes 4 and 5. For upright seams, size 1, 2 or 3 nozzles would be used depending upon the lead's thickness.

- snips

- flat dresser

- shave hook

- steel ruler.

Figure 12.18 Lead welding nozzle

Creating the welds

- It is essential that the meeting points of the sheet lead and its faces (points where the weld adheres to the sheet) are clean and any oxide removed. This cleaning process is carried out using a shave hook.

- Mark a width of about 10 mm and, using your metal straight edge as a guide, shave the surface of the metal with the shave hook.

- Cut strips to use as filler rods. This is done by cutting a thin strip of lead about 3–5 mm thick and 300 mm long. Again it needs to be shaved clean. Alternatively, lead rods up to 6 mm can be obtained from a supplier.

Now everything is prepared, you can set up the welding equipment.

Now light the blow pipe:

1 Turn on and light the acetylene first.

2 Then feed in the oxygen. A pressure of 0.14 bar (2lb/sq in) for both the oxygen and acetylene is required.

3 You need to achieve a **neutral flame** to get the best results. An incorrectly adjusted blow pipe will result in either a carburising or an oxidising flame being produced, both of which are unsuitable for lead welding.

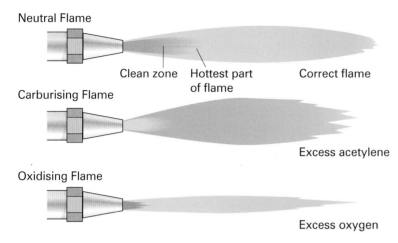

Figure 12.19 Blow pipe flames

Lead welding – good practice

✓ Flat-butted seam welds should fully penetrate through the thickness of the lead sheet.

✓ Lapped seam welds should penetrate the surface of the lead, but not through to the underside.

✓ The thickness of the seam should be between a third and a half thicker than the sheet.

✓ The width of the weld will depend on the thickness of the lead and the seam pattern. When using code 4, the minimum width of a flat-butted seam should be 10 mm.

Beware of **undercutting** – reducing the thickness of the lead at the side of the weld. This causes a weakness that can result in cracking along the weld line. The main cause of undercutting is holding the flame for too long on the vertical surface.

Safety tip

Whether jointing lead using a butted or lapped seam, you should tack the pieces together. This prevents any movement and keeps the surfaces in close contact

Figure 12.20 Examples of undercutting

Welding flat-butted seams

This seam is used for jointing pieces of flat supported lead sheet as shown.

Figure 12.21 Joining flat supported lead sheet using butted seams

Figure 12.22 Position of blow pipe

The position of the blow pipe and filler rod should be as in Figure 12.22.

- The tip of the cone of the flame should be just clear of the molten lead.

- Lead is melted off the welding rod into the weld area.

- A seam that is between one-third and one-half the thickness of the sheet is built up.

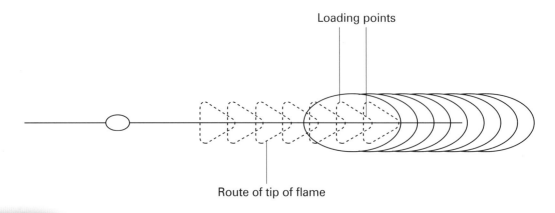

Figure 12.23 The flame is moved forwards to set the pattern of the seam

- The flame is directed into the centre of the seam and is moved forward, either in a straight line or slightly from side to side. This will set the pattern of the seam.

- The weld should just penetrate through the underside of the lead.

Welding a flat-lapped seam

This technique is an alternative to butted seams. It is often preferred when working on-site where there is a risk of fire during the welding process, because the flame will not make contact with the material beneath the lead. Even experienced plumbers will admit that this technique is slightly easier.

The illustration shows a typical example of a welded flat-lapped seam used on lead slate.

The preparation and process of welding a lapped seam is the same as for a butted seam, the only difference is that when welding thicker lead two loadings are used.

Figure 12.24 Welded flat-lapped seam

Inclined seam on a vertical face

If you can, pre-fabricate sheet lead components before fixing them. There will be occasions, however, when you may have to weld a joint in position. One such situation is an inclined seam on a vertical face.

Although this joint can be made without using filler rods – by using the overlapping edge to make the seam – it is recommended that a filler rod should be used.

Figure 12.25 Inclined seam

Fire safety when working with lead

If working on-site, always try to use lapped joints. This reduces the risk of fire which exists when making a butted seam against combustible surfaces.

On occasions, butt joints may be unavoidable. When this is the case, you should:

- wet the timber area beneath the weld

OR

- place a non-combustible material beneath the weld.

Remember

Pre-fabricate lead components before fixing wherever possible

Welding equipment and safety

The equipment

Below is a diagram of a typical welding kit.

Figure 12.26 Welding kit

FAQ

It seems absurd that we buy a roll of lead and then cut it into short lengths only to weld it together again, surely we could fix it in one long length in the first place.

Not at all, lead expands and contracts with every temperature change so there must be maximum lengths and an overlap time to allow for essential free movement to take place. If it was fixed in long lengths it would cause metal fatigue and cracking would occur.

When using a welding kit, some basic safety procedures *must* be observed.

Safety checklist

1 Gas cylinders should be stored in a ventilated area on a firm base. If possible store oxygen and acetylene separately. Empty and full bottles should also be stored separately

2 Acetylene gas bottles should be stored upright to prevent leakage of liquid

3 Oxygen cylinders are highly pressurised. They should be stored and handled carefully to prevent falling. If the valve is sheared, the bottle will shoot forward with great force

4 Keep the oxygen cylinder away from oil or grease as these materials will ignite in contact with oxygen under pressure

5 Check the condition of the hoses and fittings. If they are punctured or damaged replace them. Do not try to repair or piece them together

6 Do not allow acetylene to come into contact with copper. This produces an explosive compound

7 Make sure the area where you are welding is well ventilated

8 Erect signs or shields to warn and protect people from the process

9 Always have fire fighting equipment to hand

10 Wear protective clothing: gloves, overalls, goggles – clear are fine for lead welding

11 Make sure that hose check valves are fitted to the blow pipe and flashback arrestors to the regulators. This prevents any possible flashback on the hoses and the cylinders

12 Allow the acetylene to flow from the nozzle for a few seconds before lighting up

13 In the event of a serious flashback or fire, plunge the nozzle into water, leaving the oxygen running to avoid water entering the blowpipe

Remember

Eye protection must be worn at all times when lead welding - it is a key safety point

Fixings to sheet lead components

Abutment fixings

Flashings to abutments are fixed into position using a turned in section of lead cover flashing that is secured in position at the brickwork, blockwork or stonework joint (seam). The joints in the brickwork will therefore need to be 'raked out' to a depth of at least 25 mm prior to the sheet lead component installation. On new properties a friendly builder will often leave the joints raked out. For existing properties raking out will normally be carried out using a club hammer and chisel.

The sheet lead is secured into the brickwork joints using lead wedges.

Figure 12.27 Lead wedges

Lead wedges are strips of sheet lead between 20 to 25 mm in thickness that are folded several times to a dimension that is slightly thicker than the joint. One edge of the lead is flattened slightly to form the wedge which is then driven into the joint between the sheet lead turn-in and the upper part of the joint to secure the lead in position.

Lead wedges are normally spaced at a distance of 300 to 450 mm between each other dependent on the building material.

With standard brickwork/blockwork joints up to 18 mm the joint is normally pointed with a mastic type sealant. These sealants are either polysulphide or silicone based. If a silicone sealant is used, it must be **neutral cure,** as the acid cure variety can cause corrosion on the surface of the sheet lead. Any pointing material selected must be suitable for pointing sheet lead materials. Guidance can be sought from the material manufacturer.

Making fixings to wide joints

There are occasions when plumbers come across masonry walls such as stonework with joints wider than 18 mm, in which case a different fixing method is identified.

Here the sheet lead is turned up the back of the joint and fixed using corrosion resistant screws and washers (normally stainless steel) positioned at approximately 450 mm centres. As an alternative large headed copper or stainless steel nails may be used.

Mortar is normally used as the sealing material, but to overcome excessive cracking caused by thermal movement, a masking tape is applied as shown in Figure 12.28.

Masking tape

25 mm minimum

Figure 12.28 Making fixings to joints wider than 18 mm

Fixings to free edges of sheet lead abutment flashings

The free edges of raking (sloping), or horizontal abutment flashings, requires fixing. This is normally carried out using clips. The number of clips to be used is dependent on the degree of expose of the roof surface:

- severe exposure – every tile joint
- moderate exposure – one clip at every second tile joint
- sheltered location – one clip at every third tile joint.

The maximum clip spacing distance is normally no greater than about 500 mm.

Definition

Terne coating - stainless steel that is hot dip coated with a lead–tin alloy

Soft copper only

50

Copper or stainless steel

Figure 12.29 Application of clips to single-lap tiling

Clips used at raking (sloping) abutments may be manufactured from copper or terne-coated stainless steel whereas horizontal flashings should only be fixed using soft copper clips. The clips are usually about 50 mm wide and soft copper of 0.6 mm thickness is normally used.

Knowledge check 2

1   When producing an internal corner by lead bossing, lead is said to be _____ from the corner?

2   When producing an external corner by lead welding, what is the name of the piece of sheet lead inserted into the joint?

3   Lead welding is carried out using which type of flame?

4   Acetylene cylinders are which colour?

5   Lead wedges are produced from strips of lead approximately how wide?

Weatherings to abutments

Apron flashings at abutments

Where a pitched roof joins on to, or abuts, a wall, this should be weathered with lead sheet. The lead sheet should have a minimum upstand of 75 mm, a turn into the brickwork joint of 25 mm and the apron should be extended down the roof to at least 150 mm over the tiles or slates (this dimension should increase to 200 mm with a low pitched roof or exposed locations). Code 4 sheet lead is normally used to form an abutment flashing.

Figure 12.30 Apron flashing at abutment with lean-to roof

The lead is wedged into the brickwork joint at maximum centres of 450 mm. The maximum length of a single piece of sheet lead used with an apron flashing is 1.5 m. If the length of sheet lead is greater than this distance then one or more lap joints will need to be used. Detail is shown in Figure 12.30. If a lap joint is required, then the minimum overlap distance is 100 mm.

Abutment flashings with soakers and step flashings

a. Soakers

L = Gauge + lap + 25 mm

75
100
L

Top edge turned over tile

Gauge

Lap

b. Marking-out step flashing

85
65
1500 MAX
Water line

c. Fitting

Lead wedge

Figure 12.31 Abutment flashing with soakers and step flashings

This type of flashing is used with slated or double-lap tiled pitched roofs. The flashing includes a **soaker** and separate step flashing arrangement.

- Separate soakers (produced from code 3 sheet lead) are provided based on the tile or slate details. The formula for calculating the length of soaker is length (L) = gauge + lap + 25 mm. The minimum soaker width is 175 mm with a 75 mm upstand and a 100 mm projection onto the roof.

- The top edge of the soaker should be turned over the tile to prevent slippage occurring.

The step flashing details (produced from code 4 sheet lead as a minimum) should ensure a minimum cover over the soaker upstand of 65 mm. This is known as the water line measurement and is the minimum height of coverage to guard against capillarity. The maximum length of each piece of step flashing is 1.5 m and where lap joints are necessary, the minimum overlap should be 100 mm. The steps are secured in position using lead wedges.

Water line

Lead wedge

65

65

150-200

Edge clipped
to suit exposure

Figure 12.32 Step-and-cover flashing over single-lap tiles

Abutment step and cover flashings to single lap tiles

With this type of weathering, no soakers are used. The step and cover flashing is formed in one piece of sheet lead. The minimum width of sheet lead coil to be used is normally 300 mm. The lead should project at least 150 mm onto the tiled surface. With deeply profiled roofs, or those with a pitch below 25°, it may be necessary to project 200 mm onto the tiled roof surface. The free edges of the lead will need to be clipped to prevent wind lift.

Code 4 is the minimum thickness of sheet lead to be used for this type of step-and-cover flashing. However, code 5 may be necessary when dressing over deeply profiled tiles.

Abutment flashing with secret gutter

A secret gutter may be included in the roof design to protect against surcharging of rainwater and the risk of blockage from leaves. It may be used with a single step-and-cover flashing arrangement or a separate soaker-and-step flashing arrangement. The gutter lining pieces must not be greater than 1.5 m in length, after which lap joints will need to be used.

Continuous
(or single step
flashing)

65

15 mm
max

75

Welt

Gutter depth: 25 mm min

Figure 12.33 Abutment flashing with secret gutter

The use of a saddle at an abutment

The type of flashing to be used at a ridge to an abutment is known as a saddle. The saddle can be formed by either lead bossing or welding. The saddle should extend at least 150 mm along the ridge and at least 150 mm down the roof on each side.

Welted edge

Ridge tile

Figure 12.34 Bossed lead saddle at an abutment

Knowledge check 3

1   What is the minimum upstand dimension to a lead apron piece to an abutment?

2   When determining flashing dimensions what is the water line measurement?

3   At what minimum distance should a soaker project under a roof?

4   What is the formula for calculating the length of a soaker?

5   What is the name of the sheet lead component used at the ridge to an abutment?

Remember

Remember in your practical test, all dimensions and side flashings will depend on what is available in your centre in terms of templates. We have made up our own dimensions for this example

Chimney weathering sets

In this section we will look at how to apply the knowledge you have gained from the previous sections. The Level 2 Certificate in Plumbing requires you to competently work on installing chimney weatherings, so they will be the main focus of this section.

Installation requirements of a chimney weathering set

The job we describe here is the one that you will do in the sheet lead practical for your NVQ in Plumbing. We also show the side chimney flashings details for both slate and tile.

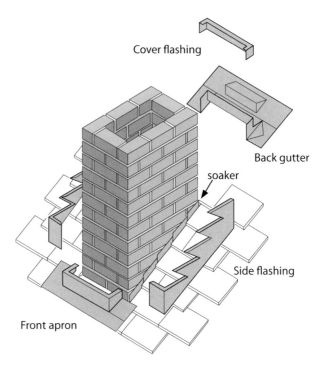

Cover flashing

Back gutter

soaker

Side flashing

Front apron

Figure 12.35 Chimney weathering

Front apron

The lead joint in the brickwork should be at least 75 mm above the surface of the tiles or slates, and an extra 25 mm added to turn into the mortar joint. The side of the apron needs to be turned at least 100 mm around the side cheek of the chimney. A lead welded front apron is normally manufactured using code 4 sheet lead. Code 4 sheet lead may be used to manufacture a lead bossed apron although code 5 is commonly used.

Figure 12.36 Front apron

The bossed part of the apron

The Lead Sheet Association recommends that the piece of lead used for the joint apron should be not less than 300 mm wide – 150 mm for the upstand against the chimney plus 150 mm for the apron over the tiles. With a low-pitched roof this apron distance over the tiles should be at least 200 mm.

The length of the piece will be the width of the chimney plus a minimum of 150 mm for each side. If the roof is covered with deeply contoured tiles increase this to 200 mm. This will form the basis of the dimensions for this example.

The bossed corner is set out and bossed as shown in a previous section on fabrication techniques.

The lead welded part of the apron

Obviously, this will be set out based on the same dimensions as the bossed section. You have already covered these marking out and fabrication techniques above.

Side flashings

There are two applications for side flashings:

- side flashings using soakers – these are used on roofs covered with slate or double lap plain tiles
- side and cover flashings (combined) – these are used where it is not possible to incorporate soakers, such as over contoured tiles.

Soakers (if required) are produced in exactly the same manner as previously described for abutment flashings. Soakers are manufactured from code 3 sheet lead whereas cover flashings are normally manufactured from code 4 sheet lead.

Setting out the step flashing

Let's assume a width of 150 mm for the step flashing. The length will be the full length of the side of the chimney. For the overall length you need to add a minimum of 75 mm for turning around the front of the chimney and over the apron.

- A water line of 65 mm is marked on the length of the lead before offering it against the brickwork.

Figure 12.37 Marking out for a lead bossed front apron

Figure 12.38 Marking out for a lead welded front apron

Side flashings

Figure 12.39 Step flashing

- The lead is placed in position at the side of the chimney and, using a folding ruler, lines are marked that correspond to the bottom of the mortar joint.

Blade of ruler lined up with joint

Allowance to turn on to front of stack

Ruler

Total width 150 mm

65 mm water line

65 mm

Figure 12.40 Setting out step flashing

- The lead is then marked out and fabricated as shown.

Setting out step flashing

Marking out step flashing

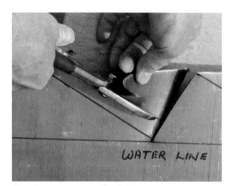

Cutting out step flashing

When installing step flashing on slates or double-lap plain tiles, the allowance for each side of the chimney is 150 mm. If the roof is covered with single-lap tiles, a measurement of 200 mm is required.

Side and cover flashings

Flashings for both sides and cover require the same techniques as setting out a simple side flashing, but now, because the lead will be extended over the roof covering to a width of not less than 150 mm, the overall width of lead will be greater.

In addition, the piece of lead that is to be turned around the joint of the chimney will have to be fabricated. This can be done by either bossing or lead welding. The details are illustrated here.

Figure 12.41 Marking lead for step flashing

Figure 12.42 Bossing and welding around the joint of a chimney

Back gutter

This forms the final part of the chimney weathering 'jig saw' and a finished gutter is shown in Figures 12.43 and 12.44.

A method is shown here of forming the back gutter by lead welding, a much easier process than bossing this complicated shape. Code 4 sheet lead is normally used for a lead welded back gutter.

Setting out

When installing a back gutter on slates or double-lap plain tiles, the allowance for each side of the chimney is 150 mm. If the roof is covered with single-lap tiles, a measurement of 200 mm is required.

Once cut, the back gutter is folded into shape and the gussets inserted and welded.

The final part of the back gutter installation is to fit a cover flashing over the upstand of the back gutter.

Figure 12.43 Welding a back gutter

Figure 12.44 Completed chimney weathering with back gutter

A minimum of 100 mm over the width of the chimney is allowed on each side for trimming around the corners, and the bottom of the flashing is left about 5–10 mm from the gutter base.

Knowledge check 4

1   What is the minimum turn-in distance of a chimney front apron into a masonry joint?

2   What is the minimum width of sheet lead coil to produce a lead bossed front apron?

3   What is the minimum upstand dimension for producing a soaker?

4   What is the maximum length of a step-and-cover flashing piece?

5   What is the name of the sheet lead component installed at the rear of the chimney?

Lead slates

The photograph shows what a lead slate looks like in position. Lead slates can be formed by bossing, but are much quicker and cheaper to fabricate by lead welding.

This type of flashing is used where a pipe penetrates a roof covering. For most domestic dwellings this is likely to be the plastic soil and vent pipe. A lead welded slate is normally produced using code 4 sheet lead.

The size of the base will vary depending on the roof covering. A typical slate for a 100 mm pipe will be 400 mm wide. The base should extend 150 mm from the front, and be not less than 100 mm under the slate. The height of the upstand should not be less than 150 mm.

Lead slate in-situ

Figure 12.45 Pipe penetrating lead slate

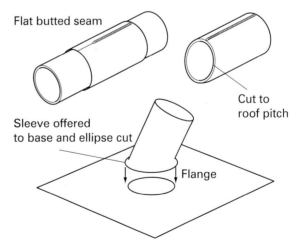

Figure 12.46 Cutting and fitting the upstand

Making the slate

Cut a piece of lead wide enough to give the height of the upstand, and as wide as the circumference of the pipe plus about 5 mm for tolerance. For a 100 mm pipe this is calculated as follows:

$$3.142 \times 100 \text{ mm} = 314 \text{ mm} + 5 \text{ mm for tolerance}$$

$$= 319 \text{ mm (round to 320 mm)}$$

The edges of the upstand are prepared for a butt weld, and then the lead is turned around a rigid pipe and butt welded.

One end of the upstand is cut to the pitch of the roof. This can be done using a bevel and taking the actual angle from the roof. Another method is to 'develop the piece' using a drawing – we will look at this technique now.

Once the upstand is cut to the required angle, the edge is dressed to form a flange. This is then placed on the base, a hole is marked and cut, and the upstand and base are prepared and welded together

> **Remember**
>
> The circumference of a circle = π D, where D = the diameter

Development of the slate piece

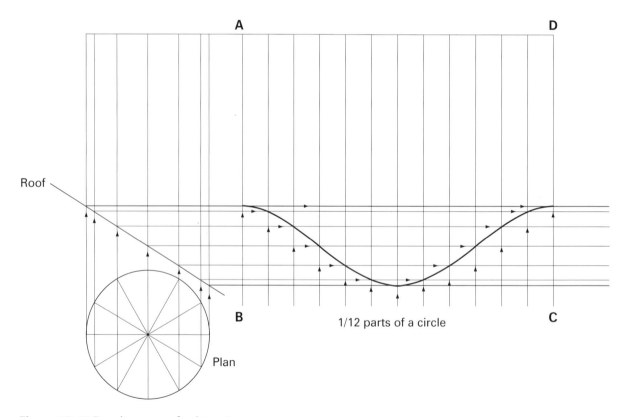

Figure 12.47 Development of a slate piece

- Draw the elevation of the slate piece
- Using a 60°–30° set square, draw plan and divide circumference into 12 equal parts
- Project points from the plan up on the elevation to give an intersection at the roof line
- Draw a rectangle ABCD, so that AB is a least equal to the height of the slate piece
- Following the arrows, project horizontal lines from the intersection at the roof line and vertical lines from dimension on AB to obtain the intersection marked by a cross
- The true shape of the slate is obtained by joining the intersecting points by hand.

Did you know?

You can try making sleeves from card for a variety of roof pitches and these can then be used as templates for various jobs

Testing sheet weatherings

When completed, a sheet lead installation needs to be tested to ensure that it is water-tight. The easiest way is to apply water to the completed installation using a hosepipe or bucket. Then check the roof internally for any leaks.

Maintenance

Maintenance does not form a large part of the work on the components that we have looked at in this chapter, but you do need to establish the basic procedures for checking systems that have been installed. Areas covered should be based on the following:

- Establish adequate safety procedures for completing work, including safe roof access and working platforms.
- Check internal roof space to establish whether there was any water penetration around the components.
- Check key components for size dimension (establish that the components were installed correctly and in particular could not give rise to capillary attraction).
- Check roof covering adjacent to components – repair/rectify as necessary.
- Check individual roof components for soundness – no splits/cracks, etc. – repair/rectify as necessary.
- Check component fixings to ensure soundness of fixing – refix (where required and re-pointed), check to establish whether additional fixings were necessary in the event of wind lift.
- Test components for leakage on completion of maintenance activity.

FAQs

What is the white staining often seen on a roof or brickwork below any chimney/leadwork?

This is white carbonate which can form on untreated lead once it is exposed to the weather. It happens as a result of not applying a patination oil to all surfaces of the lead as work progresses. This white carbonate is very unsightly and detracts from a good quality job.

Why does the correct thickness of lead used on jobs matter?

The recommendations for thicknesses of lead are based on sound historical practice and research, You must follow these LSA guidelines if you hope to achieve a good long-lasting job.

Knowledge check 5

1   A lead welded lead slate is normally manufactured from what thickness of sheet lead?

2   What is the minimum upstand piece dimension at the rear of a lead slate?

3   What is the minimum width of sheet lead required to form an upstand piece to a 150 mm pipe assuming 5 mm of additional material is required?

4   What type of joint is used to joint a lead slate upstand to the base plate?

5   What is the most likely cause of daylight being seen at the front edge of a chimney breast weathered by sheet lead?

You should now be ready to complete the multi-choice questions contained on the CD to assist with preparation for your end of unit examination on sheet lead work.

Additional information sources

Sources of additional information for this chapter include:

Lead Sheet Association – www.leadsheetassociation.org.uk
British Lead Mills – www.britishlead.co.uk

chapter 13

Environmental awareness

OVERVIEW

Environmental awareness is an understanding of why it is important to conserve energy, dispose of waste properly and prevent wastage of materials. Environmental awareness is currently very topical; the UK government is keen to promote initiatives which help to conserve energy and reduce waste.

Building Regulations have been improved and updated to ensure that buildings are more energy efficient and that building materials are more effectively used.

In this section you will look at some of the ways you can help the environment in your work as a plumber.

- **The importance of energy conservation**
  - legislation

- **Applying environmental awareness**
  - energy efficiency and heating and hot-water systems
  - other methods of improving efficiency
  - methods of reducing waste

- **Environmental hazards**
  - waste disposal

- **Customer advice**

You will have realised by now, particularly from the hot and cold water and central heating chapters, just how important energy conservation is to the plumbing industry. What you do when choosing materials and components for systems and how well you install them will have an effect on the environment.

Water Regulations, British Standard specifications and Building Regulations are in place for the public health and welfare of the consumer. They also ensure that working practices, system design and the use of materials are of the highest standards in order to maximise energy conservation.

In this chapter we will look at the legislation and some of the organisations that are involved with energy and our industry.

The importance of energy conservation

Energy conservation is aimed at reducing the CO_2 emissions into the atmosphere, which have a detrimental effect on the ozone layer, leading to global warming. This needs to be checked, and government has tackled this by bringing in legislation to make buildings better insulated and heating appliances more efficient. Outside of the building industry, vehicles are now taxed on the basis of the amount of CO_2 they produce when burning fuel. Ensuring that energy is conserved is also important in terms of ensuring that we have sufficient fuel available for use in years to come. There is a drive to move towards the use of renewable energy sources such as wind and solar for our longer term energy needs.

Energy generation (by power stations) is a major source of CO_2 emissions and there are a number of initiatives aimed at producing 'clean' energy, including:

- *Wind farms* – large windmill type structures that produce electrical power as the blades rotate.

- *Solar-powered hot-water heating systems* – large panels with a system of pipes located behind a glass panel; the water in the pipes is heated by the sun and transferred to the hot-water storage cylinder.

- *Solar photovoltaics* – photo electric cells set in panels use the sun's power to produce electricity.

- *Micro CHP* – CHP stands for combined heat and power. Micro CHP is utilising the heat produced via a heat generator such as a sophisticated boiler to also generate electricity as a by-product of the heating process. This type of technology is currently very new to the marketplace but is predicted to grow in the next few years as the drive to be

1 Solar collector
2 Hot water storage
3 Auxiliary heater

Hot water

Cold water supply

Figure 13.1 Solar heating system with indirect twin coil cylinder energy efficient increases. It is envisaged that Micro CHP appliances may be produced to cover a range of fuel types, for example gas, oil and solid fuels.

- *Ground source heat pump (GSHP)* – GSHPs transfer heat from the ground into a building to provide space heating and, in some cases, to pre-heat domestic hot water. They operate on a very similar basis to a fridge – but in reverse. For every unit of electricity used to pump the heat, 3–4 units of heat are produced. As well as ground source heat pumps, air source and water source heat pumps are also available. There are three important elements to a GSHP:

1 The ground loop – This is comprised of lengths of pipe buried in the ground, either in a borehole or a horizontal trench. The pipe is usually a closed circuit and is filled with a mixture of water and antifreeze, which is pumped round the pipe absorbing heat from the ground.

2 A heat pump – This has three main parts:
  - the evaporator, similar to the device in the cold part of your refrigerator, takes the heat from the water in the ground loop
  - the compressor moves the refrigerant round the heat pump and compresses the gaseous refrigerant to the temperature needed for the heat distribution circuit (this is what makes the noise in a fridge)
  - the condenser gives up heat to a hot water tank which feeds the distribution system (this is the hot part at the back of your refrigerator).

3 Heat distribution system – Consisting of underfloor heating or radiators for space heating and in some cases water storage for hot-water supply.

Solar panels are increasingly being used as alternative energy sources

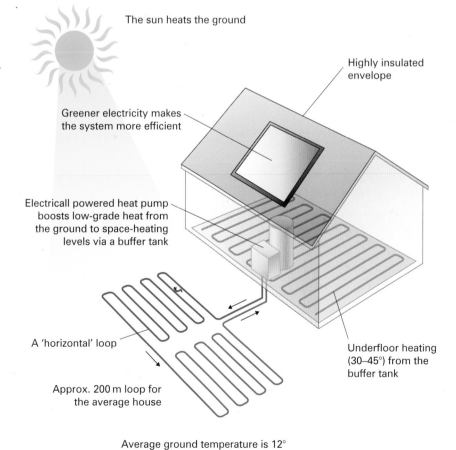

The sun heats the ground

Highly insulated envelope

Greener electricity makes the system more efficient

Electricall powered heat pump boosts low-grade heat from the ground to space-heating levels via a buffer tank

A 'horizontal' loop

Approx. 200 m loop for the average house

Underfloor heating (30–45°) from the buffer tank

Average ground temperature is 12°

Figure 13.2 Ground source heat pump

- *Biomass* – This is biological material produced from living, or recently living organisms. Biomass for energy purposes often means plant-based material (such as trees), but biomass can equally apply to both animal- and vegetable-derived material such as faecal matter from animals. In terms of wood, which is the most common biomass source that a plumber is likely to come into contact with, it takes carbon out of the atmosphere while it is growing, and returns it as it is burned. If it is managed on a sustainable basis, biomass is harvested as part of a constantly replenished crop. This is achieved either through woodland management or as part of a continuous programme of replanting with the new growth taking up CO_2 from the atmosphere at the same time as it is released by combustion of the previous harvest. There are five basic categories of biomass material:

  - *New wood* – from forestry, tree-growing activities or from wood processing
  - *Energy crops* – high yield crops grown specifically for energy applications such as pelleted wood-chips burnt in a biomass boiler
  - *Agricultural residues* – residues from agriculture harvesting or processing, such as animal waste
  - *Food waste* – from food and drink manufacture, preparation and processing
  - *Industrial waste* – from manufacturing and industrial processes.

Renewable or non-renewable fuel?

Fuels also fall into two categories:

- *Non-renewable fuels* – These are natural resource that cannot be re-made, re-grown or regenerated as quickly as they are consumed. Essentially, non-renewable fuels exist in a fixed amount that is being consumed or used up faster than it can be made by nature. The Earth's population cannot rely on the ongoing use of non-renewable energy sources in the long term as they are becoming depleted. Fossil fuels such as coal, oil and gas are non-renewable resources, as they do not naturally re-form at a rate that makes the way that they are used sustainable. This also includes the use of fossil-based fuels to produce electricity.

- *Renewable fuels* – Renewable energy effectively uses natural resources such as sunlight, wind, rain, tides and geothermal (underground) heat, which are naturally replenished at a rate that is sustainable with our rate of consumption. Renewable energy solutions include solar power, wind power, hydroelectricity and biomass fuels. Biomass wood fuels fit into the category of a renewable fuel but only when the wood is produced sustainably, as highlighted in the section on biomass fuels.

The government has also brought in measures to minimise waste, as this is also harmful to the environment, requiring energy to transport it and treat it. Recycling is a way of reducing waste by returning excess and used materials back into the production cycle. A typical example is the scrap sheet lead mentioned in Chapter 12.

Legislation

Part L1 Building Regulations

Approved Document L1: Conservation of Fuel and Power

The Building Regulations 2000 deals with the conservation of fuel and power. *Approved Document L1: Conservation of Fuel and Power* lays down requirements relating to dwellings. There are two parts covered by these regulations:

- new dwellings (L1A)
- existing dwellings (L1B).

They were brought in because it was obvious that big improvements were required to improve the energy efficiency of existing buildings. The following aspects of existing buildings now have to meet the same standards as new buildings:

- glazing
- upgrading of boilers and systems.

Inefficient boilers will require higher levels of insulation for the building.

One of the major requirements of the legislation as far as plumbing is concerned is the certification of heating and hot-water systems to show that they've been correctly installed and commissioned, and the provision of operating and maintenance instructions for users.

Why the need for change?

Everyone has probably heard about the greenhouse effect and global warming due to carbon emissions. These measures aim to:

- reduce carbon dioxide emissions from buildings by 25 per cent

For new dwellings the main changes include:

- big improvements to U values, meaning higher levels of insulation. The lower the U value, the lower the heat loss through the building fabric
- higher standards of insulation for dwellings
- certification of heating and hot water systems to show that they have been correctly installed and commissioned, and that operating instructions have been left for the user
- higher standards of design and workmanship to improve building performance, reduce gaps in insulation, thermal bridging and poor air-tightness (e.g. gaps around windows and door frames, etc.) – air-tightness testing is now mandatory for dwellings
- all new buildings to have a SAP (Standard Assessment Procedure)

Did you know?

The Domestic Heating Compliance Guide produced by the Government Department – Communities & Local Government includes full technical details of the minimum energy efficiency requirements of heating and hot water systems that must be followed

Definition

The U value – a measure of the thermal transmission of heat through the fabric of a building

Find out

Further information about the Regulations can be obtained at www.communities.gov.uk/

- minimum levels of boiler efficiency (requirements that also now apply to existing systems)

- minimum requirements for central heating control systems (requirements that also now apply to existing systems).

The Standard Assessment Procedure (SAP)

SAP is the government's 'Standard Assessment Procedure' for energy rating of dwellings; it is now a compulsory component in Part L1 of the Regulations. Every new house has to have a SAP rating.

- SAP provides a means of estimating the energy efficiency performance of a dwelling. SAP ratings are expressed on a scale of 1–100. The higher the number, the better the rating.

- SAP is also used in the calculation of the carbon index, which can be used to show that dwellings comply with Part L1.

- SAP ratings are used to predict heating and hot-water costs. These depend on the insulation and air tightness of the house and the efficiency and control of the building's systems and controls.

- The procedure for calculating SAP is shown on SAP worksheets, although in practice most people calculate SAP using a computer program approved by the Building Research Establishment. The SAP calculation is based on:
  - the size of the house
  - the insulation levels
  - ventilation levels
  - heating and hot-water systems.

- Once calculated for new dwellings, or new extensions to dwellings, the SAP rating can be submitted for Building Regulations approval and is checked by the local building control department.

SAP rating of a building is undertaken by an individual who has passed a recognised SAP assessors course. Plumbers will often use the details provided from an SAP assessment of a prospective new building in order to design an energy conscious heating or hot water system.

Most of the details with which the heating and hot water system will need to comply are contained in the *Domestic Heating Compliance Guide* supporting Part L1 of the regulations which we'll look at on page 440 in this chapter.

Did you know?

SAP ratings are similar to the miles per gallon fuel consumption figures for cars

Water Regulations

The Water Regulations, discussed in earlier chapters, are relevant to energy conservation. They are discussed in some detail at Level 3. The purpose of the Water Regulations is to:

- prevent contamination of a water supply
- prevent the waste of water
- prevent the misuse of a water supply
- prevent undue consumption of water
- prevent erroneous measurement (fiddling the meter).

It is clear that two of these areas have a direct impact on energy conservation.

Waste of water

Plumbers see this in the form of dripping taps, leaking joints and overflows on both hot and cold water systems. A dripping hot tap means that the hot water lost is replaced by cold water, which then has to be re-heated, wasting boiler energy. There are around 18 million homes in the UK: if a leaking hot tap lost an average of just one half litre of water a week in each dwelling, overall 9 million extra litres of water would have to be heated.

A dripping tap represents a waste of water and energy

In the case of cold taps and cistern overflows, wasted water has to be replaced, which on a wider scale means additional treatment and distribution by water companies, using extra electrical and mechanical energy.

Water can also be wasted through burst pipes brought about by freezing – due to a lack of insulation or pipe runs being in exposed conditions.

Undue consumption

Excessive consumption can be caused by bad design, and in particular dead legs. This results in high volumes of cold water being run off before hot water arrives at the tap. Water Regulations also make recommendations about the insulation of hot-water pipes and storage vessels, which is another way to reduce heat loss and decrease the load on the boiler.

Knowledge check 1

1. Which type of gas is mainly responsible for global warming?

2. The method of converting energy from the sun into electricity is known as _____.

3. Provide one example of a biomass type fuel.

4. Provide two examples of non-renewable fuel supplies.

5. What are the two main aims of the Water Regulations that seek to improve our use of water?

Requirements of the *Domestic Heating Compliance Guide*

Gas and Oil Systems

The *Domestic Heating Compliance Guide* lays down quite stringent energy efficiency requirements for heating systems. Here we'll look at the requirements for gas and oil systems in both new-build and existing properties.

Boilers

The minimum SEDBUK efficiency for gas boilers must be at least 86 per cent in new properties. In existing properties, it is normally expected that the appliance efficiency will be 86 per cent. However, in exceptional circumstances as indicated on completing the condensing boiler assessment procedure, the efficiency can go down to 78 per cent.

Oil boilers installed after 1 April 2007 must be of the condensing type and have a SEDBUK efficiency of 86 per cent.

System circulation

Systems installed in new properties must be fully pumped. In circumstances where a boiler is being replaced in an existing system, the system must be upgraded to a fully pumped system. If a bypass is required in the system, then it must be of the automatic type.

Hot water storage

New or replacement hot-water storage cylinders must meet the following specifications relating to the insulation and the area of the heat exchange coil:

- vented copper cylinders – BS 1566

- unvented cylinders – BS 7206

- thermal primary stores – the WHMA specification.

In addition all hot-water storage vessels must carry proper labelling to show that they are properly approved. The compliance guide also lays down a minimum specification for the insulation of central heating and hot-water pipework and the circumstances in which pipework must be insulated.

System controls

There are various requirements for each of several components of a domestic heating system.

Boiler interlock

There must be a suitable wiring arrangement so that when there is no demand for either hot water or space heating the boiler and pump are turned off.

Space heating zones

For new properties with a total usable floor area of up to 150m², the area should be divided into two separate space heating zones with independent temperature control. One zone must be the living room, and the requirement could be met by siting a room thermostat in the hall and placing TRVs on all other radiators.

For new properties with a total usable floor area over 150m², the area should be divided into two or more separate space heating zones, each with independent time and temperature control.

A new system in an existing property is required to meet these requirements, although in the case of a replacement boiler in an existing system, it is acceptable for there to be just one heating zone.

Water heating zone

All dwellings must have a separate water heating zone.

Time control of space heating and hot water

A system in a new property or a full replacement system in an existing property must have independent time control of the space heating and hot water circuits, for example by use of a programmer.

Did you know?

The space heating zone in a central heating system is the circuit feeding the radiators

In the case of a cylinder being replaced in an existing system, if a separate time control is not provided to both circuits, then it is permissible to have one time clock controlling the operation of both circuits together.

Temperature control of space heating

Separate temperature control of each zone can be provided by

Figure 13.3 Thermo-mechanical hot-water valve

- room thermostats or programmable room thermostats in all zones

- a room or programmable room thermostat in the main zone and individual TRVs on radiators in the other zones

- a combination of both of the above

- for existing systems in which the cylinder is being installed and there is a semi-gravity system then it is acceptable for a thermo-mechanical thermostat to be installed as a minimum.

Temperature control of hot water

A cylinder must have a thermostat controlling a motorised valve. For new properties with a total usable floor area over 150 m², there must be two water heating circuits, each with their own time and temperature control. For existing systems in which the cylinder is being installed and there is a semi-gravity system, it is acceptable for a thermo-mechanical thermostat to be installed as a minimum.

Commissioning

There are very specific guidelines for commissioning a system.

- The system should be thoroughly flushed on completion of the installation.

- A corrosion inhibitor must be added to the system.

- The system must be commissioned in-line with manufacturer requirements.

- The property owner must be fully instructed in the effective operation of the system and all user operating instructions must be left on-site.

- A commissioning record must be left on-site.

The installation of a boiler, central-heating or hot-water system is identified under the Building Regulations as work on a controlled service or fitting. Therefore, such an installation requires that either a notice be given to the local authority before the work begins, or a compliance certificate be issued on completion of the work by an installer who is a member of a competent persons scheme.

Solid fuel systems

Boilers and heating appliances must meet the minimum efficiency requirements as defined in the *Compliance Guide*. The appliances are normally approved by HETAS. The level of sophistication of the control system is identified by the manufacturer as a control system that is safe to use in conjunction with the appliance.

Typically, the following holds true for solid fuel systems:

- There must be thermostatic control of the boiler for all appliances, excluding those that will have an open fire.

- Simple appliances such as open fireback boilers and room heaters will usually be fed from a semi-gravity system in order for them to work safely, in which case boiler interlock and temperature control of the hot-water circuit cannot be provided. However, the requirements for control of the space heating circuit are the same as for gas and oil systems in terms of zoning requirements, time and temperature control.

- Less sophisticated independent boilers, such as gravity feed boilers, are subject to the same requirements as room heaters.

- Fully automatic solid fuel boilers, such as pelleted woodchip boilers burning biomass type fuel, are available if the manufacturer indicates that it may be possible to fully pump these, incorporating full boiler interlock and controls as per gas and oil systems.

- The system should be commissioned according to the same requirements as gas and oil systems.

Other methods of improving efficiency

We have covered the following procedures earlier in the book, but they are particularly relevant to efficiency:

- All systems pipework should be insulated and installed to comply with Water Regulations.

- Systems and components should be serviced and maintained regularly to:
  - ensure they are working to design specification
  - they are not wasting water.

- Systems should be designed to:
  - keep dead legs as short as possible
  - avoid over capacity, i.e. heating high volumes of hot water storage that may not be required.

Methods of reducing waste when plumbing

It is important that you get into the habit of keeping waste material to a minimum when you are carrying out plumbing work. Here is a checklist.

Waste reduction checklist

1 If it is economically possible, try to repair an appliance or component rather than replace it. New components have to be made, and the manufacturing process uses energy. Old appliances have to be transported and disposed of (although replacing an inefficient boiler would be a good move)

2 Take time to carefully measure and set out pipework for bends. This will reduce the amount of wasted pipe when cutting it to length

3 When using capillary integral solder ring fittings do not use additional wire solder on the joint

4 Treat screws and other fixings like money – too often they are left all over a job

5 Be extra careful when fitting sanitary ware, a broken one has to be disposed of and replaced

6 This also applies to the storage of sanitary ware and other damageable goods

7 Take care of your tools. Defective tools have to be replaced – and so do stolen ones

8 Think about how you use water when you are working on a system. Have a thorough check around all the fittings to make sure they have all been soldered. You do not want to be filling and draining systems unnecessarily

9 Do not overdo the use of jointing compounds, fluxes and other materials

10 Take care when taking floorboards up; avoid having to replace damaged boards with new ones

On the job: Efficiency regulations

Jordan is a second year apprentice working under the supervision of Anthony, who has been with the company quite a few years. They have nearly finished installing a heating system in an occupied terraced house. Jordan decides to use some initiative and begins to insulate the hot water primaries in the airing cupboard.

'Don't bother with that', shouts Anthony. 'Waste of time and effort.' Jordan is about to tell Anthony about the new regulations and the importance of energy efficiency he learned about at college when Anthony continues, 'You apprentices are all the same, get on with something worthwhile.'

The job is completed (minus insulation) and while Jordan is putting the tools in the van the customer thanks them and gives them a £10 tip each for doing such a good job. Jordan feels guilty because the job is not as complete as it should be.

1   What should Jordan do now?

2   What would you do?

3   Is the insulation important?

Environmental hazards

The main environmental hazard as far as plumbing is concerned is asbestos, dealt with in Chapter 2 *Health and Safety*. A popular pastime on some of the larger sites is burning waste material. All sorts of things get thrown into a site fire: plastics, paint cans and so on. This activity is not environmentally friendly and is usually against site rules. Particular attention is drawn to the burning of plastics and polystyrene which give off dangerous fumes.

Waste disposal

There are special arrangements for the disposal of some types of asbestos. On larger sites, skips are usually provided for waste. These, once full, are taken to licensed sites for disposal. Do not overfill the skip, and do not put in items that are dangerous, e.g. flammable material.

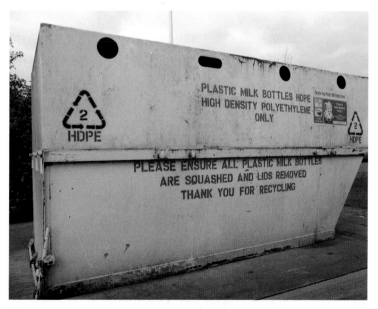

Dispose of construction waste responsibly

On smaller jobs you might still use a skip, for the disposal of an old bathroom suite for example. If you try to take 'industrial waste' to your local tip, you are likely to be turned away or have to pay for its disposal. Remember that the disposal of refrigerators is covered by legislation and they must be dealt with at specialist sites. Do not be tempted to let the customer just dump them in the skip.

Plumbers work with:

- copper
- brass
- lead
- cast iron and low carbon steel.

All of these have a value at the scrap merchant – think about recycling before you think about throwing away.

Customer advice

It is good business practice, as well as a requirement of the L1 Building Regulations: *Approved Document (Central Heating Systems)*, to provide information to the users on the control and operation of a new system installation.

You will probably find that a high percentage of callouts are from customers who do not know how to work their system controls properly. Here is a list of what we think should be done for the customer:

- Provide a company folder containing:
  - emergency contact number for burst pipes, etc.
  - general advice line number – you might be able to sort a problem out over the phone
  - customer guidance leaflets that most manufacturers produce; also *go through them with the customer*.

- Label the various fittings or components that the customer may need to use in an emergency, e.g. stop valve, gate valve, etc.

- Show them, or tell them, where the service valves are, and what they do.

- Walk them around the system: show them the various components; tell them what they can touch and what they must not touch.

- Explain what the components do in simple terms, e.g. 'This is the cylinder thermostat, and, when the water in this cylinder gets to that pre-set temperature, it shuts off the heat to the cylinder; you don't have to alter that setting'.

- When you are finished, give your customer the chance to ask questions.

- Leave the manufacturers' installation and user instructions on the job for future reference purposes. Ensure that a full commissioning record is completed for the work – it's not only good sense, it's also the law.

FAQ

Why does avoiding dead-legs in hot-water systems have an impact on the environment?

In two ways: first by saving water which is run off to waste while waiting for hot water to arrive, and second by saving energy on heating the water in the first place. The hot water that is left in the pipe when we close the tap cools down in a very short time and is wasted.

Surely a little bit of insulation to hot-water pipes in an airing cupboard will not make that much difference to the environment?

Well, every little does help. If we were to multiply all the heat losses on a national basis it would be considerable, so you see, it will make a difference.

Knowledge check 2

1   Which document details the minimum energy efficiency specification for a heating system in a dwelling?

2   What is the SEDBUK rating of a new gas boiler to be installed in an existing property?

3   A new building will require two or more space heating zones when it has a floor area of _____.

4   What is the minimum energy efficiency measure that must be applied to the hot-water circuit when replacing a hot-water cylinder in an existing semi-gravity heating system?

5   Provide two examples of where a plumber could reduce the impact of their work on the environment.

Additional information sources

Sources of additional information for this chapter include:

Department of the Environment, Food and Rural Affairs (DEFRA) –www.defra.gov.uk

Communities and Local Government – www.communities.gov.uk

chapter 14

Employment rights and responsibilities

OVERVIEW

This chapter supports the Employment Rights and Responsibilities unit of the plumbing technical certificate programme. The unit is an additional unit to the mainstream plumbing qualifications that must be completed by all apprentices completing a Government-funded apprenticeship programme. The chapter includes:

- **Key employment legislation**
  - Employment Rights Act 1996
  - Employment Relations Act 2004
  - Employment Act 2002
  - other legislation impacting on employment

- **Protection against discrimination**
  - discrimination on the grounds of race
  - sexual discrimination
  - discrimination on the grounds of religion or belief
  - discrimination on the grounds of sexual orientation
  - discrimination against disability
  - age discrimination

- **Basic employment rights**
  - statement of employment
  - entitlement to paid leave
  - working hours
  - termination of employment
  - wages and statutory sick pay
  - maternity leave
  - paternity leave
  - pensions

- **The plumbing industry**

- **Career opportunities in plumbing**

Key employment legislation

There are a number of items of legislation designed to protect workers' rights. The main laws affecting employment are:

- the Employment Rights Act 1996
- the Employment Relations Act 2004
- the Employment Act 2002.

Employment Rights Act 1996

The main features of the act are:

- *Wages rules* – the Act determines methods of wage payment and establishes you with the right not to suffer from unauthorised wage deductions. It offers you 'guaranteed payment' – the guarantee of wages even where an employer is unable to supply you with work during your hours of employment. It has requirements laid down concerning Sunday work and requires your employer to provide written statements of employment.

- *Time off work* – one of the Act's most important aspects concerns law involving time off work. You are entitled to certain protections involving having time off that may be owed to ante-natal care, training or public duties. The Act outlines procedures involving suspension due to maternity leave or medical problems. The rights and obligations of individuals seeking maternity leave are covered in particular detail.

- *Dismissal and redundancy* – another topic key to the Act involves requirements related to dismissal and redundancy. It is stated that unfair dismissal – where an employer has insufficient evidence to back up their formal reason for dismissing an employee – gives the employee a right to compensation and it gives an indication of the circumstances in which unfair dismissal has occurred. The Act also provides a guarantee of payment in respect of redundancy.

- *Dispute resolution* – the Act identifies the process by which you can resolve disputes related to any violation of your rights as an employee. Indeed, you have the right to go to an employment tribunal if you have a dispute with your employer over a range of issues.

Employment Relations Act 2004

This Act was first introduced in 1999 and was further modified in 2004. The Act established a number of new rights at work covering the following areas:

- trade union recognition
- industrial action ballots
- unfair dismissal of strikers

- maternity leave
- parental leave
- time off for dependants
- employment tribunal awards
- the right to be accompanied in disciplinary and grievance hearings
- part-time work – equality with full-time work.

Employment Act 2002

The Act is a wide-ranging employment package, covering:

- work and parents
- dispute resolution in the workplace
- improvements to employment tribunal procedures
- provisions to introduce the Fixed Term Work Directive
- a new right to time off work for union learning representatives.

Other items of legislation impacting on employment

- *Equal Pay Act* – the Act requires that men and women be given equal pay for equal work in the same establishment. The jobs need not be identical, but they must be substantially equal. It is job content, not job title, that determines whether jobs are substantially equal.

- *Data Protection Act* – the Act provides a common-sense set of rules which prevent the misuse of your personal information without stopping it being used for legitimate reasons.

  The details of the Data Protection Act are quite complex. The following are eight main rules that are the key points of the Act. These require personal information to be:

  - fairly and lawfully processed
  - processed for limited purposes
  - adequate, relevant and not excessive
  - accurate
  - not kept longer than necessary
  - processed in accordance with your rights
  - kept secure
  - not transferred abroad without adequate protection.

Did you know?

The fixed-term work directive is European legislation designed to ensure that workers on fixed-term job contracts, for example 12 months duration, have the same employment rights as permanent staff members

Did you know?

An employer is legally required to have Employers' Liability Insurance as a safeguard against employee accident and illness whilst at work.

Organisations using personal information must comply with these requirements and, moreover, are required to register under the Act. The Act provides stronger protection for sensitive information about your ethnic origins, political opinions, religious beliefs, trade union membership, health, sexual life and any criminal history.

- *National Minimum Wage Act* – this is a legal right which covers almost all workers in the UK. It became law in1999 to prevent unduly low pay. There are three rates of National Minimum Wage (NMW). They are based on recommendations of an independent body, the Low Pay Commission.

  - The main (adult) rate applies to workers aged 22 and over.

  - The development rate applies to 18- to 21-year-olds.

  - The third rate applies to 16- to 17-year-old workers who are above school leaving age.

Apprentices under the age of 19 are not entitled to the NMW and nor are apprentices over 19 in their first year. The vast majority of adult workers in the UK are entitled to be paid the NMW provided they are not genuinely self-employed, including:

- part-time workers
- casual labourers
- agency workers
- piece workers, including homeworkers.

The main exemptions from the requirements of the Act are:

- the genuinely self-employed
- voluntary workers
- workers who are based permanently outside the UK.

Protection against discrimination

There are a number of items of legislation which are designed to provide protection against discrimination in the workplace.

Discrimination on the grounds of race

It's unlawful to discriminate against someone, either directly or indirectly, on the grounds of race, colour, nationality – including citizenship – or ethnic origins, under the Race Relations Act 1976. Racial harassment is defined as a form of discrimination.

Discrimination on the grounds of race can be either direct or indirect:

- Direct discrimination may include treating somebody less favourably on the grounds of their race, colour, ethnicity or national origin.

Did you know?

Discrimination on the grounds of someone's accent does not count under the Race Relations Act

- Indirect discrimination can occur when an employer applies an apparently general rule that in practice disadvantages people of a particular race, colour, ethnicity or national origin and that cannot be justified.

This form of discrimination applies to the way employers recruit, train, promote and select people for dismissal on the grounds of redundancy, or, after they have left, if an employer refuses to provide a reference.

Sexual discrimination

As with other forms of discrimination, the legislation recognises both direct and indirect discrimination:

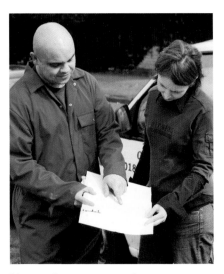
Men and women must be treated equally at work

- Direct discrimination occurs when an employer treats somebody less favourably on the grounds of their sex.

- Indirect discrimination occurs when an employer applies an apparently general rule which in practice disadvantages one sex and which cannot be justified. It includes the way employers recruit, train, promote and dismiss people. Former employees have the right to bring action against an employer if he/she discriminates against them after they have left.

The penalties for sex discrimination can be high, for both organisations and individuals, since there is no limit on compensation.

There are no length-of-service or age requirements in bringing a claim and claimants who are employees do not need to have left employment.

The law is explicit that less favourable treatment of women on the grounds of pregnancy, or maternity leave, counts as unlawful sex discrimination.

The law makes sexual harassment – and harassment on the ground of sex – explicitly unlawful in employment or vocational training. Sexual harassment can include insensitive jokes, displays of sexually explicit material, sexual innuendos or lewd comments or gestures.

Discrimination on the grounds of religion or belief

Discrimination against an employee, or job candidate, on the grounds of their religion or belief became unlawful from December 2003 under the Employment Equality (Religion or Belief) Regulations.

The regulations apply to discrimination on the grounds of religion or religious belief. They cover discrimination on the grounds of perceived as well as actual religion or belief and the belief or religion of someone with whom the person works with.

Did you know?

The Sex Discrimination Act was introduced in 1975

Remember

There are limited employer exemptions from the Regulations on the grounds of religious belief e.g. the appointment of a church minister where an applicant is restricted to a particular faith as an occupational requirement

As with other forms of discrimination, the legislation recognises both direct and indirect discrimination:

- Direct discrimination may occur in areas such as recruitment, selection, training, promotion, selection for redundancy or dismissal when someone is treated less favourably than another because of their actual or perceived religion or beliefs.

- Indirect discrimination occurs when an employer applies a provision, or practice which disadvantages people of a particular religion or of particular religious beliefs and which cannot be justified.

The regulations also outlaw:

- harassment – unwanted conduct that violates people's dignity or creates an intimidating or offensive atmosphere

- victimisation – treating people less favourably because of something they have done under or in connection with the regulations, for example made a formal complaint of discrimination, or given evidence, in a tribunal case.

There are no length-of-service, or age requirements, in bringing a claim and where the claimant is an employee, they do not need to have left employment.

Discrimination on the grounds of sexual orientation

Discrimination against an employee or job candidate on the grounds of their sexual orientation became unlawful from December 2003.

The Employment Equality (Sexual Orientation) Regulations apply to discrimination on grounds of orientation towards persons of the same sex (lesbians or gays), the opposite sex (heterosexuals) and the same and opposite sex (bisexuals). They cover discrimination on the grounds of perceived, as well as actual, sexual orientation and the sexual orientation of someone with whom the person associates or works.

As with other forms of discrimination, the legislation recognises both direct and indirect discrimination:

- Direct discrimination may occur in areas such as recruitment, selection, training, promotion, selection for redundancy or dismissal or in awarding employment related benefits when someone is treated less favourably than another because they are, or are thought to be, lesbian, gay, heterosexual or bisexual.

- Indirect discrimination occurs when an employer applies a provision, criterion or practice which disadvantages people of a particular sexual orientation and which cannot be justified.

From December 2005, indirect discrimination could also occur where a civil partner is treated less favourably than a married employee on a range of employment rights and employee benefits including the right to:

- make a request for flexible working

- statutory paternity leave and pay for a newborn child or newly adopted child

- vocational training

- access to a benefit that is given to a married employee, such as a pension or private health care.

Discrimination against disability

It is unlawful for any employer, regardless of size, with the exception of the armed forces, to directly discriminate against a disabled person because they are disabled. Also they must not discriminate against a disabled person for a reason related to their disability, unless this can be justified.

What counts as a disability?

The Disability Discrimination Act 1995, which was further updated in 2005, defines disability as a physical, or mental impairment, which has a substantial and long-term adverse effect on the ability to carry out normal day-to-day activities. This does not include drug or alcohol abuse or a tendency to start fires, steal or physically abuse others, amongst other exclusions.

What employers must not do – an employer mustn't treat a disabled person less favourably than other members of staff. For instance, they cannot refuse to interview, employ, train or promote a disabled person, simply because of their disability. Since 1st October 2004, harassment on the grounds of disability has also been explicitly outlawed.

What employers must do – employers have a duty to make reasonable adjustments to enable a disabled person to work or continue working.

Penalties for discrimination

Where an employment tribunal finds that disability discrimination has occurred, penalties can be high, since there is no limit on compensation.

There are no length-of-service, or age requirements, in bringing a claim and where the claimant is an employee, he or she does not need to have left employment.

Age discrimination

The age discrimination regulations apply to all employers in the private and public sectors, vocational training providers, trade unions, professional organisations, employer organizations, etc. They cover employees of any age and other workers, office holders, partners of firms and others.

The regulations affect all areas of employment including:

- recruitment, terms and conditions, promotion and transfers
- the provision of training
- provision of benefits
- dismissal
- retirement

Did you know?

The Employment Equality (Age) Regulations came into force on 1 October 2006

- occupational pensions.

The legislation provides for a default retirement age of 65. Compulsory retirement ages below 65 are unlawful unless objectively justified. In addition, employees have the right to request to work beyond that age. Employers have a duty to consider such requests.

Knowledge check 1

1 When was the Employment Relations Act first introduced?

2 Which piece of legislation would be breached if a female was treated less favourably during a job interview than a male?

3 Unfair penalties imposed on an employee by his/her employer following an industrial tribunal would be regarded as _____.

4 What is the default retirement age under the age discrimination regulations?

5 Which piece of legislation lays down the circumstances in which an employer must recognise a trade union in representing the employee's interests in a plumbing business?

Basic employment rights

Under the requirements of employment legislation you are entitled to a number of basic employment rights. The following summarises the key points of those rights.

Statement of employment

Previously known as contracts and conditions of employment, employment law requires that you receive a statement of employment within two months of starting work, if that employment lasts for a period of more than one month. The statement may be accompanied by a staff handbook which contains a detailed description of each of the procedures laid down in the statement. The statement should usually cover the following:

- the name of your employer and your name

- the date your employment started

- your job title and a summary of your duties

- the period of employment, stating whether it is a permanent position

- the place of work

- how much you will be paid, how often, and the method of payment; it should also include information such as travel allowances and any deductions from pay

Did you know?

An employer must enter into negotiation with you before making any amendment to your statement of employment

Remember

The disciplinary procedure lays down the steps that your employer must take to deal with issues such as poor work performance, illegal activities, etc. related to your employment

- hours of work
- holiday entitlement
- procedures for dealing with absence from work through illness, or for other reasons, and how to notify the employer if absent
- details of pension scheme if applicable
- details of how to terminate employment (for example, length of notice required by both you and your employer)
- disciplinary rules and procedures (these are usually contained in a separate document such as a staff handbook)
- grievance procedures (which again could be contained in a staff handbook).

Remember

The grievance procedure lays down the steps that you must follow to raise issues of concern that you may have related to your employment

Did you know?

Your employer is not obliged to allow you to carry over paid leave from one year to another

Entitlement to paid leave

On starting a job you qualify immediately for holiday leave and pay. You have a statutory entitlement to 4.8 weeks of paid leave a year. This figure will rise to 5.6 weeks on 1 April 2009. Your employer may include public holidays as part of the annual leave entitlement and may set the times when you take holidays such as during the Christmas shutdown. If your employment ends and you have leave due you have a right to be paid for the leave time due and not taken.

Time off work may also be permitted to carry out duties, such as:

- dealing with emergencies involving dependents (close family)
- jury service
- trade union duties.

The entitlement to time off for most of these activities is usually on an unpaid basis unless your employment statement says otherwise. Activities such as jury service do provide reimbursement for loss of earnings. In addition employees aged 16 and 17, who did not reach a certain standard of education at school, have the right to reasonable time off with pay while working to study for a qualification that will help them reach that standard.

Did you know?

The standard working week for plumbing operatives as laid down by the JIB is 37.5 hours

Working hours

Requirements for working hours are laid down in the Working Time Regulations. Workers aged 18 or over cannot be forced to work for more than 48 hours a week on average. However, most workers can agree in writing to work longer than the 48 hour limit. The agreement must be signed by the worker. Workers can cancel the opt-out agreement whenever they want, although they must give their employer at least seven days' notice. For young workers (under 18) the maximum working week is 40 hours and each working day must be no longer than 8 hours.

Termination of employment

In order to dismiss you your employer must provide good grounds for the dismissal. If you have been in employment for 12 months or more then a claim of unfair dismissal can be made to an employment tribunal if the grounds are deemed to be unfair. There are exceptions, however, and a claim for unfair dismissal may be made with less than 12 months service if on the grounds of discrimination, pregnancy, parental responsibilities, etc. The notice period in terms of a dismissal should be a minimum of the following:

- one week if employed between one month and two years

- an additional week's notice for every continuous year of employment between two and twelve years

- minimum twelve weeks notice if in continuous employment for 12 years or more.

Wages and statutory sick pay

An employer is required to provide an itemised payslip showing:

- the gross wages earned (before deductions)

- the amounts of and reasons for any deductions, for example National Insurance

- the net, or take home, pay.

An employer is legally required to keep records of the payments made, including any deductions made, this includes details of any statutory maternity or paternity leave payment.

The rate of pay in smaller companies is often negotiated between employer and employee. Legally the minimum rate of pay must be in line with the NMW. In larger companies pay rates, and indeed working conditions, may often be determined collectively between an employer and a trade union.

The JIB for PMES produces national standard terms and conditions of employment for plumbing operatives. These are often regarded as the minimum pay and conditions for plumbing operatives. In terms of pay the JIB identifies specific rates of pay for apprentices based on the year of apprenticeship. Pay for qualified plumbers is based on grading against the following definitions:

- trained plumber – Level 2 qualified with post-qualification experience

- advanced plumber – Level 3 qualified with post-qualification experience

- technical plumber – Level 3 qualified with significant post-qualification experience and significant job responsibilities.

Employees have protection in law against unauthorised deductions from their pay.

Did you know?

The Advisory, Conciliation and Arbitration Service (ACAS) provides essential support in employment relation matters including resolving disputes

Sickness Pay

The procedures for reporting in sick will normally be laid out in the statement of employment or staff handbook. In most cases you will be required to self-certify sickness for periods of up to seven days. Periods over seven days normally require a medical certificate to be provided from a doctor. Your employer will expect you to notify him/her of your inability to work due to sickness as soon as possible and to keep them updated on progress if the sickness period becomes longer.

If you are off sick for four consecutive days or more (assuming a doctor's certificate is provided) you are entitled to Statutory Sick Pay (SSP). SSP is paid by the employer in the same way that wages are paid. Records of absences lasting over four days and of payments made must be kept. The employer is responsible for payment of SSP for up to 28 weeks.

Redundancy pay

Employees have the right to a lump-sum redundancy payment if they are dismissed because of redundancy. The amount is related to the employee's age, length of continuous service with the employer, and weekly pay up to a maximum. The employer must also provide a written statement showing how the payment has been calculated at, or before, the time it is paid.

Maternity leave

- All pregnant employees are entitled to paid time off for antenatal care.

- All pregnant employees are entitled to 52 weeks' maternity leave – 26 weeks' Ordinary Maternity Leave (OML) and 26 weeks' Additional Maternity Leave (AML).

- Pregnant employees who meet qualifying conditions based on their length of service and average earnings are entitled to up to 39 weeks SMP, which is paid by their employers and mostly or completely refunded by the Government.

- Women who are not entitled to SMP but meet qualifying conditions based on their recent employment and earnings may claim up to 39 weeks' Maternity Allowance, paid direct by Jobcentre Plus.

- Employers may make reasonable contact with a woman on maternity leave for a number of reasons, such as to discuss arrangements for her return to work.

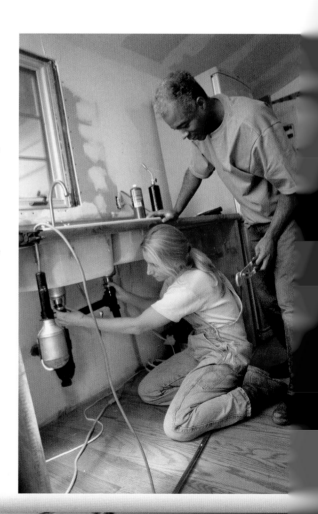

Remember

Qualification for Statutory Maternity Pay (SMP) is based on a continuous period of at least 26 weeks employment ending with the 15th week before the expected week of childbirth

- Employees may undertake up to ten 'keeping in touch days' during their maternity leave – allowing work under their contract of employment – by agreement with the employer.

- Employees who wish to return to work either earlier, or later, than agreed with the employer should provide eight weeks' notice, unless the employer agrees to less notice being given.

- Employees have a right to return to the same job after maternity leave. There may be some exceptions to this if the employee takes more than 26 weeks' maternity leave and if it is not reasonably practicable for the employer to hold her job open, but she must still be offered a job that is suitable for her and the terms and conditions must be no less favourable.

- Pregnant employees, and those on maternity leave, are protected under sex discrimination legislation which outlaws unfair treatment, including dismissal, on grounds of their sex, pregnancy or maternity leave.

Paternity leave

Employees must satisfy the specific conditions in order to qualify for paternity leave. They must:

- have or expect to have responsibility for the child's upbringing

- be the biological father of the child or the mother's husband or partner

- have worked continuously for their employer for 26 weeks ending with the 15th week before the baby is due.

Employers can ask their employees to provide a self-certificate as evidence that they meet these eligibility conditions.

Eligible employees can choose to take either one week or two consecutive weeks' paternity leave (not odd days).

They can choose to start their leave:

- from the date of the child's birth (whether this is earlier or later than expected)

- from a chosen number of days or weeks after the date of the child's birth (whether this is earlier or later than expected)

- from a chosen date later than the first day of the week in which the baby is expected to be born.

Leave can start on any day of the week on or following the child's birth but must be completed:

- within 56 days of the actual date of birth of the child

- if the child is born early, within the period from the actual date of birth up to 56 days after the first day of the expected week of birth.

Only one period of leave is available to employees regardless of whether more than one child is born as the result of the same pregnancy.

During their paternity leave, most employees are entitled to Statutory Paternity Pay (SPP) from their employers.

Statutory Paternity Pay is paid by employers for either one or two consecutive weeks as the employee has chosen.

Pensions

Any pension provision available to you will be laid down in your statement of employment. If a company scheme is not available to you then, as a minimum, organisations employing five or more people are obliged to offer access to a stakeholder pension scheme (from 8 October 2001) if they do not already offer pension provision.

The plumbing industry

The plumbing industry is a part of:

- The Mechanical Engineering Services (MES) sector, which also includes the following industries:
  - heating and ventilating
  - refrigeration and air conditioning
  - gas fitting.
- The MES sector forms part of the wider Building Services Engineering industry containing:
  - plumbing
  - heating and ventilating
  - refrigeration and air conditioning
  - electrical installation

The key plumbing industry organisations and their role in the industry are detailed in Chapter 1 on pages 4–7. An understanding of these organisations forms a part of the unit in Employment rights and responsibilities.

Career opportunities in plumbing

The main career opportunities open to you in the plumbing industry are described in Chapter 1 on pages 14–15. An understanding of these opportunities forms a part of the unit in Employment rights and responsibilities.

Knowledge check 2

1   A statement of employment must be provided within what period of time of commencing work with a new company?

2   What is the maximum period of paternity leave that may be taken?

3   What is the maximum period of maternity leave that may be taken?

4   An employer will normally require a doctor's certificate to be provided after a period of sickness lasting more than how many days?

5   What grades of qualified plumber are recognised under the JIB for PMES grading scheme?

Additional information sources

Sources of additional information for this chapter include:

- Department for Business Enterprise and Regulatory Reform – providing guidance on employment matters – www.berr.gov.uk/employment

- Business Link – providing guidance on employment matters – www.businesslink.gov.uk

- ACAS – providing guidance on personnel requirements in companies – www.acas.co.uk

Glossary

abutment flashing used where a pitched roof needs weathering against a brick wall

accident book a document where detail of all accidents is recorded no matter how minor

adhesion the force of attraction between molecules and another material they are in contact with, e.g. water in a glass, tyres on a road

allen key an hexagonal tool designed to fit into socketed screws to allow them to be turned

alloy a type of metal made from two or more other metals

alternating current type of electrical current as used in domestic properties

ampere a flow of one coulomb in one second, i.e. the quantity or amount of electricity that flows every second

annealing the process of softening something using heat, e.g. copper

anode positively charged electrode

APHC Association of Plumbing and Heating Contractors

Apprenticeship in Plumbing formal agreement between trainee and employer to confirm period of training and expected outcomes

ascending spray the water discharges below the top of the appliance, i.e. directly into its bowl

atmospheric pressure the pressure in the environment created by the force of gravity

atom smallest particle of a chemical element

bending spring spring used either internally or externally to aid the manual bending of copper tubing over the knee

bi-metallic strip strip made of two different metals bonded together which have different coefficients of expansion, thereby causing the strip to bend when it is heated

bonding term used to describe electrical earthing arrangement

bossing working lead sheeting into shape using a range of mallets and dressers

BSP British Standard Pipe

butt joint method of joining sheet lead where the edges are welded up against each other

capillary action the process by which a liquid is drawn up the surface of a solid material

catalyst a substance that increases the rate of a chemical reaction while remaining unchanged itself

cathode negatively charged electrode

CE mark mark denoting that an appliance conforms to European Standards

centralised storage system stores heated water ready to be supplied to a number of outlets

ceramics   fired pottery

chamfering   smoothing the cut edge of a pipe by slightly bevelling or angling it

chasing out   the process of cutting a recess into plaster or blockwork for a socket or cable

check valve   a device used to protect the water supply from any contaminants that may get into it

circuit breaker   a safety device to a fuse that can be set to interrupt an electric current in certain situations

cohesion   the tendency for molecules to stick together

compound   a mixture of two or more elements

compression ring   the ring used to form a pressure-tight seal in a compression joint

conduction   the transfer of heat energy through a material

conductor   a material that allows the transfer of electricity

conduit   a preformed PVC channel that covers cable runs and protects cables against damage

convection   the transfer of heat by means of the movement of a locally heated fluid substance (usually air or water)

corrosion   degradation of a material caused by air, water, acids, alkalis, chemicals or electrolytic action

coulomb   a unit of measurement of electrons

cross-flow   occurs when two branches are located opposite each other allowing water flowing from one branch to affect the pressure in the other branch

crutch head   the handle of a stop tap

current   flow of electrically charged particles

curtilage   an area attached to a dwelling house and forming one enclosure within it

deformation   when a material will not return to its original stage following the application of force

density   the relative lightness or heaviness of solid materials

dezincification   the breakdown of zinc in brass fittings by the electrolytic corrosion between zinc and copper

differential   relating to the difference between two opposing forces

direct current   electrical term given to current flow as from a battery

earth   to run an electric current safely away to earth, or ground

earth equipotential bonding   an earthing system that allows a fault or electrical leak on any part of a system to go to earth, rendering the system safe

earth fault loop   total earth path around the installation and back to the transformer

electrical conductivity   how well or poorly a material will conduct electricity

electricity   the flow of electrons through a conductor

electrolyte   a solution or molten substance that conducts electricity

electromagnetism   the magnetic force produced by an electrical current

electro-motive force   producing an electric current

electron   elementary particle with a negative electric charge

fascia   a piece of wood or plastic fixed to the rafters just below the roof allowing a fixing point for guttering

ferrous   metals that contain iron

ferrule   a metal fitment used for mains connection

first fix — fixing of pipework to floor joists, boards and sheets, stud partitions, roof trusses and timbers, etc., sometimes before a building is watertight

flexible push-fit plumbing systems — term given to system of pipes and fittings using push-fit joints

fluid — a substance such as a gas or liquid that lacks definite shape and is capable of flowing

foul drainage — anything discharged from a sanitary appliance – WC, bath, basin, sink, etc.

fuse — a small unit containing a strip of wire that will melt when the current exceeds a certain level, thereby breaking the circuit

fusion welding — the process of joining two pieces of lead by melting the two edges of the lead together (called the parent metal) while a filler rod of lead is added. Also the process of jointing some waste and soil pipework

gas — air-like, not solid or liquid

glass — hard, brittle, usually transparent substance made by fusing sand (silica) with soda and lime

gravity — force of attraction dependent on the relative mass of the bodies in question

gravity circulation — movement of water around a property which relies on the fact that cold water is heavier than hot water: gravity therefore exerts a stronger pull on the cold water, allowing the hot water to rise through the system

gripper fixings — laid underneath carpets, usually at the edges, to hold the carpet firmly in place

hand bender — small portable machine for bending copper tube

hertz — the number of cycles of AC that are produced every second

immersion heater — an electric element fitted inside the hot-water storage vessel, controlled by a switch and thermostat

impounding reservoir — a man-made reservoir

installation — fitting a new system

instantaneous hot-water system — heats water on demand

insulator — an item to stop transfer of electricity, also used to describe material for reducing flow of heat

JIB for PMES — Joint Industry Board for PMES

kinetic — movement

lap joint — method of joining sheet lead, where the sheets initially overlap

liquid — offering no resistance to change of shape, not solid or gaseous

localised system — heats water at the point at which it is required

main earthing terminal — as found at electrical point of entry/main distribution board

maintenance — carrying out any necessary repairs to worn or broken parts; keeping appliances and systems in good working order

making good — repairing and finishing off the brickwork, blockwork and concrete

malleability — the ability to be worked without fracture

mandatory units — the ones candidates must do

manipulative joint — a joint where you work or form the end of the tube

marring — the migration of polymers between polystyrene and a PVC cable sheath, potentially leading to damage to the PVC cable

micro-businesses — either sole traders or businesses employing 1–4 people

milestones — key targets that you need to achieve at specified times within your apprenticeship

Mohs' scale — hardness scale, developed in 1812 and named after the German mineralogist Friedrich Mohs (1773–1839)

neutron — elementary particle that has no electric charge

nucleus	the central part and the main mass of an atom
observed assessment	carrying out a set task with an assessor in attendance to confirm competence
olive	the ring that is compressed onto a pipe in a compression fitting
open vented	a system that has a vent pipe always open to atmosphere
orbit	circulate around another body
overflow pipe	discharges excess water safely
packing-gland nut	nut used to compress packing to make valve spindles watertight
patina	a protective barrier which forms on metals
PAT tests	checking for safety and keeping maintenance records of all portable equipment to ensure it is in safe working order
permeation	the possibility of microscopic particles of oxygen entering the water supply through the external wall of the pipe
plant	large, usually heavy machinery used on building sites, such as cranes, excavators, specialised earthmoving equipment, forklift trucks and power access equipment
polybutylene	a product of the polymerisation of ethers
potential difference	energy used up by one coulomb between two points in a circuit
properties	physical and working attributes of materials
proton	elementary particle with a positive electric charge
proving unit	low-voltage, inverted DC testing device
radial circuit	starts at the consumer unit and connects to every point on the circuit, terminating at the last one
radiation	the transfer of heat from a hot body to a cooler one without the presence of a material medium (other than air), by means of 'heat' waves
residual current device	a highly sensitive unit that measures changes in the electrical current between different electrical conductors in a system and automatically disconnects the circuit if a small change occurs
resistance	the degree to which a material or device resists the flow of an electric current
ring circuit	connects the consumer unit and every point on the circuit, and returns to the consumer unit
rungs	the horizontal 'steps' of a ladder
sacrificial anode	a metal that is relatively more easily destroyed such as magnesium; it will be destroyed by corrosion first, therefore giving longer life to other metals
secondary circulation	a method of installing hot-water pipework whereby a return pipe is fitted near the top of the cylinder, to maximise energy efficiency
second fix	the installation of bathrooms, radiators, staircases, doors, kitchen units, architraves and skirtings after a building has been made watertight
Sector Skills Council	links with government and looks after the training and development needs of their specific area, e.g. SummitSkills
SEDBUK rating	a measure of the seasonal efficiency of a boiler installed in typical domestic conditions in the UK
service	regular checking of a system to ensure it is still working properly
simulated assessment	assessment that takes place at the centre

soaker — a piece of lead fitted between the slates or tiles and then turned up and fitted to the side of the chimney

soffit — the underhang of the roof, above the fascia

solid — having stable shape, not liquid or gaseous

solvent power — ability to dissolve other substances

sparge pipe — a flush pipe mounted on the face of a urinal drilled with a number of holes, used to wash the face of the slab and the channel

specific heat capacity — the amount of heat required to raise 1 kg of a material by 1 degree Celsius

stand bender — a larger type of bending machine that stands on legs/tripod

'stat' — a widely used abbreviation for 'thermostat'

sterilisation — purification by boiling or dosing the water supply with chlorine or chlorine-ammonia mix

stiles — the upright bars of a ladder, into which the rungs are set

stillson — type of wrench for use on steel pipe and fittings

stocks and dies — stocks are the body and handle of the tool, dies are the actual cutter

storage hot-water system — stored heated water ready to be supplied to one or more outlets

stratification — the formation of layers of water in a hot-water storage vessel

SummitSkills — the sector skills industry organisation dedicated to improving skills in building services, such as plumbing, by training and development

supplementary evidence — additional proof that you carried out the work

surface tension — the way in which water molecules cling together to form a 'skin'

surface water — the term used for water collected via the rainwater system

swaging tool — a specialist tool for 'opening out' the end of a pipe for manipulative fittings

swarf — debris produced from cutting action

termination — end point

terne coating — stainless steel that is hot dip coated with a lead-tin alloy

thermal — heat

thermal conductivity — how well or poorly a material will conduct heat

thermal insulators — wooden, ceramic and plastic materials that are poor conductors of heat

thermistor — a device that changes electrical resistance in response to a temperature change

thermoplastic — a plastic that softens when heated

thermosetting plastic — a plastic that keeps its shape after initial forming

throating — when a pipe's diameter has been reduced at the bend by the roller, guide and former being too tight

TN-S system — electrical installation system in which exposed metalwork is connected to the main earthing terminal and in which there is a separate earth conductor

toolbox talk — a briefing given by management/supervisory staff on subjects such as safety, customer care, etc.

TRVs — thermostatic radiator valves

undercutting — reducing the thickness of lead sheeting where two pieces are welded together

U value — a measure of the thermal transmission of heat through the fabric of a building

venturi a device that causes a differential pressure in a pipe to activate a switch such as on a gas water heater

vitreous china ceramic produced from a solution of clay and water, called slip, or casting clay, which contains ball clay, china clay, sand, fixing agent and water

voltage electro-motive force, measured in volts

warning pipe alerts occupants to a problem with float valves

water cycle water's journey from the sky to your tap

water governor a pressure-reducing valve

water mains the network of pipes that supply wholesome water to domestic and commercial properties

water table the natural level of water under the earth

water undertaker the legal term for the water companies that supply domestic water

welding the jointing of sheet lead materials by the application of heat to the two surfaces to be jointed

wholesome water good-quality water provided for human consumption

work at heights usually described as work taking place ata a height of 2 m or more, above ground level

workplace evidence record formal written evidence of your work carried out on-site

Index